MANAGING
HUMAN
RESOURCES

A Challenge to
Urban Governments

Volume 13, URBAN AFFAIRS ANNUAL REVIEWS

MANAGING HUMAN RESOURCES

A Challenge to Urban Governments

Edited by
CHARLES H. LEVINE

Volume 13, URBAN AFFAIRS ANNUAL REVIEWS

 SAGE PUBLICATIONS / BEVERLY HILLS / LONDON

For information address:

SAGE PUBLICATIONS, INC.
275 South Beverly Drive
Beverly Hills, California 90212

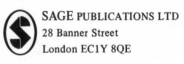

SAGE PUBLICATIONS LTD
28 Banner Street
London EC1Y 8QE

Printed in the United States of America

International Standard Book Number 0-8039-0741-9 (cloth)
International Standard Book Number 0-8039-0938-1 (paper)

Library of Congress Catalog Card No. 77-79869

FIRST PRINTING

CONTENTS

*To the Memory of
Clifford L. Kaufman (1940-1977),
talented political scientist,
committed student of urban problems,
and good friend*

Introduction:
Managing the Human Resources of
City Governments in the Post-Reform Era

CHARLES H. LEVINE

☐ THE IMPORTANCE OF MUNICIPAL EMPLOYEES to the effective functioning of city governments has been, until very recently, an often acknowledged but understudied dimension of urban governance. After years of relative neglect the transformation of municipal personnel systems into more complex and expensive arrangements that affect the quality and quantity of service delivery, the fiscal health of cities, and the viability of elected and appointed officials have reawakened scholarly interest. This collection of essays reflects the heightened interest in studying municipal employment as human resource systems. The issues and forces discussed in the contributions outline the new convergence of political forces and administrative imperatives involved in the transformation of public personnel administration from a static technical activity to a dynamic political one. The essays taken as a whole suggest that simple, apolitical, and purely local-based models of city government personnel management are inappropriate for either understanding or designing the human resource systems of city governments. Collectively, they beg for a new synthesis that accounts for the new roles, demands, complexities, techniques, and intergovernmental relationships that have come to dominate human resource management in the post-reform era.

The main body of scholarship that dominated public personnel administration throughout most of this century was embodied in the "merit system" which emerged as a reaction to abuses in government management at all levels of government from the "spoils system" and machine politics. The positive and negative functions and consequences of machine politics are well known; public employment was used for patronage in exchange for votes, campaign contributions, and loyal obedience so that machines could centralize control over governmental decision making. The spoils system produced graft and corruption, but it also allowed lower social class groups to participate in politics and gain benefits from government and it provided a mechanism to overcome the political fragmentation caused by the weak powers granted mayors in most city charters.[1] Even though the balance between the negative and positive consequences of machine politics continues to be subjected to scholarly scrutiny, these efforts seem to be largely academic, at least for human resource management, as urban political machines and political patronage systems continue to wither away and national manpower programs mandate local government adoption of civil service systems based on merit principles.[2]

The municipal reformers sought "good government" by breaking the power of the machines to control electoral competition through their control over government decisions on licensing, franchising, contracts, services, and law enforcement in addition to employment.[3] The reformers proposed a variety of electoral and decision-making reforms to assure electoral competition, accountability, and the neutrality of city employees: the initiative, referendum, and recall; the direct primary and proportional representation; nonpartisan elections, the council-manager form, and at-large elections. Although the civil service system and its "merit concepts" had earlier origins in the Pendleton Act of 1883 (and the earliest municipalities and counties to adopt civil service systems— Albany, New York, and Cook County, Illinois—have been strongholds of the old machine politics), the central tenets of the merit system became enmeshed in the doctrine of civic reform (Nigro and Nigro, 1976).

The merit system embodies a set of provisions, institutions, and procedures aimed at developing and maintaining a professional cadre of civil servants who are neutral and protected from partisan political pressures (see Stahl, 1976). It covers recruitment, selection, promotion, compensation, discipline, training, and political prohibitions. In many places civil service reform produced exactly the consequences intended by its advocates; in others, it has been charged with fostering unresponsiveness, unnecessary paperwork, legalistic complexity, and with denying employees their constitutional rights as citizens (see Sayre, 1948; Newland, 1976). Critics have charged it has produced less rather than more effective government and have called for further reforms to strengthen executive authority over the bureaucracy and to increase the powers of line managers over their employees.[4]

From roughly the end of World War II to the early 1960s, the merit principles were supplemented with procedures and techniques taken from changes in federal personnel practices aimed at making personnel systems "more efficient, better related to management needs, and more concerned with the employee as a person" (Nigro and Nigro, 1976:10). Some of these changes merely represented the "catching up" of the public sector with the industrial relations and personnel management movements that had swept through the private sector years before; others stemmed from the pragmatism of government management during World War II where the necessity of fast action overwhelmed bureaucratic proceduralism. During that same period, municipal employment doubled in size and labor costs rose to become the single largest expense item of city government. In many communities, weak-mayor arrangements were strengthened and in some cities council-manager arrangements were replaced with strong-mayor systems in order to make the administration of city services more responsive to voters. During this period, the controversy over spoils versus merit gave way to the debate over neutrality versus responsiveness as the fulcrum of local public personnel politics (Newland, 1976).

The struggle for black political power in the 1960s tended to divert attention away from the quieter shift in political power in the postwar period to the bureaucracies of city government. Even though white-dominated bureaucracies like police departments and welfare agencies were often the focus of black protest, the underlying sources and extent of bureaucratic power were frequently obscured beneath the rhetoric and dynamics of racial conflict. It took the protests of white middle-class neighborhood groups in the late 1960s and early 1970s over issues such as the discontinuation of services, the location of highways and public housing, and more recently the closing of neighborhood schools to expose how the power of bureaucracies, rooted in professionalization and unionization, had supplanted both political machines and reform movements as the dominant forces in American urban politics.

What had occurred during the post-World War II period was noticeable only after race relations became "normalized" in the seventies. Quietly, we had entered a new era of city politics—a post-reform era—in which public employees were the central actors.[5] In the new era, the principal antagonists were no longer "good government" reformers and machine politicians, urban renewal entrepreneurs and neighborhood groups, or even blacks and whites. Instead, the battles that have come to dominate city politics involve "program politics" in which bureaucrats struggle with one another for scarce resources; "contract politics," in which taxpayers and elected officials struggle with employee unions over collective bargaining settlements; and "implementation politics," in which local governments and federal agencies struggle over local compliance with federal laws and program guidelines. At the heart of each of these new political struggles lies the municipal public employee and his union; their power is the new constraint on local decision making.

The post-reform era has posed new challenges for city managers, mayors, and other city officials charged with the responsibility for maintaining and enhancing confidence in city government, financial solvency, quality services, and administrative accountability. What has evolved to meet these needs is a "third-generation system of knowledge, concepts, and skills" for municipal personnel management (Crouch, 1976:10). These challenges and the new techniques that have emerged to help cope with them are not mere "add-ons" to the corpus of merit system procedures and personnel management practices inherited from previous "generations."[6] Instead, we have entered a new era with new sets of issues, participants, and power relationships in which human resource management has evolved into a multifunctional and operationally fragmented arrangement in most communities. The unitary conception of public personnel administration developed during the reform era has collapsed under the complexity of differentiated role demands on city employees and their demands for power and resources from city governments. The result is a conceptual fragmentation that reflects the operational fragmentation of the human resource system itself.

The essays in this volume are organized into five sections which reflect the major developments of the post-reform era; that is, changing roles, affirmative action, collective bargaining, human resource development techniques, and national public employment programming. The contributions follow a variety of formats and theoretical orientations. Included are review essays, surveys, case studies, conceptual frameworks, aggregate analyses, and speculative arguments. This variety simply mirrors the complexity and variety of municipal public employment, the emergence of new issues in the field, and the reinterpretation of old issues as the orthodoxy of the reform movement continues to erode away under the combined pressures of political reality, management necessity, and scholarly curiosity (see Thompson, 1975). It is the hope and intention of this volume to add further impetus to the growing scholarly interest in the structure and management of the public sector, particularly the human resources of city governments. It is the long range hope that this scholarly interest might eventually contribute to improving methods and practices.

What emerges from these essays is a clear consensus that no theory, descriptive or prescriptive, of municipal human resource management will encompass the appropriate mix of variables unless it accounts for the overlay of personnel procedures and functions (control, staff, or service) that have evolved over the years and the new pressures and stresses that have emerged during the post-reform period. The present fragmentation in approaching the study of the human resources of city governments may be troubling to some who might prefer a more neatly ordered paradigm (see Rosenbloom, 1973; Milward, 1977); yet when compared to the narrow ideological and technical orientation that dominated public personnel administration during the reform era, the variety of approaches taken in this *Urban Affairs Annual Review* should be a refreshing change.

THE CONTRIBUTIONS

Change in municipal employment has been felt nowhere more strongly than in the roles of public employees and in the definition of their appropriate and legitimate rights and obligations. In the post-reform era, technological and social advances have combined with the democratization of government employment through judicial decrees and union pressure to make public employee roles more differentiated, multifaceted, and more subject to redefinition and renegotiation. This theme is prominent in David T. Stanley's contribution, "The Ambiguous Role of the Urban Public Employee" in which he argues that municipal employees' roles are "necessarily and permanently ambiguous" because they have come to occupy four main roles in the human resource systems of city governments: (1) as public servants who provide service or protection; (2) as clients or beneficiaries who are provided employment through their public service job; (3) as politicians who are in effect party representatives in city government; and (4) as managerial participants, who through their union, may not only influence compensation, but also may effect the way public resources are divided and public services are rendered. Stanley concludes that as co-managers public employees have developed a new stake in the viability of city government far beyond that of the traditional public employment relationship and, as a result, the financial distress of some city governments becomes their distress as well.

At the heart of the ambiguity over the appropriate role(s) for public employees lies the growth of professionalization which grew out of the more fundamental 20th century development in Western societies toward work rationalization, specialization, differentiation, and the rewarding of technological expertise. In American cities, professionalism took on added impetus as it became a central tenet of the municipal reform movement. Harlan Hahn, in his contribution, "Alternative Paths to Professionalization: The Development of Municipal Personnel," carefully examines the link between professionalism and the reform model of personnel and its roots as a reaction to the excesses of the boss-dominated political machines of the 19th and early 20th centuries. Hahn argues that by dismantling patronage the close service-oriented "direct-aid" link between municipal employees and their clients—where votes were exchanged for service—was disrupted. Without electoral accountability there became a tendency among public employees to use the image of neutral competence and detachment as insulation from public scrutiny. To overcome this problem, Hahn suggests an alternative conception of "professionalism" in which public employees are encouraged to become client-centered by increasing their awareness of the "life-and-death" implications of their actions for their clients. To illustrate, Hahn analyzes the historical evolution of the role of law enforcement agencies and suggests that they might recapture some of their earlier valuable pre-reform roles as "community service agencies" by combining the best of both the "reform" and "machine" traditions.

While the professionalization and unionization of city employees were changing their roles gradually through adjustments inside city governments, equally important changes were emanating from outside city government in the form of judicial decisions interpreting the constitutional status of public employees. The impact of these decisions on municipal employment are discussed in David H. Rosenbloom's essay, "The Public Employee in Court: Implications for Urban Government." By historically tracing the judicial reinterpretation of the public employee relationship, Rosenbloom explains how the constitutional status of public employees has been transformed since the 1950s. He devotes special attention to some recent and emerging court cases involving such thorny constitutional issues as freedom of expression and association, procedural due process, and equal protection of the law.

Rosenbloom concludes that of all the issues, "deconstitutionalization"— deciding that the federal courts are not an appropriate forum to review routine personnel decisions—is an idea whose time has come because it "may be seen as redressing the balance of power between public employer and employee by depriving the latter of the added advantage of being able to obtain in court what cannot be gotten at the bargaining table."

Some of the most fundamental changes in public employment produced by the courts have involved decisions guaranteeing equal employment opportunity and encouraging affirmative action programs for the hiring and promotion of minorities and women. But affirmative action has collided head-on with the older reform traditions of professionalism and the "merit system," newer technological imperatives for advanced education and training, and union demands that seniority be the major determinant of promotion and retention. These contradictory pressures have yet to be resolved by local governments and represent a continuing source of strain in human resource management that from time to time spills over into open political conflict.

These issues are treated in depth in the contributions by Frank J. Thompson, "Institutional Barriers to Equity in Local Government Employment: Implications for Federal Policy Derived from Attitudinal Data," and Robert W. Backoff and Hal G. Rainey, "Technology, Professionalization, Affirmative Action, and the Merit System." Together they point to the difficulties faced by affirmative action programs confronting opponents with legitimate but contradictory counterclaims.

The resolution of these counterclaims is examined in Thompson's interpretation of responses from a survey of urban personnel specialists. His findings give credence to critics of government bureaucracy who argue that professionalization is the enemy of affirmative action. In his survey, Thompson found widespread opposition to such affirmative action instruments as hiring targets and timetables and only a very slight positive relationship between the professionalization of the personnel specialist and attitudes favorable to disadvantaged job seekers. Thompson concludes his analysis by exploring federal policy alternatives for reducing local institutional barriers to hiring the disadvantaged. He argues that despite the complexities involved, a program that would link "payment to performance" would be necessary to provide the incentive for local governments to voluntarily reduce barriers.

Backoff and Rainey are concerned with the interrelationships of several factors which have major implications for urban personnel management, but which are seldom considered jointly: affirmative action, merit systems, professionalization, and technological advancements. They present a model of the urban personnel management system and set forth the main norms or pressures exerted by the major factors in the system. They then posit the potential impacts of these factors—and particularly of conflicts among them—on various functions, decisions, and relationships within the system. They conclude by considering the possible adaptive responses of city governments to these contradictory pressures.

Of all the transformations in the post-reform era, the widespread and rapid unionization of municipal employees has been the most perplexing and troublesome for city officials. The effects of increased unionization on urban management, the problems it has brought, and the opportunities it offers are explored in the contribution by Felix A. Nigro and Lloyd G. Nigro, "Public Sector Unionism." Throughout their essay, Nigro and Nigro point out important distinctions between collective bargaining in the public and private sectors and focus on how collective bargaining has changed the relationship between public employees and city officials. They see unionization and bilateral bargaining as a viable substitute to the "old style and demonstrably ineffective unilateral paternalism" of public labor relations before unionization. After reviewing evidence concerning the alleged inordinate power of municipal unions in city politics, they conclude that the evidence does not support the argument that collective bargaining inherently causes the financial distress of governments. Also, they argue that while some unions have abused their new power, unionization and the participation of employees in decision making through bilateral bargaining have produced important and underappreciated side benefits in terms of improved employee morale and valuable inputs into policymaking.

A somewhat different but complementary approach to examine municipal collective bargaining is taken in the essay, "Collective Bargaining in Municipal Governments: An Interorganizational Perspective," by Charles H. Levine, James L. Perry, and John J. DeMarco. Their basic purpose is to build a framework rooted in interorganizational theory for organizing the findings and important questions of the literature on municipal collective bargaining. By adopting an interorganizational perspective, their framework opens the analysis of collective bargaining to include actors and dimensions besides the two negotiating organizations and the dynamics of their interactions. The framework is therefore able to account for the behavior and role of third parties and state governments in directly and indirectly manipulating constraints and agendas that shape settlement outcomes.

The new complexity of the post-reform era brought on by new political and economic demands on city government has spurred the development of new management techniques to help managers and city officials cope with these pressures. These techniques augmented the existing procedural apparatus developed during the reform era to administer the merit system and the supervisory techniques that were developed and adopted by city governments after World War II. Three clusters of these new techniques are discussed in this collection: applied behavioral science, manpower forecasting and planning, and productivity measurement.

In their essay, "Applying Behavioral Science to Urban Management," William B. Eddy and Thomas P. Murphy explore the uses of behavioral science to change and improve organizational and individual effectiveness. They argue that because of changes in the context of urban management (brought on by instability, complexity, the openness of public organizations to external pressure, the new responsibility of city government for "soft" problems involving social and psychological relations, and increased employee participation in decision making) new approaches to urban management rooted in applied behavioral science are needed since the traditional view of the administrator's role does not provide adequate tools for handling these real world problems.

To illustrate their main point, Eddy and Murphy present three case examples where applied behavioral science techniques were valuable aids to solving difficult urban management problems: organization development in Kansas City, Missouri, change agentry in Dayton, Ohio, and management by objectives in Chesapeake, Virginia. While these examples and overall developments in urban governance seem to support a favorable forecast for the future utilization of applied behavioral science techniques in urban management, Eddy and Murphy conclude their essay by observing that the use and effectiveness of these approaches are nevertheless constrained by the political realities of situations and the ability and willingness of urban managers to have a clearer understanding of the techniques and their usefulness in solving problems.

Another set of urban human resource management techniques is discussed in Eugene B. McGregor's essay, "Issues and Problems in Applying Quantitative Analysis to Public Sector Human Resource Management." McGregor concentrates on three "intellectual problems" related to human resource management: forecasting, "depicting over time likely configurations of supply and demand for labor"; planning, "formulating and selecting alternative strategies to meet expected human resource needs"; and programming, "translating into reality the selected strategy."

Most of McGregor's essay is devoted to the difficult conceptual and quantitative problems of building a model to forecast the demand and supply of labor in a public agency under the assumption that without adequate forecasting, planning, and programming are severely constrained. Even though his model appears an attractive and scientific way to determine human resource capability, McGregor cautions that because the human resource "needs" of public agencies are determined by political processes and values which permit some functions to grow and others to decline irrespective of "objective analysis," one must be careful in the use of forecasting techniques for long-run projections where assumptions based on short-term incremental budget adjustments will not hold.

The other side of the human resource equation, the forecasting of supplies, involves determining the availability of appropriately trained and experienced people which McGregor contends depends ultimately on models of the career mobility of personnel. To illustrate the overall utility of this conceptualization, McGregor presents two situations, one involving expanded demand for services in a city with a growing economic base and another involving a contracting organization faced with declining ceilings, and applies probability and Markov chain techniques to each. McGregor then concludes by demonstrating how forecasting techniques can be extended to planning and programming activities, particularly those involving changes in urban service delivery patterns and priorities.

The adoption of most management techniques is predicated on the assumption that they will somehow contribute to the attainment of higher levels of employee productivity and organizational output. To determine whether or not a technique has its intended effects, some measures of productivity and output are therefore necessary. But in the public sector the measurement of productivity and output is usually difficult and in many cases it is impossible, and this reality makes claims about the effectiveness of various management techniques based on improved productivity and output results suspect. This paradox—between the need for greater productivity and the impossibility of its measurement—is the central theme of Roy W. Bahl and Jesse Burkhead's contribution, "Productivity and the Measurement of Public Output."

Bahl and Burkhead argue that expenditure control, "accompanied by rhetoric on the issues of efficiency and productivity," has emerged from the fiscal crisis of New York City as the major policy variable in urban management. But they assert that much of the interest in productivity, based on a simplistic relationship between inputs and outputs, is misguided and "conceptually flawed" and "such measurement may be wholly misleading in terms of answering the question of whether the citizen is getting more or less for a tax dollar."

Bahl and Burkhead especially criticize the tendency to "finesse the output measurement problem" by a variety of assumptions in models or through the use of proxy measures which have limited utility for making management decisions for expenditure control. Because of the conceptual and measurement difficulties confronted in productivity and public output research, Bahl and Burkhead recommend that instead of continuing on this research path, research focusing on public sector labor markets might be a more auspicious way to gain "an understanding of the way expenditures grow, whether they should be controlled and how they may be controlled."

Another striking change in urban government in the post-reform era has involved the emergence of links between federal policy and local government. Of particular interest is the new role of city governments as implementors of programs legislated, designed, funded, and "steered" in Washington. Public employment programs in particular demonstrate how this new link between federal and local governments has vastly complicated local human resource politics and ensnarled the implementation of national programs in webs of locally based political entanglements.

These matters are discussed in the contributions by Randall B. Ripley and Donald C. Baumer, "National Politics, Local Politics, and Public Service Employment" and Charles W. Washington, "Public Employment Policy-In-Use: The Atlanta Experience with CETA." Taken together, both essays point to the advantages and limits of using local units of government to implement national policy and demonstrate how programs designed in Washington become modified when "in-use" at the local level.

Ripley and Baumer begin their essay by framing public service employment within the broader ideological debate between Democrats and Republicans over the proper management of the economy and unemployment in particular. They point out that the issue becomes further complicated because of the variety of public employment programming options and the different purposes they serve. To illustrate and analyze the links between national policy and local implementation, Ripley and Baumer focus on the Public Service Employment (PSE) provisions (Titles II and VI) of the Comprehensive Employment and Training Act (CETA) of 1973, as amended, because CETA implementation decisions are made at the local level while the national and regional offices of the Department of Labor assume a "steering role."

From their evidence Ripley and Baumer conclude that even though there are strong trends toward homogeneity of CETA programming across localities, considerable variability exists suggesting that CETA can be used as both a local political tool and a means to achieve national employment goals. But their data also suggests that local political pressures and limited job opportunities make the implementation of national policy aimed at enhancing the employability of the disadvantaged, women, and minorities in permanent unsubsidized employment difficult. They conclude by raising a number of open and important questions for further research including the critical political question: "Under what political conditions can both elected officials and professional staff members resist the seemingly strong pressures to guard city or county payrolls against shrinkage by employing mostly white males in the better jobs and females and minority persons and economically disadvantaged persons in seemingly "dead end" jobs?"

Washington's study of CETA implementation in Atlanta provides an in-depth complementary analysis to the Ripley and Baumer study. To understand how

CETA as a national espoused public policy (NEPP) is converted during implementation into a public policy-in-use (PPIU), Washington develops a framework
that allows for different sets of goals and objectives to emerge and be interjected
during different stages in the formulation and implementation process through
rule modification (PPRM), discretion modification (PPDM), and local initiative
(locally espoused public policy; LEPP). Using this framework, Washington goes
into great depth to explicate how the conflicting objectives of CETA (to
improve the fiscal stability of local budgets, to improve the quality of skills
possessed by the disadvantaged through work experience, and to employ the
unemployed and underemployed through public service employment and emergency jobs) "require numerous administrative complexities to assure that the
antirecessionary effects of public service employment are realized." Like Ripley
and Baumer, Washington concludes that local jurisdictions do possess considerable flexibility in how they may use PSE and emergency jobs funds, but the
Atlanta case also "illustrates how a jurisdiction's administrative procedures and
mechanisms can be designed to assure that jobs are additive and integrated into
the governmental structure."

SOME CONCLUDING THOUGHTS

The conceptual disunity and ambiguity created by the rapid change and
complexification in the human resource systems of city governments make
stating anything theoretically significant and general about the subject difficult.
Despite the reluctance of students of the field to repeat the mistakes of the past
by forcing premature closure on its theoretical development, there is agreement
that human resource management needs to be better structured conceptually—if
only tentatively and roughly to account for the emergence of the overlaps and
contradictions which make its study and practice so perplexing (Newland,
1976:532).

One cut at the conceptual confusion focuses on four distinct but interrelated
arenas of human resource management that have emerged in the 1970s to meet
the responsibilities confronting city governments.[7] This perspective recognizes
that the operational fragmentation of human resource management has arisen
because of the need to incorporate new programs and practices in city government; for example, in the past decade most large and medium size cities
organized special units to accommodate new functions like collective bargaining,
affirmative action, and public employment programs. Because of the layering of
responsibilities and operational fragmentation, human resource management has
become open to outside pressure at multiple points facilitating the development
of multilateral conflict between line managers, personnel specialists, interest
groups, and union representatives with an almost infinite array of coalitional

alignments possible. In this scheme, the organization of human resources is seen to encompass functional responsibilities to: (1) assure employee rights and enforce rules of conduct and procedure; (2) develop, marshall, and maintain human resources to carry out public programs; (3) adapt organizational size and skill levels and reallocate scarce manpower as program priorities change; and (4) participate in national manpower programs to foster the economic and social development of minorities and combat unemployment generally.

Each of these four human resource management responsibilities involve distinct but sometimes overlapping arenas of politics.[8] The *juridical arena* includes actors who legally structure the public employment relationship and enforce civil service rules. In municipal collective bargaining, for example, this arena includes the actors who structure state laws and the courts who interpret and enforce them. The legacy of the municipal reform movement left many cities with elaborate administrative machinery and detailed codes for applying and enforcing the "merit system." These rules and procedures tend to tightly constrain and formalize the political bargaining and maneuvering in this arena. The *human resource management arena* involves employee-management relations and all the practices involved in inducing contributions from employees including job design, wages, fringe benefits, working conditions, job security, career development, and training. In an era dominated by collective bargaining, each of these elements involves detailed elaboration and complex negotiation as issues such as seniority, retirement options, and parity confound issues which before collective bargaining were treated as subject to management discretion alone. The *program priority arena* involves the manpower implications of changes in the priorities or funding levels of city government agencies. Changes in policy create new agencies, force reorganizations, and sometimes require retrenchments and reductions-in-force. Before any shift in manpower complements can occur, however, city officials must confront the power of entrenched bureaucrats most of whom consider themselves professionals with limited flexibility to undergo retraining or to assume new jobs. Finally, the *national manpower policy arena* includes those policymaking and policy implementing actors concerned with the character of the national manpower pool, unemployment levels, affirmative action, and their relationship to national and local needs. Most of the political maneuvering in this arena takes place in Washington, beyond the reach of local political forces. Nevertheless, local political interests intrude in this arena through the mechanism of cooperative lobbying groups and through the use of the discretion granted local authorities in the implementation of national programs.

With at least four major human resource subsystems to contend with, the management of a city government's human resource capacity is bound to be difficult. Unlike a neatly arranged hierarchical structure in which both top management and first line employees understand and agree on decision premises, the human resource system is complicated by the presence of differentiated and diffused decision points. Because of this fragmentation, human resource management requires coping with multiple demands in a variety of decision-making arenas. In this context, decision making entails partisan mutual adjustment and disjointed incrementalism (see Lindblom, 1965).

Over the past two decades, each of these four subsystems or arenas have become more elaborated and interact with elements of one another more frequently. There has been a temptation to view the whole human resource system as hopelessly complex and discount any claims that a method can be developed to integrate its management as "wishful thinking." Pessimists have even predicted the long-run paralysis of some city governments because of their inability to prevent the system from breaking apart under the pressure of conflicting demands and rapid change. In fact, nothing could be further from the reality of the situation. Rather than fragmenting under environmental pressure, the interdependence of the subsystems is increasing the intertwining of the whole system leading to a more unitary but more complicated "organized social complexity."[9]

For the time being this new arrangement seems to be defying conceptualization, so a "rough cut" proxy like the four arena structure proposed here has utility. The new convergence of issues and actors demands understanding and this can be enhanced by keeping in mind the basic responsibilities of human resource management in city governments. By using the four arenas as starting points we can account for the changing roles, techniques, demands and opportunities that have emerged to confound human resource management. This compromise will have to suffice while we wait for the development of a fully operational theory of city management as it has emerged in the post-reform era.

NOTES

1. The literature on the positive and negative consequences of the political machine is enormous. Four of the most valuable contributions are Riordon (1963), Merton (1945), Lowi (1967), and Wolfinger (1972).

2. See for example, Scott (1969) and Shefter (1976).

3. For a classic statement of how machines controlled government decisions over an incredibly wide range of issues, see Steffens (1931).

4. For a discussion of the Nixon Administration's plan to overcome the recalcitrance of parts of the bureaucracy, see Nathan (1975). For a model for accomplishing it, see the "Federal Political Personnel Manual. The 'Malek Manual'," (1976).

5. I credit Clarence Stone and Robert Whelan for first labeling the present period of urban politics as "the post-reform era."

6. While the idea that these developments are "add-ons" comes from Newland (1976), I interpret their meaning and significance differently. See also Shafritz (1975).

7. This approach has been developed at greater length in Levine and Nigro (1975).

8. The logic that political configurations follow from policy problems and the use of the term "arena" comes from Lowi (1969).

9. See LaPorte (1975) for a fuller explication of the concept of "organized social complexity" and the challenge it presents for policymaking and public management.

REFERENCES

CROUCH, W.W. (1976). "Local government personnel administration: The setting." In W.W. Crouch (ed.), Local Government Personnel Administration. Washington, D.C.: International City Management Association.

Federal Political Personnel Manual. The 'Malek Manual' (1976). Bureaucrat, 4(January):429-507.

LaPORTE, T.R. (1975). Organized social complexity. Princeton, N.J.: Princeton University Press.

LEVINE, C.H., and NIGRO, L.G. (1975). "The public personnel system: Can juridical administration and manpower management coexist?" Public Administration Review, 35(January/February):98-106.

LINDBLOM, C.E. (1965). The intelligence of democracy. New York: Free Press.

LOWI, T.J. (1967). "Machine politics—old and new." Public Interest, 4(fall):84-92.

——— (1969). The end of liberalism. New York: W.W. Norton.

MERTON, R.K. (1945). "Some functions of the political machine." Pp. 72-82 of "Social theory and social structure." American Journal of Sociology.

MILWARD, H.B. (1977). "Politics, personnel and public policy." Public Administration Review, (forthcoming).

NATHAN, R.P. (1975). The plot that failed, Nixon and the administrative presidency. New York: John Wiley.

NEWLAND, C.A. (1976). "Public personnel administration: Legalistic reforms vs. effectiveness, efficiency, and economy." Public Administration Review, 5(September/October):529-537.

NIGRO, F.A., and NIGRO, L.G. (1976). The new public personnel administration. Itasca, Ill.: Peacock.

RIORDON, W.L. (1963). Plunkitt of Tammany Hall. New York: Dutton.

ROSENBLOOM, D.H. (1973). "Public personnel administration and politics: Toward a new public personnel administration." Midwest Review of Public Administration, 7(April): 98-110.

SAYRE, W.S. (1948). "The triumph of techniques over purpose." Public Administration Review, 8(spring):134-137.

SCOTT, J.C. (1969). "Corruption, machine politics and political change " American Political Science Review, 63(December):1142-1159.

SHAFRITZ, J.M. (1975). Public personnel management. New York: Praeger.

SHEFTER, M. (1976). "The emergence of the political machine: An alternative view." In W. Hawley et al., Theoretical perspectives on urban politics. Englewood Cliffs, N.J.: Prentice-Hall.

STAHL, O.G. (1976). Public personnel administration (7th ed.). New York: Harper & Row.

STEFFENS, L. (1931). Autobiography. New York: Harcourt, Brace and World.

THOMPSON, F.J. (1975). Personnel policy in the city. Berkeley: University of California Press.

WOLFINGER, R.E. (1972). "Why political machines have not withered away and other revisionist thoughts." Journal of Politics, (May):365-398.

2

The Ambiguous Role of the
Urban Public Employee

DAVID T. STANLEY

□ IS JOHN JONES, CITY EMPLOYEE, primarily a public servant, a political tactician, a beneficiary, or a comanager? He may seem to be all four, or even more, both to himself and to others.

Identity crises have been popular in recent years, but urban public employees do not have them, at least not real crises in a make-or-break, live-or-die sense. They do have an identity *problem*, a large and complex problem. Their status, their roles, are subject to various interpretations by themselves, their bosses, their critics. This ambiguity is confusing, but is it serious? It may be argued that public employees who have a muddled view of who they are and what they are doing (and whose superiors also have a muddled view) may perform less effectively than they should. They may waste time and energy debating objectives, rights, and processes or doing things for the wrong reasons. On the other hand, they may not be muddled at all but perfectly clear about their purposes and duties but may play different roles to get more advantage or satisfaction out of a situation.

This paper presents several different ways of looking at urban employees. The analysis shows how these various perceptions are evident in the everyday administration of cities. Much of the ambiguity, as we shall see, is unavoidable, but there are ways of reducing its undesirable effects.

In order to bring out these varied perspectives we must make several assumptions about the urban governments in which the employees work: (1) their personnel practices are based partly on patronage, partly on civil service; (2) they recognize unions and use collective bargaining to determine pay, benefits, and working conditions; (3) they participate in federally assisted manpower programs; (4) they are passably managed, as local governments go; and (5) they are in increasing fiscal difficulty.

MULTIPLE EMPLOYEES, MULTIPLE BOSSES

Any categorization of human activity is an oversimplification, so we shall start by listing only four main roles played by urban employees. Call them conceptual models if you want to be more academic about this.

1. *The Public Servant.* The custodian who sweeps or mops the corridors of the city hall is rendering a public service. The citizens, acting through their mayor and the director of the building management department, hired him to sweep and mop, to do this reasonably well at minimum cost, and to cause no trouble. This perception is based on the idea that the purpose of the government is to provide service or protection.

2. *The Client or Beneficiary.* The custodian may owe his job to somebody. He may be a friend of the father of the councilman from his ward—a political patronage employee. Alternatively, he may be a beneficiary of a public man-power program intended to help members of underemployed groups. His presence in either case reflects the idea that one purpose of government is to provide employment.

3. *The Politician.* The employee may be so active in politics that he is in effect a party representative in the city government. He may collect political contributions from other employees, urge them to vote, assist citizens aggrieved by some city action, or intervene with the city in favor of some businessmen. (It may take someone above the mop-jockey level to do *all* of these things.)

4. *The Managerial Participant.* The employee may have another role with political implications. He may influence the way public resources are divided and the ways public services are rendered. He does this as a member or (to make the example more emphatic) an officer of a union or association. This model emphasizes political pluralism as well as labor management relations. Employee organizations compete with other citizen pressure groups to get advantages and benefits from government activities. Their officers are just as much politicians serving a constituency as are councilmen who are responsive to certain commercial interests. Like the council member, the union officer has a share in management. He is on some occasions almost a comanager because many a decision is negotiated rather than made unilaterally by management.

The man with the mop may play all of these parts on the cluttered stage of public employment. The complexities and interrelationships in the situation will become evident as we examine below the various ways in which local governments deal with employees.

WHO IS MANAGEMENT?

It would be a mistake to assume that this multiple employee is working for a unitary boss. On the contrary, he has many superiors and may on occasion have trouble finding out which one to deal with. He knows who assigns his work and approves his leave. However, he may have trouble finding out who can give him a promotion to a more responsible job because several layers of management may be involved. If he wants a cost-of-living pay raise, or if he wants overtime to be paid at double time instead of time-and-a-half, the mayor and council may have to decide. If he wants a more liberal pension plan, the state legislature may have to act.

Such multiple bossism has been a major cause of distress to employee union leaders. They have seen bargained pay agreements overturned by legislative bodies. They are uncertain whether operating division heads can commit the city government to working conditions and schedules. They may have grievance decisions reversed by courts.

This diffuseness of management encourages, indeed aggravates, the ambiguity of the city employee's position. In effect he works for one boss, files claims with another, negotiates politically with a third, and may face a fourth across the bargaining table.

FILLING JOBS

If every local government job were filled on the basis of merit, the identity problem would not be solved, but it would be easier. In fact, however, a large percentage of urban employees are chosen because of some sort of political or personal preferment or patronage. Statistics about what proportion of jobs are filled in this way are hard to come by, but an informal estimate by a federal government source says that some three out of five state and local government jobs are covered by merit systems. The rest are not all pure patronage. Our man with the mop is a patronage case if he got his job because he knew somebody, not because of his skill. He can be replaced if he loses the favor of his sponsor or if someone more deserving comes along.

Above the broom and mop level there can be similar motivations. A politically hired assistant to a department head is there primarily because of his connections. Of course, he must be literate, diligent, and accommodating enough to do the work satisfactorily, but his major loyalty is to his patron. He is more of a beneficiary than a public servant. He may also be a politician in the sense that he meets with constituents, helps in working out deals, and answers complaints, and he may be a liaison between party officials and the city government. It is a part of his job to see that the city bureaucracy is dependably responsive to partisan political leadership.

MERIT SYSTEMS

Civil service or merit systems are based on the public servant concept. The idea is that citizens compete for public employment on the basis of their qualifications for the work to be done. It is assumed that there is more than one applicant for every job and that the government will hire the best qualified. A presumably independent civil service commission is typically established to enforce the competence and purity of the appointment system. Once appointed, employees are promoted on the basis of merit, trained to do their jobs more efficiently, compensated according to the relative difficulty and responsibility of their work, and assured of tenure if they do their work satisfactorily and behave decently. The purity of the concept, however, has become tarnished by other considerations.

Take qualification requirements. Police officers, for example, must be tall, husky men so that they can cope with unruly citizens. This requirement, perfectly defensible in the past, has fallen before the onslaughts of the female and the short, particularly Mexican-American and Puerto Rican applicants. The main force of the drive is equal opportunity for all groups—a fair grab for government jobs by particular groups.

Fairness in job opportunity has also affected civil service testing. Written examinations tended to favor applicants with high verbal abilities and to handicap those who were less comfortable with books and pens. Courts have now required that employment tests be clearly related to the work done on the job (e.g., *Griggs* versus *Duke Power Co.*, a case applicable to the public sector).

In both of these examples the idea of making jobs available to more types of citizens has altered merit system policies. The public servants now employed as a result of these changes may be viewed in part as beneficiaries.

PUBLIC EMPLOYMENT PROGRAMS

Manpower programs for underprivileged socioeconomic groups also illustrate the supremacy of the beneficiary concept over that of the public servant. Public employment programs make governments "employers of last resort" for citizens who cannot find work elsewhere. Their lack of skills and, in some cases, lack of satisfactory work habits, necessitate extra efforts in supervision and training. Although they do such useful work as typing records or mowing public lawns, they are primarily clients being trained in the hope that they can move into the regular work force.

In some city governments that are in fiscal trouble—and more will be said about them later—such employees are becoming more public servants than beneficiaries. The fact that "summer youth" and "CETA" (Comprehensive Employment and Training Act) employees can be used in laboring and clerical jobs enables the cities to reduce their regular staffs for similar jobs and still get the work done. In Cleveland, for example, a large park system (41 properties, 2,600 acres) was being managed by only 34 regular employees because of the work done by about 150 CETA workers. This naturally leads to competitiveness between the two groups. In New York City a regular employees' union went to court to keep regulars from being laid off as long as there were "CETAs" doing similar work.

Another quirk in this already-odd situation was the metamorphosis of "regulars" into CETAs. This happened in Detroit and Cleveland, where CETA funds were used to reemploy laid-off policemen, trash collectors, and other employees. Such workers simply resumed their usual duties at federal expense. Nevertheless, they had really changed from public servants to beneficiaries. Arguments arose as to whether they were proper beneficiaries, considering the intent of the federal manpower laws. One way to look at it was negative: they had proved their employability for regular work and therefore should *not* be paid from funds intended to make citizens more employable. The other side of the argument was that the law was partly intended to relieve unemployment, of which they too were victims. U.S. Department of Labor officials have stiffened their opposition to the use of CETA to make beneficiaries out of regular workers, and Congress has formally recorded its disapproval (P.L. 94-369, sec. 201(5)).

PROMOTION

Most public employees want to be promoted, and role conflicts are again abundant in the process of deciding who will get the better-paid, more responsible jobs. The purest merit idea, choosing the employee who can do the job best, still requires a human and therefore fallible judgment. The judgment is limited and guided by civil service procedures based on the public servant concept. Selection is limited to people with certain qualifications; tests of skill or knowledge are required; employees within the department are given an "edge" over others. Such requirements do help in the choice of the "best" candidate. In part, however, they reflect the public employee's desire to hang on and get ahead. The employee unions have been strong supporters of limiting competition to insiders and of using selection methods that favor people familiar with the work. The unions have been particularly effective in some cities in making seniority (among employees with the basic qualifications) the key criterion for promotion. This standard is little used for promotions to supervisory positions but is more likely to be applied in advancing laborers to truck drivers, or truck drivers to bulldozer operators.

These policies that give advantage to the insiders and the senior employees add to our confusion over who the urban public employee is. If he is clearly chosen by a wise management from a wide field of candidates because he can do the job better than anyone else, he is being labeled a superior public servant. Yet if he is chosen because he is the senior person on a list of adequate candidates, this reflects a policy insisted upon by the union and agreed to by management. The employee promoted is still a public servant, and he can hardly be regarded as a beneficiary. He is rather a member of a group that participates in management in such a way that the group is benefited. If the employee is promoted as a result neither of civil service competition nor of a negotiated seniority clause but as a result of political or personal favoritism he is of course a beneficiary.

GRIEVANCES AND ADVERSE ACTIONS

Once placed in a job, whether by hiring or promotion, the employee may have some experiences that cause him to challenge decisions by management. Let us discuss them in two groups: (1) job-related grievances and (2) adverse actions, such as suspensions, demotions, or discharges.

Working life includes frequent grievances. The employee is moved to a work location he does not like, denied an opportunity to earn overtime, or assigned distasteful duties. Most grievances are resolved by discussion with the foreman and perhaps also with the next higher supervisor. We are concerned here with those that are still not settled by such discussions. An employee who has a political sponsor runs off to him, the sponsor intervenes, and the matter is settled—thus making the employee a client or beneficiary. If there is a civil service grievance procedure, the employee has a right to appeal to a quasi-independent appeals body and perhaps ultimately to the civil service commission itself. In this type of proceeding the employee is considered a possibly wronged public servant whose problem needs to be considered and adjusted on its merits.

Unions, however, are not inclined to accept civil service grievance procedures as the final word. To them the civil service appeals body may be independent of the management of the employee's department, but it is still part of the overall management of the government. The unions have insisted on, and in many cases gained, the right to appeal grievances to an arbitrator or arbitration board, in whose selection the union has an equal voice with management. This use of third-party arbitration is further evidence of the public employee in his role of managerial participant. Indeed, the union may participate at the initial stages too, with a shop steward or business agent representing the employee's interests in discussions with his supervisors.

We come second to the most painful form of grievances, those in which the employee is deprived of his job or is reduced in pay or rank. When the employee contests a suspension, demotion, or discharge he does so through the union-negotiated grievance procedure in some cities. In others he does so through a civil service appeal, either because there is no negotiated grievance procedure or because the civil service appeal is superior by law. In the latter instance the union finds itself fighting adverse actions through civil service channels and other (usually lesser) grievances through the negotiated procedure leading to arbitration. However, the fact that the negotiated procedures have begun to supplant civil service appeals suggests the rise of the managerial participant concept to the detriment of the public servant concept.

Nothing has been said so far about urban employees behaving as partisan politicians to remedy grievances. This occurs also, in two ways. First, the union may lobby with the mayor or city council to correct some grievance that affects a number of employees (as, for example, denial of overtime pay). Such a matter could be resolved administratively, but the union has chosen to handle it politically, possibly because this is a more effective method or because the union leadership wants to impress its members. Second, a job action or strike may be started. In New York City in the summer of 1975, sanitation men struck to protest announced layoffs, and school teachers struck over working conditions, particularly the amount of paid nonteaching time in a work week. These were political acts in that they were steps taken by citizen groups trying to change the way in which public resources were allocated. Although both unions are managerial coparticipants also, they took their grievances out of the managerial realm when they took to the streets. Their actions drew the attention of the media (first page and prime time) and of top city officials (the mayor himself visiting sanitation truck garages). The sanitation men achieved no change in their layoff situation, but the teachers did reach a compromise between their demands and the city's position.

It may surprise some readers to find strikes referred to as political acts rather than as economic or labor-management events. Strikes in the private sector are economic struggles between management and labor. In governments, however, strikes are one way that unions have of influencing public opinion, the legislative body, and the chief executive in making decisions on what public services are rendered and how much shall be spent on them. Strikes are illegal in most states, but this fact has in general not moved them from the category of familiar forms of political action to that of criminal deeds. Striking unions in such states are far more likely to be surrendered to, bought off, compromised with, or waited into defeat than prosecuted.

WORK MANAGEMENT

The discussion of grievances has already involved us in the matter of who manages the work. The entire area of what a city government should do, by what methods, at what expense, and with how many personnel traditionally belongs to management acting for citizens (Stanley, 1972:21-25, 89-111). Management's rights, as distinguished from matters subject to collective bargaining, may be set forth in laws or in union agreements, or even not stated explicitly but generally understood.

The boundary between management's right to administer and employees' right to influence management is fuzzy. Employees have always influenced management informally. Now, however, public employee unions have a formal share in the process—further testimony to the comanagerial role of the employee. There are several classes of examples.

1. Decisions on the subcontracting of city work to private firms (e.g., for trash pickup, building cleaning, or pavement patching) may be forbidden by union agreements. If not prohibited, such decisions may require prior consultation with employee groups.

2. Management and the union may agree on how much work an employee shall be required to do (e.g., check 60 welfare cases, supervise 80 probationers) or on how many employees shall be assigned to a given task or piece of equipment (four men on a hook-and-ladder, three on a garbage truck) or on prior consultation on installation of labor-saving equipment.

3. Assignments to overtime work, to shifts, to locations, or to desirable types of equipment or clientele may be decided not by management but by bidding by employees, with awards being made on the basis of seniority.

4. Provision of safety equipment, work clothing, or wash-up facilities may be a matter for collective negotiation rather than for unilateral decision by city officials.

In all such types of decisions (and more types still could be named) the urban employee is a public servant if he simply has to accept what management decides. He is a comanager (or, more accurately, his union is) if the matters are negotiated. He is a politician if he tries to influence management decisions by pressuring the city council or by going on strike.

PRODUCTIVITY PROGRAMS

In recent years, special productivity improvement programs have emphasized and further complicated the role of the public employee in work management. The rationale for such programs is obvious. As government costs rise, particularly personnel costs, it is incumbent upon government managers to deliver maximum service for every dollar and every work hour spent. There have been a variety of efforts in city governments to deploy fire departments more efficiently, to collect more trash with the same or smaller sanitation forces, to utilize police officers more productively, and to get school teachers to spend more time in actual instruction (New York, 1972, 1973; National Commission on Productivity and Work Quality, 1974; Committee for Economic Development, 1976).

These efforts would seem to add to the urban employee's identity problem. As public servant, he can hardly oppose more economy and efficiency in service. As a union member, however, he naturally opposes productivity measures that make him work harder or that will result in layoffs or demotions. He also demands extra pay for more productive work—for example, working with a higher capacity trash truck, processing more vouchers, or patrolling more blocks.

In short, he puts his interests as a recipient ahead of his responsibility as a coconserver of the taxpayers' dollar.

PAY AND FRINGES

Before the days of merit systems and classification and pay plans, urban employees were political beneficiaries. Each one was paid what the political leadership decided he was worth. Once they began to be regarded as public servants, they became public servants looking out for themselves. Civil service commissions or personnel offices made formal surveys of pay rates in the local area (or took informal looks) and made recommendations to the city council accordingly. Since the personnel staff and the city department heads wanted raises too, they tried to see that the local government's pay rates and fringe benefits did not lag behind those of other employers. With the arrival of collective bargaining, however, pay setting is "a whole new ballgame." The unions make demands every year or every two or three years, management responds, and agreement is ultimately reached. The result is that city employee pay rates have caught up, kept up, and in some cities have even gone ahead of rates paid by other employers (Stanley, 1972, chapter 4; Perloff, 1971). Logic suggests, and some studies show, that collective bargaining has lifted public employees' salaries higher than they would have gone if there had been no unions (Fogel and Lewin, 1974). Also unions have probably kept city managements from being as stingy as they wanted—or perhaps needed—to be.

THE URBAN EMPLOYEE IN HARD TIMES

Fiscal stringency in local governments has narrowed the dimensions of both management discretion and collective bargaining. Both in the late 1960s and in the mid- to late-1970s, city governments have struggled to balance their budgets as costs rose more rapidly than revenues (Stanley, 1972, chapter 6, esp. pp. 112-15; Stanley, 1976). The cities in the worst trouble were the large, old cities of the East and Midwest with static or declining economies, loss of industrial and commercial employment, diminishing populations, and no opportunity to expand. Among these, New York City was a uniquely severe case because of its comprehensive combination of services, including a university system, a hospital system, and a huge welfare burden, and a long history of unsound financial management. New York could not pay its bondholders without massive state and federal help.

The fiscal stress in such cities in 1975 resulted in unusual pressure on the employee unions. This is a natural result because, in any city, personal service costs (pay and benefits) run from 60 to 80% or even more of the operating budget. In St. Louis, unions agreed to forego any pay raise for fiscal year 1975-1976 in order to avoid layoffs. Unions in New York City also agreed to a no-raise policy, but there were heavy layoffs anyhow. There were payless days off for Detroit employees—again in lieu of layoffs. Buffalo employees were to get a 3.5% "bonus" pay increase at a time when public sector settlements in New York State were averaging 7%.

The ambiguity of the urban employees' status is well illustrated by these examples. As public servants, they had reason to feel exploited or victimized. They did their work but were given no pay increases at a time when other employed people were getting them and when the cost of living was going up about 9%. As beneficiaries, claimants, lobbyists competing for public goodies, they got nothing but retention of their jobs, and not always that. Most importantly, as comanagers, they were a party to the decision that hurt them. No matter which of these hats they were wearing, there were unhappy faces beneath.

The net effect of the pressure for retrenchment was to emphasize the urban employee's role as a public servant and to make him less effective in his other roles. Fiscal crises strengthened the hand of management as tough bargainer, job abolisher, and antifeatherbedder. The unions tried hard to respond. They made strong public statements urging pay increases and opposing job reductions. They struck for more money (unsuccessfully in San Francisco and with modest success in Cleveland), struck to protest an increase in duty requirements (New York City schoolteachers), and struck to protest layoffs (New York sanitation men). They objected to city work being done by CETA employees (Buffalo and New York). On the whole, however, the fiscal stringencies were overpowering, and the employees lost strength both politically and managerially.

POLITICAL ACTIVITY

More needs to be said about the urban employee as politician, particularly off the job. In a frankly political administration in which patronage is a way of life, employees are identified as adherents of the politicians in power and are expected to vote right, to contribute to campaigns, and, depending on individual interest and capacity, to work on party matters. Even in such administrations there will be some executives and many workers hired and retained because of their skills and knowledge needed in public service—physicians, fire fighters, and auditors, to name only a few.

Merit systems make a greater effort, however, to separate service from politics. Part of the separation is by law. The Hatch Act, applying not only to employees of the federal government but also to state and local employees paid from federal funds, forbids partisan political activity. As unions came on the scene, they first respected such restrictions, recognizing that adherence to one political group would jeopardize their future if that group went out of power. Of course they did not stay out of politics entirely. Their leaders passed the word informally to vote for candidates who would treat the unions generously. In some cities the public employee unions joined with private sector unions in political action groups (Stanley 1972:2).

As time went on, pressure grew, much of it from public service unions, to permit overt political activity. Essentially they emphasized a point already recognized by courts for a number of purposes, that a citizen should not lose the rights inherent in his citizenship simply because he becomes a government employee (Rosenbloom, 1971). To forbid him to run for office or to campaign for others makes him, in effect, a second-class political citizen. The Congress listened to this line of reasoning and voted liberalizing amendments to the Hatch Act, but the legislation was vetoed by President Gerald Ford, and an effort to override the veto was unsuccessful. Opponents of the liberalization argued that the entire civil service could be politicized, recalled patronage excesses of the Nixon Administration, and predicted widespread shakedowns of employees for political contributions.

The urban public employee may or may not want to go to political precinct meetings or help run campaigns. Some employees resent the prohibition, others welcome its shelter. Most employees, unless they actually owe their jobs to a politician, will resent pressure on them to contribute to political war chests or to do political chores. They certainly should not be either rewarded or penalized by their job superiors because of what they do at the party clubhouse.

The most serious conflict between the employee's role as politician and his role as public servant occurs in jobs in which the employee makes or influences government decisions that benefit certain groups. Much of politics is a matter of deals and favors, and it is clearly bad public business to make public decisions on the basis of political favoritism. Is a street the first repaired because it needs it most, because it carries the heaviest traffic, or because it goes near the party chairman's house? Is a city clinic placed according to citizens' needs, or is it built on land sold to the city by a heavy party contributor? Are purchases of concrete, fuel, vehicles, furniture, or office supplies decided on a price-quality basis or on a political basis? If such decisions are made by avowedly political officials, the mayor and council members, they are accountable to the voters for what they have done. If, however, such decisions are in effect made by presumably objective, service-minded employees, such employees should be kept from political involvement.

THE SELF-SELECTED ROLE

Suppose, instead of analyzing and speculating on the role of the urban public employee, we asked the employee, "What is your work?" or "What do you do?"

To begin with, nobody would say "I am an urban public employee." A few might say "I work for the City of St. Louis" (or Peoria or whatever). They would probably do so only if they held some vague, hard-to-understand job like assistant to the comptroller, rather than a more commonly used kind of occupational identification. The employee is much more likely to identify himself primarily by profession, trade, or skill and secondarily by employer. Thus: "I'm a police officer" or "I'm an attorney for the City" or "I'm a lab technician—work in Bellevue Hospital" or "I'm a secretary in City Hall." A short-term employee, whether professional, technical, or clerical, is more likely to identify with his skill than with his urban government.

As to the other roles that we have been discussing, the employee is likely to identify himself with them when he needs them or has to spend time on them. He thinks of himself as a Democrat if the campaign is on his mind or if he needs help from an influential politician of that party or if he is asked to perform a political chore or cough up a political dollar (or fifty). He thinks of himself as a unionist if he is an officer or steward, if the union is representing him in a grievance proceeding, if he is eagerly awaiting the outcome of pay negotiations, or if he is picketing city hall. Role identification, then, varies with the person and the situation.

CLUES FOR MANAGEMENT

A complex and fuzzy picture has been painted of the identity of the urban public employee. So what? Who cares, and how important is it? The main point of this is that urban officials and those who second-guess them (legislators, editorial writers, opposition politicians, and barroom political scientists) need to be aware of the complexities and to avoid simplistic assertions and recommendations. They will get into unnecessary difficulty if they speak and act as if all (or most) urban employees are either political hacks, aggressive unionists, wheedling beneficiaries, or dedicated and able public servants.

Management officials can and should start with the thought that employees are working human beings. The employees need (to simplify the findings of social psychologists) recognition; security, both financial and emotional; a sense of belonging; and a feeling of success and progress. It is hard to gratify such needs for all employees all the time, but continual awareness of the needs is essential.

After that, a more analytical approach is required. Personnel management is very different in highly political and nonpolitical situations, in unionized and nonunionized governments, and with "summer youth" employees compared to regular members of the work force. Employees respond to different motivations, and different styles of supervision if they perceive their roles differently, but in all cases they will put their self-interest first.

The role of urban public employees is necessarily and permanently ambiguous for the reasons that we have developed. A few general guidelines will minimize confusion and maximize good performance in this unclear situation:

(1) As far as possible, separate "political" jobs from "merit" jobs. They need different handling, and both categories will be better for it.
(2) Within the merit system, use simple, practical, job-relevant selection methods.
(3) In unionized situations, use a professionally competent labor relations staff member and bargain hard with the unions.
(4) Make special provision for supervision, training, and discipline of beneficiaries of public employment programs.

Nevertheless there are no simple ways to cope with the multiple human, political, and organizational intricacies. It will be helpful to adopt the attitude that, in the language of Harlan Cleveland, "Crises are normal, tensions are promising, and complexity is fun."

REFERENCES

Committee for Economic Development (1976). Improving productivity in state and local government. New York: Author.
EHRENBERG, R.G. (1973). "Municipal government structure, unionization, and the wages of fire fighters," Industrial and Labor Relations Review, 27(October):36-48.
FOGEL, W., and LEWIN, D. (1974). "Wage determination in the public sector." Industrial and Labor Relations Review, 27(April):410-431.
LIPSKY, D.B., and DROTNING, J.E. (1973). "The influence of collective bargaining on teachers' salaries in New York State." Industrial and Labor Relations Review, 27(October):18-35.
National Commission on Productivity and Work Quality (1974). So, Mr. Mayor, You Want to Improve Productivity. . . . Washington, D.C.: U.S. Government Printing Office.
New York, City of (1972). Productivity program (goals and timetables). New York: Author.
––– (1973). Second quarter progress reports. New York: Author.
PERLOFF, H.S. (1971). "Comparing municipal salaries with industry and federal pay." Monthly Labor Review, 94(October):46-50.
ROSENBLOOM, D.H. (1971). Federal service and the Constitution: The development of the public employee relationship. Ithaca, N.Y.: Cornell University Press.
STANLEY, D.T. (1972). Managing local government under union pressure. Washington, D.C.: Brookings Institution.
––– (1976). Running short, cutting down: Five cities in financial distress. Washington: Brookings Institution.

3

Alternative Paths to "Professionalization": The Development of Municipal Personnel

HARLAN HAHN

☐ PERHAPS THE MOST CRITICAL BATTLE in the history of American cities was the conflict at the beginning of the 20th century between the defenders of the old-style political "machine" and the advocates of civic "reform." By seeking to substitute administrative competence and impartiality for the corruption and favoritism that had flourished during the era of the political "boss," the "reformers" secured the adoption of many proposals that left a lasting imprint upon the structure of city governments throughout the country. The demise of the machine and the success of the reformers had a powerful impact not only upon the characteristics of public polities but also upon the nature of municipal personnel.

Although the "reform" movement encompassed many proposals—including nonpartisan elections, at-large representation, and the city manager plan—perhaps the basic difference between the machine and reform measures was reflected in the selection of municipal employees. While administrative offices traditionally had been filled by patronage dispensed by party bosses, the reformers attempted to undermine the foundations of the machine by introducing a civil service system through which personnel would be chosen on the basis of ability and qualifications. In addition, the reform movement sought to secure the insulation and autonomy of civil servants from partisan or public pressures. Although both the machine and reform programs appeared to place strong reliance upon the role of personnel in the accomplishment of policy goals, they

[37]

seemed to contain different implications concerning the characteristics of public employees and the relationship between bureaucrats and city residents. In fact, the triumph of the reform movement—and the collapse of the political machine—probably has had more effect upon the subsequent development of government personnel than any other influence.

Since the historic struggle between the machines and the reform effort, perhaps the principal factor affecting civil servants has been the increasing trend toward "professionalization." In large measure, this movement was launched by the reformers themselves. The concept of professionalism, as defined primarily by the criteria of training and qualifications, became a major tenet—and perhaps the most important legacy—of the reform era.

This chapter attempts to examine the implications of the machine and reform models of municipal administration concerning the development of professionalism among city personnel. Major attention is devoted to the human dimensions of urban administration rather than to the managerial aspects of this subject. Thus, emphasis is placed not only upon the relationship between supervisors and employees within the bureaucracy but also upon the relationship between civil servants and the clients whom they serve.

This essay, which is designed to be suggestive rather than definitive, is divided into several sections. In the first portion, the personnel implications of the machine and reform models are explored. The second section compares and contrasts an alternative definition of professionalism with the concept inherited from the reform movement. Third, an attempt is made to examine the impact of different notions of professionalism upon the activities of a municipal agency. The police are the focus of this example. And, finally, an effort is made to evaluate the ideological and programmatic implications of both definitions of professionalism as well as the machine and the reform models.

THE MACHINE AND THE REFORM MODELS

According to classic descriptions, the old-style urban machine was associated with some of the most reprehensible aspects of political power. In fact, even the value-laden words machine and boss have never been replaced by more neutral terms in the conventional vocabulary of social scientists. During the era of the machine, several political bosses amassed personal fortunes at public expense. There was widespread bribery, graft, intimidation, and election fraud. As a result, corruption became virtually synonymous with the machine in the minds of both researchers and municipal reformers.

Yet, like most institutions, the machine had desirable as well as undesirable characteristics. As Merton (1957:60-76) pointed out, the machine—and especially the system of political patronage—fulfilled many important "latent functions" such as regulating competition for jobs, providing economic gains for a localized constituency, and creating alternative channels for social mobility. Even more fundamentally, the machine demonstrated that government was not solely an institution for the resolution of abstract disputes between distant or remote interests in society; in fact, it "humanized" government. While the ostensible purpose of the machine was the acquisition of votes to sustain its power, the machine also revealed that government could be a positive force in the amelioration of personal needs.

The substantive assistance provided by the machines naturally yielded many disappointments. But it would be foolhardy either to ascribe the blame exclusively to the machine or to ignore the importance of the aid that was dispensed. During the era of the machine, resources were limited; and many ethnic groups— as well as other Americans—still believed that unlimited opportunities to share in the rewards of the industrial revolution awaited them in the private sector of the economy. A bucket of coal during the holiday season, or an occasional public job, hardly seemed like munificent government benefits; but they did appear to be a critical and sustaining factor in the lives of many urban residents during the 19th century.

The primary thrust of the services offered by the old-style machine seemed to be directed at the solution of immediate problems rather than to be large-scale attempts to improve the plight of deprived areas of the city. Citizens sought the aid of machine workers primarily when they encountered specific problems, and the responses they received often were based upon personal—or partisan—considerations rather than administrative regulations. But perhaps the most persuasive rationale for the machine was suggested by Martin Lomasny, a political boss in Boston, who responded to muckraker Lincoln Steffens' proposal to destroy the machine by saying:

I think that there's got to be in every ward somebody that any bloke can come to—no matter what he's done—and get help. Help, you understand; none of your law and your justice, but help. (Steffens, 1931:618)

The type of aid that Lomasny—and many other urban residents—were seeking was not based upon abstract considerations of equity and fairness; instead, it reflected a simple desire for assistance, regardless of procedural requirements or the technical provisions of the law. In many respects, the ability to secure help when it was needed may have been as important to rank-and-file citizens as extensive government programs.

Perhaps the most important component of the machine was the relationship that it implied between municipal personnel and their clients. Although patronage employees frequently acted on the basis of friendship or partisan loyalties, their activities seemed to focus on direct aid rather than an attempt to determine whether or not clients would be eligible for the categories of assistance provided by a specific agency. When an urban resident in the 19th century came to a party worker with a problem, the total resources of the machine could be mobilized to achieve a comprehensive solution.

The primary qualification possessed by municipal employees who owed their jobs to the party machine was their knowledge of the people whom they served. Perhaps the classic analysis of the talents of a party worker, and of the formation of a political organization, was provided by George Washington Plunkitt (Riordon, 1963:8-9):

> After goin' through the apprenticeship of the business while I was a boy by workin' around the district headquarters and hustlin' about the polls on election day, I set out when I cast my first vote to win fame and money in New York City politics. . . . Did I get up a book on municipal government and show it to the leader? I wasn't such a fool. What I did was to get some marketable goods before goin' to the leaders. . . . I had a cousin, a young man who didn't take any particular interest in politics. I went to him and said: "Tommy, I'm going to be a politician, and I want to get a followin'; can I count on you?" He said: "Sure, George." That's how I started in business. I get a marketable commodity—one vote. Then I went to the district leader and told him I could command two votes on election day, Tommy's and my own. He smiled on me and told me to go ahead.

Like many other workers for the party machine, Plunkitt launched his career primarily through his ability to relate to other citizens and to gain their confidence. The skills rewarded by the machine seemed to place greater emphasis upon sensitivity to the concern of others and the capacity to establish rapport than upon technical expertise or accomplishments.

Many municipal employees during the era of the machine lived in the neighborhoods that they served. They were familiar not only with the general conditions of the community but also with the circumstances of individual clients. They brought to their jobs a cluster of values, traditions, and experiences that made them almost indistinguishable from the remainder of the population. In many respects, therefore, the position of patronage employees in the 19th century probably was similar to the status of community organizers in the 20th century. They were required to possess an intimate knowledge of the needs and problems of the neighborhood, and they occupied a role which placed them on the same level as those who sought their assistance.

Since the job performance of patronage appointees was closely tied to the needs of the machine, those employees and their clients were able to meet on terms of relative equality. Workers for the machine obviously had something to offer in the form of services and assistance; but members of the public also possessed something which was vital to the maintenance of the machine, i.e., votes. Any failure by the employee to fulfill the needs of the client might have an effect upon the ability to gain needed electoral support. And the inability to deliver votes could have an impact upon the job security of the employee. When city residents presented a problem to machine workers in the 19th century, therefore, they approached those workers as *coequals.*

The need for electoral support also may have restrained municipal employees from abusing their authority. Under the machine, allies might have received more assistance than declared opponents, but it is doubtful that potential or uncommitted voters were totally ignored in the distribution of government benefits. Most political organizations that depend upon the electorate have not purposely sought to alienate important segments of the public. Contemporary bureaucracies have attempted to establish extensive rules and regulations to prevent administrators from acting in an arbitrary or capricious manner. But the constraints imposed by impartial guidelines may have less force than the threat of electoral displeasure as a method of promoting administrative accountability. In their dual roles as municipal employees and party workers, patronage appointees were dependent upon both the local government and the party for their jobs. To the extent that political parties were more sensitive to the will of the electorate than administrative bureaucracies, therefore, this arrangement provided an important means of holding the personnel of city government responsible to the general public.

The reciprocal dependence of the party machine and the electorate also may have had some important implications for the relationship between party leaders and city personnel. Perhaps the principal difficulty for the leader of the machine was the problem of making appointments. Party bosses probably were vulnerable to the axiom that "ten applicants for a patronage position usually produce one friend and nine enemies." The machine was not able to avail itself of the convenient rationale provided by credentials and qualifications in explaining its appointment decisions. Although political patronage might appear to discourage party competition, the inescapable necessity of disappointing many job-seekers may have created its own mechanism for swelling the ranks of the opposition party.

Reliance upon the good will of the party boss undoubtedly created tension and insecurity for patronage workers. Appointees had to fear both the ire of the boss and the threat that their party might lose in the next election. Yet, the scanty available evidence suggested that not only did party leaders demand unswerving allegiances from their workers, but they also displayed considerable loyalty to patronage personnel. As long as employees remained faithful to the machine, party bosses appeared willing to sustain and to reward those who had compiled extensive records of service.

The personnel decisions made by the leaders of the party machine seemed to reflect a Jacksonian belief that all persons were intrinsically capable of serving in government positions. In part, this view may have been based upon the fact that most of the services provided by local government during the 19th century were not highly complicated or technical. Cities at that time were primarily engaged in building the water, power, sewer, and street systems that were essential to future urban development;[1] and many of those jobs did not require extensive training or credentials. The intricate body of information that controls the administration of contemporary urban policies had not yet emerged, and most municipal functions could be performed as easily by unskilled immigrants as by Ivy League graduates.

In addition, many party leaders felt that patronage appointees should be able to prove their value through dedication, hard work, and the capacity to establish and maintain good relationships with the people whom they served. Those traits could be readily tested in the crucible of service to the party organization. Under the machine model, therefore, party loyalty and service were considered as important for city employees as personal abilities and qualifications.

The advent of the reform movement injected a new philosophy into the administration of local government. Since most government posts were filled by competitive examinations rather than by political appointments, the introduction of the civil service system seemed to impose more limits upon the opportunity to secure public employment than did the party machine. Civil service exams stressed personal or intellectual qualities, which seemed to be innate or inherent properties, rather than attributes such as party service, which were theoretically available to everyone. By replacing one principle, namely, party service, with another set of criteria, i.e., personal abilities, as a basis for the recruitment of city personnel, the reformers had a major impact upon the nature of municipal bureaucracies.

The emergence of the civil service system did not, of course, mean that all patronage was eliminated. As Stuart Chase (quoted in Chapin, 1935:40) once observed:

> The astute political leader . . . often plays upon the sensibilities of his more idealistic and dynamic colleagues, who really motivate the reform movement, by insisting that the persons appointed to subordinate executive positions should be "in sympathy with the administration's point of view." This is nothing more than a clever cloak to disguise the ordinary stock variety of partisan politics in the interest of patronage.

Many "reformed" governments simply elevated patronage to higher-level positions which were available to the educated and the influential rather than to workers questing for social mobility.

The adoption of reform measures, however, did produce an important change in the orientations of most city personnel. As increasing emphasis was placed upon abstract criteria in the administration of policy, less attention was devoted to the satisfaction or dissatisfaction of clients. As a result, familiarity with administrative rules overshadowed familiarity with the needs of clients as the principal concern of municipal employees.

By substituting generalized rules and regulations for personal needs as a standard in the allocation of public resources, the reform movement had a major effect upon the delivery of government services and benefits. Since city employees owed their primary allegiances to administrative agencies rather than political parties, they seemed preoccupied by the task of distributing specific types of assistance rather than by the effort to identify a range of government programs which might be adopted to solve the problems of clients. In addition, a host of regulations were adopted to protect the tenure and the perquisites of public jobholders. Civil servants were insulated from public pressures allegedly to shield them from the whim of partisan politicians and to enable them to perform their duties impartially. The growing distinction between administrative responsibilities and electoral politics appeared to make government functions less comprehensive and more impersonal.

Personnel policies established under the reform plan also seemed to require different talents of public employees. Many of the skills necessary for effective political organization no longer were relevant to the jobs defined by reformed governments. The increased importance attached to individual knowledge and intelligence appeared to diminish the value of personal empathy and rapport. While the civil service system may have promoted bureaucratic training and expertise, it did not stress capacities for personal interaction or provision of the "help" which had been a hallmark of the political machine.

Perhaps the most important consequence of municipal reform, however, was evident in the relationship between city personnel and their clients. Because civil servants were not dependent upon voters to maintain their jobs, urban residents possessed a serious disadvantage vis-à-vis administrators which the average citizen was unable to redress. Bureaucrats were not recruited on the basis of their social or political contacts with the community, and they adopted practices that enhanced their separation from the remainder of the population. By virtue of their qualifications and familiarity with government regulations, public employees no longer occupied an equal relationship with their clients. Civil servants assumed an increasingly dominant or superior position, while clients were forced to accept a subordinate or inferior status.

The innovations promoted by the reform movement altered the relationships not only between employees and the electorate but also between civil servants and public officials. Whereas the primary interest of elected representatives was the satisfaction of their constituents, the principal goal of city personnel was compliance with established bureaucratic procedures. Since familiarity with the technical provisions of administrative regulations was the special province of civil servants, politicians often found themselves no match for skilled administrators in the effort to supervise the operation of executive agencies.

At least a partial solution to this dilemma was provided by the demands of administrators for increased resources and highly qualified personnel. By endorsing the need to secure more competent and trained employees, elected officials were able to focus more attention on the administration rather than the drafting of public policy. Successful programs still could be hailed as testimony to the merits of policy design, but failures could be ascribed to the need for improved personnel. Thus, elected representatives were allowed to accept credit for some legislation, and they were provided with a ready excuse for other laws that did not fulfill their objectives. This position also had several important advantages for bureaucrats. Not only were they granted a defense against blame or criticism, but they were also given a justification for increasing their status and for expanding their influence.

THE DEVELOPMENT OF PROFESSIONALIZATION

Perhaps most importantly, the argument for increasing the quality of personnel was highly compatible with the professionalization of civil servants. Municipal employees were not content simply to serve as the impartial and passive administrators of government programs. Instead, they sought to leave their imprint upon the administration of public policy by developing a body of information about their occupational roles which could be imparted to others. Prevailing concepts of professional practices became the primary basis for the education of prospective employees, for the evaluation of personnel practices, and for recommendations to policy makers. This approach allowed civil servants to carve a distinctive niche for themselves in the administrative process, while preserving the image of neutral competence and detachment. As a result, professionalization appeared to enhance the insulation of administrative personnel from the general public.

The growing trend toward professionalization, which was spawned by the reform movement, placed increased importance upon the role of personnel in municipal administration. In many respects, the individual qualifications of bureaucrats became even more crucial in determining the impact of government programs than the substantive regulations or benefits that they administered. Although the old-style political machines did sustain their operations by relying heavily upon the social skills of patronage employees to satisfy the public, the doctrine of professionalism seemed to reflect a belief that urban problems could be solved primarily through the talent and expertise of government personnel.

The assumption that massive social changes could be accomplished through the intensive efforts of administrators rather than through large public expenditures perpetuated a myth that few elected officials were prepared to expose. By focusing upon the need for professional personnel rather than increased policy benefits, both politicians and taxpayers were relieved of a major burden. Public officials usually have been more willing to assume the obligations of salary increases for government employees than to support the costs of direct aid to the clients of local bureaucrats.

The movement toward professionalization has enabled bureaucrats to fulfill several important interests simultaneously. By making personnel the focal point of government programs, administrators are able to enhance their own position in the political process without necessarily assuming responsibility for policy failures. The inability to solve many pressing social problems is not widely interpreted as a fault of government employees; on the contrary, it is frequently cited as indicating the need for more talented and expert civil servants. Thus, the increasing crime rate is regarded as reflecting a need for better-trained policy officers; mounting welfare rolls imply a demand for highly skilled social workers; and deteriorating student performances on standardized examinations denote a need for better teachers.

The difficulty with this analysis, of course, is that it cannot be sustained indefinitely. The inevitable calculus of politics demands that government agencies ultimately be held responsible for the fulfillment of policy goals which they define for themselves. If a primary emphasis is placed upon personnel, elected officials eventually may demand to know how improvements in the quality of government employees have contributed to the attainment of important social and political objectives. Yet, few clear empirical relationships have been found between the training or qualifications of personnel and the solution of complex social problems. As a result, the failure to achieve major goals might cause professional civil servants to become crucial targets of political criticism and opposition.

Fundamentally, the concept of professionalization in government employment was shaped by the principles of the reform movement. Since the civil service system was based upon the assumption that personal capabilities were critical in the administration of policy, those traits also became the standards of conduct of public personnel. In most discussions of professionalism, the general definition of the term has focused on training and qualifications. While training usually has been demonstrated by educational credentials, qualifications have been exemplified by a host of characteristics including the passage of examinations, experience, and a reputation for effectiveness in the occupation.

In addition, most concepts of professionalism stress altruism and autonomy, or self-regulation. As Lieberman (1970:57-58) noted:

> The professional maintains his neutrality ... by insisting as a fundamental proposition that his motivation is altruistic. ... The professional asserts that the skills he has to offer are actually or potentially beneficial to all. ... To achieve a basis for applied altruism, the professional sets standards for himself, preaches professional ethics, limits the degree of competition with his fellows to which he is willing to submit, and disavows advertising. And because his is a specialized task which by definition no one else can fully comprehend, the professional demands autonomy in order to enforce and preserve his standards.

In part, the difficulty of developing a precise definition of professionalism[2] probably is related to the demand for "autonomy," or the right to establish standards of conduct and to discipline members of the profession without substantial interference. Many professions not only seek to be self-regulating but also are basically self-defining. And, to critics who express doubts about the right of an occupation to regulate activities which only it can define, professionals often reply that they are acting altruistically or in the interest of their clients. While some citizens might be willing to tolerate such tautological reasoning in certain private occupations, others might question its propriety in government positions which incorporate principles of accountability to the general public.

There is, however, an alternative definition of professionalism. A professional also might be defined as a person who occupies an actual or potential "life-and-death" relationship with a client. The most obvious examples of this relationship can be derived from the fields of medicine and law, in which responsibility for the health of a patient and for the personal fate or fortune of a client are placed exclusively in the hands of physicians and attorneys, respectively. But this definition also can be used as a relative measure to assess the professionalization of other occupations. A teacher who encourages or discourages the scholastic aspirations of a student, for example, can have a substantial—and unknown—effect upon the ability of the student to pursue a decent livelihood and to derive satisfaction and meaning from life. A social worker who determines the eligibility of applicants for welfare obviously has a major impact upon the daily existence of clients and their personal well-being. And the police officer who makes a decision to arrest—or not to arrest—a criminal suspect is in a crucial position to shape the remainder of that person's life. According to the alternative definition, professionals could be defined by the extent to which they are in a position to affect the fate or destiny of clients whom they serve. Although training and qualifications may be of some value in this endeavor, the definition based upon the nature of the relationship with a client seems to connote a different orientation for government personnel than the conventional understanding of professionalism.

The alternative definition of professionalism is essentially a *client-oriented* definition in which primary emphasis is placed on the problems and needs of clients. Employees who are aware—and imbued with the knowledge—that their decisions have potentially life-and-death consequences might be required to focus their attention upon clients and to concentrate upon efforts to improve their welfare. On the other hand, the definition of professionalism which emphasizes training and qualifications does not indicate a specific role for the client in the determination of professional behavior. Instead of being client-oriented, bureaucrats who adhere to conventional understandings of professionalism are apt to be oriented principally toward their administrative superiors and toward their peers in the profession. The interactions implied by the two concepts of professionalism, therefore, seem to flow in opposite directions. Whereas the alternative definition is designed to create a horizontal and equivalent relationship between clients and professionals, the traditional definition appears to suggest a vertical relationship between professionals and supervisors or others who establish the norms of the profession.

Although the alternative definition of professionalism does not imply that all of the demands of clients should be approved, decisions stemming from the two definitions conceivably might have different results. In many circumstances, bureaucrats are forced to make a choice between the needs or desires of clients and the requirements of administrative rules. Under the prevailing doctrines of professionalism, the performance of employees often seem to be measured less by the effect of their decisions upon clients than by the extent to which they comply with commonly accepted understandings of appropriate administrative behavior. This tendency produces an emphasis on procedures to the neglect of policy impact. By contrast, the alternative definition encourages administrators to consider not only the general provisions of government regulations but also the unique and personal problems of specific clients. Professionals would become "experts about their clients" as well as experts about the policies they administer. By shifting the emphasis in the definition of "professionalism," we might accomplish significant and far-reaching changes in the implementation of public policy.

The definition of professionalism which stresses the life-and-death implications of a relationship with a client obviously imposes complex and burdensome obligations. Administrators would be required to gain a thorough understanding not only of the details of laws and regulations but also of the individual characteristics and circumstances of their clients. In the latter capacity, bureaucrats would need to acquire knowledge of the personality, motivation, and aspirations of the people with whom they work in order to determine the outcome of a case. This is an awesome responsibility which is almost certain to yield failures and disappointments as well as successes and satisfaction. Unlike the traditional definition of professionalism, however, this orientation seems to suggest a clear association between personnel characteristics, the handling of individual cases, and the resolution of broad social problems. On the basis of their experience and familiarity with their clients, for example, teachers might be able to identify the rewards and punishments that would be most likely to promote the academic achievement of individual students; social workers might uncover the types of support that would enable welfare recipients to become productive and self-sustaining members of the community; and officers of the criminal justice system might determine whether or not the punitive enforcement of the law would be most effective in the rehabilitation of a criminal suspect. By focusing on the life-and-death nature of their relationships with clients, employees who embrace this concept of professionalism could develop a new approach to their jobs which might assist them in identifying incentives for individual clients that would contribute to the attainment of general policy objectives.

The definition of professionalism which emphasizes training and qualifications seems to be a natural outgrowth of the reform movement and the introduction of the civil service system. Whether or not the perpetuation of the model of the old-style machine would have spawned a different definition of professionalization is, of course, a matter of historical conjecture. Perhaps the continuation of the machine might have yielded a system of favoritism and corruption which would be antithetical to any of the contemporary understandings of professionalism. Conceivably, the machine model might have produced a different concept of professionalism, or it may have evolved a notion of professionalism substantially similar to the alternative definition suggested in this essay. What is clear, however, is that both the machine model and the alternative definition of professionalism proposed here seem to imply a need for different institutional arrangements and for the identification of different types of skills and abilities among public personnel. Perhaps this point can be illustrated by examining the evolution of one type of government service and by suggesting some alternative ways in which this profession might have developed. As a municipal agency that underwent the transition from machine to reformed governments and involves an occupation that is experiencing the process of professionalization, police departments seem to constitute a particularly appropriate example for this analysis.

ALTERNATIVE PROFESSIONALIZATION: THE CASE OF THE POLICE

Although police departments were not established in most major American cities until the middle of the 19th century, they rapidly fell under the influence of political parties. In fact, during the era when patronage controlled the recruitment of policemen as well as other government employees, the police seemed to become an extension of the local machine. In this capacity, they began to serve the values of the machine, which, in the words of Lomasny's eloquent justification, often entailed the provision of "help" rather than the application of "law" and "justice." Police officers not only acquired extensive latitude in the exercise of their duties but also performed responsibilities that exceeded a strict definition of law enforcement duties.

Although historical records are difficult to evaluate and easy to romanticize, available evidence suggests that police forces in the 19th century often maintained a cordial and cooperative relationship with the community. Police departments at that time were not embroiled in the bitter controversies with ethnic minorities which have characterized many subsequent eras, and numerous directives from police chiefs to their officers stressed the need to preserve a favorable image with the public. Many policemen lived in the neighborhoods that they patrolled, and they frequently used their knowledge of the neighborhood—and its residents—in deciding whether to invoke a harsh or a lenient enforcement of the law. Basing their decisions on their intimate familiarity with a suspect, policemen often chose either a journey home or a trip to jail as the most appropriate method of avoiding a repetition of the offense. Although police behavior in the 19th century was tainted by charges of corruption and favoritism, law enforcement officers frequently sought to prevent crime by utilizing their discretion in a manner which reflected a close relationship with individual offenders rather than the technical requirements of the law.

In addition, police departments assumed numerous responsibilities that seemed unrelated to the commission of crimes. During the era of party machines, policemen became involved in a broad range of matters such as family problems, the discipline of children, and personal couseling as well as the provision of economic, legal, medical, and other forms of assistance to the community. As one of the most conspicuous and accessible representatives of local government, police officers were the natural recipients of public demands upon the political process. Since policemen depended upon political patronage for their jobs, they performed many services which were essential to the survival of the machine.

With the emergence of civil service reforms, however, increasing emphasis was placed upon a narrow definition of police roles and upon the reduction of crime as the principal goal of law enforcement agencies. Police training and recruitment stressed skills that were considered essential for coping with crime, and many law enforcement officers felt that they were developing a standard of professionalized practices which would enable them to win the battle against crime. This trend was accentuated not only by technological innovations such as the automobile and two-way communications systems but also by a change in the attitudes and characteristics of police personnel. After some law enforcement agencies had waged a courageous effort to implement an almost unenforceable Prohibition law, the growth of morale and the ravages of the Great Depression combined to attract a new type of recruit to many police forces. In 1940, for example, more than half the recruits who entered the New York City Police Department were college graduates (Niederhoffer, 1967:16). Although this pattern was partially reversed by the "veteran's preference" provisions of civil service regulations after the Second World War, many public officials became convinced that improvements in the training and qualifications of police personnel offered the most effective means of combatting crime.

As increased record-keeping revealed a dramatic increase in the crime rate during the second half of the 20th century, public opinion surveys disclosed that a growing number of Americans were becoming alarmed by this development. Policy makers responded by enacting various measures, including the Omnibus Crime Control and State Strats Act of 1968, to upgrade the quality and the status of police personnel. Perhaps the major difficulty, however, was that no one could establish a clear relationship between the training or qualifications of police officers and their capacity to produce a decrease in the crime rate. Although a clear association exists between the speed with which police officers arrive at the scene of a crime and their ability to arrest a suspect, few other empirical connections have been discovered between any of the police procedures developed in the process of professionalization and the reduction of crime.

Contrary to the impression conveyed by detective stories and radio or television dramas, police officers seldom are able to "solve a crime" or even to identify a suspect as a result of their own independent effort. When a culprit is caught, this outcome usually occurs because someone told a police officer "who did it." Most arrests are based upon information supplied by the victim or by a witness, friend, relative, or acquaintance of the victim. Furthermore, as many police officers have testified, even the ability to arrest suspects may have little impact upon the problem of crime. Since many serious crimes are committed by recidivists or habitual offenders, neither the identification of suspects nor their subsequent conviction and punishment might contribute to a decrease in the rate of crime.

Most of the short-term "solutions" to crime, which were inspired by conventional definitions of professionalization, have not had a demonstrable effect upon the problem. Improved personnel, weapons, and procedures have not produced either a decrease in crime or an increase in the willingness of private citizens to assist law enforcement officers in the performance of their awesome responsibilities. In fact, many such practices have provoked charges that the police have abused their discretion. The growing professionalization of law enforcement personnel probably has intensified rather than diminished the reluctance of the public to cooperate with the police.

But these facts—discouraging as they appear—seem to suggest another approach to the solution of crime as a major social problem. Since the evidence indicates that arrests usually are made on the basis of information furnished by the public, increased emphasis might be placed upon methods of promoting public trust and confidence in law enforcement agencies. By adopting a long-range strategy based upon the alternative definition of professionalism, which reflects the life-and-death nature of a relationship with another person, police departments might be able to improve their public image, to mitigate a major source of criticism, and—possibly—to develop an approach that could have an impact upon the reduction of crime.

Ironically, as surveys revealed that citizens were becoming increasingly worried about crime, most of the calls made by the public to police departments have involved requests for assistance with other problems. Urban residents have continued to demand many of the services which trace their origins to police departments during the machine era. In fact, the three types of calls most commonly received by the police have entailed requests for emergency medical aid, intervention in marital disputes, and assistance with problems concerning children (Cumming et al., 1965). While police officers spend the largest share of their active-duty time responding to these calls, they have frequently regarded such requests as an unwarranted—and unwelcome—intrusion upon their primary goal of combatting crime (Cummins, 1971).

Obviously, the role of the police in such personal problems as sickness or injury, family arguments, and the guidance of children may imply the same type of life-and-death relationship as the decision to arrest—or not to arrest—a criminal suspect. Although many law enforcement officers tend to deemphasize the importance of these functions, they represent a valuable contribution to the community regardless of their relationship to the effort to curb crime.

But there is another possible by-product of service functions which deserves attention. Perhaps this point was illustrated most effectively by an exchange which occurred during a seminar of the New York state chiefs of police in 1971. As a sheriff from a small upstate community was complaining about the fact that his department traditionally had provided ambulance services for local residents, the chief of police from one of the largest cities in New York injected, "Hell, in my city, we don't just take people to the hospital. We send them a 'get-well card' the next day."

The aid of police officers in a critical life-and-death situation is a form of assistance that is not likely to be forgotten. Although service responsibilities seldom are mentioned in discussions of crime, an increased emphasis upon this role might indicate an approach which would enable the police to shed their punitive image and to gain the confidence of the community. As citizens display increasing apprehensiveness about police power and express a reluctance to "get involved" in criminal incidents, the provision of services seems to comprise a firmer basis for cooperation than the threat of punishment. The public gratitude and support generated by a growing awareness of service activities could provide police officers with a valuable resource in their efforts to secure the information that they need in order to fulfill their law enforcement responsibilities.

In fact, it even may be suggested that police departments should be retitled—and reorganized—as community service agencies. Rather than stressing the ostensible goal of reducing crime, such agencies might emphasize the predominate purpose of providing services. Whereas the former function denotes the public image of police departments, the latter activity seems to reflect what they actually do as measured both by public demands and by the allocation of time. As a result, such agencies could be organized into medical assistance, domestic problems, children's, and law enforcement divisions rather than burglary, vice, and homicide squads, which encompass only a small fraction of police duties. As the only agency of government that continually circulates throughout the community on a 24-hour-a-day basis, the police are probably in a position to perform such services more effectively than any other government agency. Although the combination of law enforcement and service responsibilities in a single agency might arouse the fear of invasions of privacy, police departments have performed these tasks simultaneously for a long period of time without apparent difficulty. In addition, this danger could be minimized by legal safeguards and by a redefinition of the criteria for the evaluation of personnel.

Perhaps the most important implication of the proposal for community service agencies is the impact that it might have upon the conduct of employees. Civil servants in such agencies would be rated for promotions and salary increases both by their performance of law enforcement duties and by the effectiveness with which they responded to requests for assistance. Since both roles involve a potentially life-and-death relationship with another person, service functions would no longer be subordinate to law enforcement activities in the priorities of police departments. This change might contribute to the increased comprehensiveness and quality of police activities in the community.

The use of the alternative definition of professionalism as a basis for the assessment of personnel action also offers a means of supervising police discretion, which is one of the most vexing and critical problems in the field of law enforcement. Unlike the classic model of administrative organizations, Wilson (1968:24) has noted that discretion in police departments tends to increase as one moves down the hierarchy. Police administrators have few means of maintaining scrutiny or control over patrol officers in the field, and few guidelines have been formulated to assist officers in dealing with the complex problems that they encounter in the community. The exercise of discretion has been a basic cause of public criticism of the law enforcement policy.

Because it emphasizes the ultimate goal of rehabilitation, the concept of professionalism which stresses a life-and-death relationship with clients might offer a more useful standard for the exercise of discretion than an exclusive emphasis on training and abilities. On the basis of their knowledge of individual offenders as well as the circumstances of a case, police officers could reserve the right to invoke the power of arrest and to make recommendations concerning appropriate forms of punishment. Although arrest and imprisonment might be the most feasible method of preventing some persons from repeating a crime, this ojective might be served most effectively for other suspects by a stern warning or by some other mode of punishment. By instilling the realization that they are in a position to shape the remainder of a suspect's life, the alternative definition of professionalism could stimulate the development of guidelines and information which might assist law enforcement officers in the exercise of their discretion. Police officers would assume responsibility for the effect of their decisions both upon other people's lives and upon the resolution of a major social problem, and they would be given an opportunity to participate directly in the effort to reduce crime by imposing sanctions specifically designed to prevent future violations of the law.

Although some might fear that the alternative definition of professionalism might produce increased inequalities in law enforcement, this proposal actually could have the opposite effect. Prevailing policies in most communities reflect the underenforcement rather than the full or equal enforcement of the law. In many circumstances, police officers make the dicision to arrest—or not to arrest—a suspect primarily on the basis of such superficial characteristics as the suspect's color, appearance, dress, or demeanor. The alternative concept of professionalism would replace those criteria with standards representing an awareness of the effect of the decision upon the suspect's life, the police officer's familiarity with individual suspects, and their professional judgment about the most efficacious method of discouraging suspects from repeating the crime. Although some might also object that a close personal association with criminal suspects might be a less savory relationship than contact with other types of clients, perhaps the pinnacle of professionalism is represented by the ability to serve the needs of all human beings regardless of their social status or personal reputations.

Combined with the restructuring of police departments as community service agencies, therefore, the application of the alternative definition of professionalism to the conduct of law enforcement officers seems to provide an innovative means of fulfilling critical social values. By emphasizing the potential effect of their actions upon other people's lives, police administrators would be provided with a foundation not only for developing the public cooperation that is essential to the identification of suspects but also for supervising police discretion and for approaching the ultimate objective of reducing crime.

SOME CONCLUDING OBSERVATIONS

Perhaps the most counterproductive trait inherited from the conventional definition of professionalism has been reflected in the basis upon which government personnel attempt to relate to their clients. Since the reform era, civil servants often have relied principally upon authority to secure compliance with their directives. In some occupations, such as police work, authority has been accompanied by the potential use of physical force, but in other professions it has been exerted primarily through status and expertise. Rather than using the rewards or punishments available in the programs that they administer, public employees frequently have employed their training and qualifications—as well as the power of their offices—as a fundamental means of managing their interactions with clients.

This orientation has seemed to produce a relationship of dominance and subordination between bureaucrats and clients which exemplifies some of the least desirable features of contemporary liberal attitudes in America. Many of the welfare policies enacted in the new post-New Deal era have seemed to imply a belief that important policy objectives can be achieved through the intervention of so-called professionals in the lives of clients to improve the client's position in society. Administrators have been encouraged to believe that their training and experience can assist them in determining proper methods of intervening and advising clients. Yet this assumption not only failed to reflect an empirical connection between professional qualifications and the capacity to improve the welfare of other people, but it also contained some implicit values which may be even more destructive.

The difficulty with this approach, of course, is that it can easily lead to an invidious form of paternalism. Obviously any persons who, on the basis of their training, expertise, or qualifications, are encouraged to believe that they are able to make enlightened decisions about the lives of others may find it difficult to restrain the impulse to exercise direction or control over those lives. Professionals who feel that they are capable of acting altruistically often may think that they are in a position to make decisions for their clients. When this belief is coupled with a strong emphasis upon authority, the temptation to tell others how they should lead their lives may be almost irresistible.

Conventional definitions of professionalism often seem to promote traits which could have a corrupting influence upon public employees. In addition to its potential abuse as a thin disguise for self-interest, paternalism can be even more stifling and oppressive than overt opposition. The training and experience of most professionals may encourage them to be sympathetic to the needs of clients, but it seldom permits them to be empathetic. By adopting a new standard of professionalism which emphasized their life-and-death relationships with clients, government personnel might gain a valuable resource in the effort to avert the dangers inherent in their position.

The use of the definition which emphasizes a life-and-death relationship with clients cannot eliminate all of the potential evils of professionalism, but it might ameliorate some of them. The alternative definition does not remove the need for altruism among public servants; but, by demonstrating the life-and-death consequences of administrative behavior, it proposes a rigorous standard for appraising the actions of government personnel. Similarly, the alternative definition does not seem to embody the same type of coequal relationship with clients that might have emerged from the machine model of administration. As long as bureaucrats are not dependent upon the electorate for their jobs, it may be impossible to sustain interactions based upon the exchange of mutually desired resources. Yet, the emphasis on the life-and-death implications of administrative decisions might enable clients to regain some of the influence which has shifted to civil servants. Because administrators would be compelled to consider their clients—as well as professional peers and supervisors—as a basic point of reference, major progress could be made in the effort to restore equilibrium to the relationship between clients and public employees.

Perhaps the most significant consequences of the life-and-death definition of professionalism might become evident in the methods that civil servants use to gain the respect of their clients. As the analysis of police problems has indicated, authority may be a less reliable basis for sustained public cooperation and support than providing needed assistance. By focusing attention on the performance of servants rather than the exercise of authority, the change from the conventional to the alternative definition of professionalism could yield major improvements in public satisfaction with the conduct of government personnel.

In addition, the alternative definition of professionalism and many of the characteristics of the machine appear to suggest a need to restructure governmental institutions. Prevailing definitions of professionalism, based upon training and qualifications, seem to be highly compatible with a strict division of labor and the development of specialized skills. A community services agency which includes and expands all of the current activities of police departments, for example, offers a model of organization which might be used in planning a new form of bureaucracy designed to focus on the needs and problems of clients.

The alternative concept of professionalism also could produce a method of increasing the accountability of public employees. Whereas the conventional understanding encompasses notions of self-regulation and self-definition, the alternative definition introduces the possibility of utilizing the satisfaction or dissatisfaction of clients both as a basis for developing a more precise delineation of professional activities and as a means of governing the conduct of personnel. As an expression of the sentiments of that portion of the community which is most directly concerned with the administration of government programs, the input of clients might be a particularly valuable resource in the evaluation of professional civil servants. Since public policies are presumably designed to serve the needs of clients, the emphasis of the capacity of bureaucrats to affect the destiny of other people would seem to comprise an appropriate point of departure for the assessment of personnel decisions.

Finally, there seems to be a need for major changes in the recruitment of administrative personnel. According to the alternative definition of professionalism, sensitivity and rapport may be characteristics that rival or outrank the significance of educational credentials or specialized expertise. Such qualities, of course, suggest the importance of developing incentives which might attract persons who are concerned about their communities as well as their personal advancement. As Lipsky (1973:114) has pointed out, "Higher salaries . . . have not previously resulted in the recruitment of significantly more sensitive or skillful people, although it has been the (somewhat self-serving) recommendation for bureaucratic improvement for many years." Efforts must be made to identify employees who not only understand the intricacies of public policy but also derive enjoyment and satisfaction from working with clients.

Obviously, the effort to improve the performance of municipal personnel is a difficult and imposing challenge. Although few citizens would tolerate a return to the corruption and favoritism that emerged in the 19th century, many of the concepts implicit in the old-style machine reflect important social values that should not be overlooked or ignored. By utilizing the best of both traditions and by adopting a standard of professionalism that places increased emphasis upon the needs and problems of clients, bureaucracies might be equipped to meet the problems confronting them.

NOTES

1. In fact, some authorities such as Steffens (1931) have contended that business interests seeking municipal contracts—rather than ethnic machines—may have been the primary sources of corruption during this era.

2. Many discussions of professionalism have been based upon a loose definition of the term. In fact, some studies have suggested that the word "profession" could be applied to the pursuit of almost any full-time vocation to distinguish it from an "amateur" or part-time endeavor. See Wilkensy (1964).

REFERENCES

CHAPIN, F.S. (1935). Contemporary American institutions: A sociological analysis. New York: Harper.

CUMMING, E., CUMMING, I., and EDELL, L. (1965). "Policeman as philosopher, guide and friend." Social Problems, 12(winter):276-286.

CUMMINS, M. (1971). "Police and service work." Pp. 279-290 in H. Hahn (ed.), Police in urban society. Beverly Hills, Calif.: Sage.

LIEBERMAN, J.K. (1970). The tyranny of the experts. New York: Walker.

LIPSKY, M. (1973). "Street level bureaucracy and the analysis of urban reform." Pp. 103-115 in G. Frederickson (ed.), Neighborhood control in the 1970s. New York: Chandler.

MERTON, R.K. (1957). Social theory and social structure. New York: Free Press of Glencoe.

NIEDERHOFFER, A. (1967). Behind the shield: The police in urban society. Garden City, N.Y.: Doubleday.

RIORDON, W.L. (1963). Plunkitt of Tammany Hall. New York: E.P. Dutton.

STEFFENS, L. (1931). The autobiography of Lincoln Steffens. New York: Harcourt, Brace.

WILENSKY, H.L. (1964). "Professionalization of everyone?" American Journal of Sociology, 70(September):137-152.

WILSON, J.Q. (1968). Varieties of police behavior. Cambridge, Mass.: Harvard University Press.

4

The Public Employee in Court:
Implications for Urban Government

DAVID H. ROSENBLOOM

☐ DURING THE 1960S AND EARLY 1970S, public personnel administration became a matter of increasing concern to governments in the United States. Because the "core" of the public employment problem is most pronounced on the urban level, cities bore the brunt of the far-reaching changes that were occurring in the public sector. Public employees were becoming more militant; collective bargaining emerged as a genuine constraint upon urban governments; and some cities, such as New York, were driven to the brink of bankruptcy partly as a result of labor settlements with their employees. Moreover, strikes by public employees became a common disruption of urban life. In some instances—the New York City teachers' strike, for example—they were an outgrowth of power struggles over the content of public policy. Thus, not only city budgets but also vestiges of "sovereignty" became bargainable items. At the same time, a related but less apparent threat to the cities' ability to deal effectively with or control their employees was emerging: the constitutional status of public employees was undergoing a remarkable series of transformations. During the 1950s, constitutional protections for public employees became a reality for the first time in the nation's history. By the end of the next decade, these had been extended so far that public employees came close to having a "property right" in their jobs, thereby forcing public employers to adapt to a dramatically new situation. No longer could it be taken for granted that governments could control the speech, association, and general behavior of their

*AUTHOR'S NOTE: *I am indebted to Susan Brannigan and Peter Carl Flood for their assistance in the preparation of this essay.*

employees. Toward the second half of the 1970s, however, the U.S. Supreme Court appeared anxious to limit some of the freedoms and protections afforded to public employees by the newer constitutional approaches. These changing constitutional doctrines affecting the status of public employees are of considerable importance to students and practitioners of administration, as are an analysis of persistent problem areas and the implications of current judicial reasoning for urban government.

CHANGING CONSTITUTIONAL APPROACHES

THE DOCTRINE OF PRIVILEGE

Historically, there have been four major approaches to the constitutional status of public employees. (Rosenbloom, 1971a; Rosenbloom and Gille, 1975). First, the doctrine of privilege dominated judicial thinking from the Founding Period until the 1950s. This approach relied upon the presumption that, because holding a government job is a voluntarily accepted privilege rather than a right or a coerced obligation, the government could place any restrictions it saw fit upon public employees, including those infringing upon their ordinarily held constitutional rights as citizens. In the words of one observer (Dotson, 1955:77):

> Its central tenet is that office is held at the pleasure of the government. Its general effect is that the government may impose upon the public employee any requirement it sees fit as conditional to employment. From the point of view of the state, public employment is maintained as an indulgence; from the position of the citizen, his job is a grant concerning which he has no independent rights.

The rationale for this approach is still best conveyed by Justice Oliver Wendell Holmes's oft-quoted statement that "The petitioner may have a constitutional right to talk politics, but he has no constitutional right to be a policeman" *(McAuliffe* v. *New Bedford,* 1892:220). In *Bailey* v. *Richardson* (1950:58, 59; affirmed 1951), a case arising out of the federal loyalty-security program established at the beginning of the Cold War, it was held, and subsequently accepted by the Supreme Court, that "due process of law is not applicable unless one is being deprived of something to which he has a right" and that therefore "the plain hard fact is that so far as the Constitution is concerned there is no prohibition against the dismissal of Government employees because of their political beliefs, activities or affiliations." Consequently, public employees could be (and were) dismissed for such things as favoring racial integration, reading Tom Paine or even the *New York Times,* and not attending church services, as well as for engaging in a host of unconventional or nonconformist activities.

Could such a judicial approach withstand the great increases in public employment that would occur in the 1960s, the development of new concepts of civil rights and civil liberties under the Warren Court and the growing dependency of citizens outside of public employment upon such governmental "privileges" as welfare payments?

THE DOCTRINE OF SUBSTANTIAL INTEREST

The doctrine of privilege had several undesirable consequences. Aside from the fact that it deprived an increasing number of citizens of some basic constitutional rights, it also had debilitating effects on the public service as a whole. This was most evident at the national level, but the same tendencies were found in state and local governments as well. First, it made for timid civil servants who were unwilling to realize their full potential as contributors to the making of public policy. It has been maintained that "Government would come to a standstill if our 'closet statesmen' in the civil service suddenly started doing only what they were told" (Storing, 1964:152) and consequently avoided advocating proposals of their own. But a study (Jahoda, 1955:111) of civil servants in Washington, D.C., in 1951 indicated the existence of a code of behavior under which it was deemed that

> you should not discuss the admission of Red China to the U.N.; you should not advocate interracial equality; you should not mix with people unless you know them very well; if you want to read the *Nation* you should not take it to the office; . . . you should take certain books off your private bookshelves. . . .

In other words, at the very time when governments and citizens in the United States were in increasingly greater need of better, more forceful administration, public servants were becoming reluctant to provide a stimulus toward dealing more effectively with the nation's many social, economic, and political problems. Moreover, the overwhelming majority were afraid to try to enlighten the electorate by speaking out on the crucial administrative and political issues of the day.

Second, not only were those in the public service deterred by the doctrine of privilege and attendant infringements upon their constitutional rights as citizens from engaging in activities which could have been beneficial to the political community at large, but potential applicants were also repelled by the lack of freedom associated with the public service. In the words of Robert Ramspeck, once Chairman of the U.S. Civil Service Commission,

> Today, the Federal Government affects the lives [sic] of every human being in the United States. Therefore, we need better people today, better qualified people, more dedicated people, in the Federal Service than we ever needed before. And we cannot get them if you are going to deal with them on the basis of suspicion, and delve into their private lives, because if there is anything the average American cherishes it is his right of freedom

of action, and his right to privacy. [U.S. Senate, Judiciary Committee, 1967:3]

The same tendencies would be even more pronounced at the state and local levels, where public service jobs have generally carried less prestige than has federal employment.

As these liabilities of the doctrine of privilege became more apparent, as the McCarthy era faded further into the political background, and as the courts pushed forward the frontiers of civil rights and liberties, a new approach concerning the constitutional status of public employees emerged. This outlook, sometimes called the "doctrine of substantial interest," rejects the very notion that a distinction between privileges and rights is of relevance to the constitutional protections afforded to either public employees or other citizens. It begins "with the premise that a state [or the federal government] cannot condition an individual's privilege of public employment on his non-participation in conduct which, under the Constitution, is protected from direct interference by the state" *(Gilmore* v. *James,* 1967:91). Moreover, "whenever there is a substantial interest, other than employment by the state, involved in the discharge of a public employee, he can be removed neither on arbitrary grounds nor without a procedure calculated to determine whether legitimate grounds do exist" *(Birnbaum* v. *Trussell,* 1966:678). Although the doctrine of substantial interest emerged largely out of lower court decisions, the Supreme Court embraced it on several occasions. For example, in *Board of Regents* v. *Roth* (1972:571), the Court stated that it "fully and finally rejected the wooden distinction between 'rights' and 'privileges' that once seemed to govern the applicability of procedural due process rights," and in *Sugarman* v. *Dougall* (1973:644) it reiterated that "this court now has rejected the concept that constitutional rights turn upon whether a governmental benefit is characterized as a 'right' or as a 'privilege.' " As a result of this approach, the constitutional rights of ordinary citizens and those of public employees became similar, although not identical, for the first time in the history of the nation. For instance, in *Kiiskila* v. *Nichols* (1970:749) it was reasoned that

A citizen's right to engage in protected expression or debate is substantially unaffected by the fact that he is also an employee of the government, and as a general rule, he cannot be deprived of his employment merely because he exercises those rights. This is so because dismissal from government employment, like criminal sanctions or damages, may inhibit the propensity of a citizen to exercise his right to freedom of speech and association.

The demise of the doctrine of privilege and the emergence of the substantial interest approach had far-reaching and almost immediate consequences for public services in the United States. Public employees were "liberated," and they celebrated this fact in many ways (Rosenbloom, 1971b). They began to speak out on the major issues of the day, including those with which they had special familiarity as a result of their jobs. Thus, in the federal service there were several antiwar petitions and statements issued by employees of the various departments, including those of State and Defense. Civil servants in the Department of Justice and others spoke out in the area of civil rights policy. Sometimes these activities took the form of protests, and on at least one occasion a department head was literally chased by some 300 employees seeking to present him with a petition concerning their grievances and demands for the furtherance of racial equality (*Washington Post*, 1970:B1). Such activities gave rise to a variety of more or less formal employee organizations dedicated to changing the nature of both public employment and the society at large. One such group, Federal Employees for a Democratic Society (FEDS), stressed the desirability of participatory bureaucracy in which hierarchy would be reduced to a minimum, rank and file employees would participate in the making of public policy to a greater extent, and employees would be free to refuse to perform work that violated their consciences (Hershey, 1973:51-63). Although less well documented, similar activities took place at the state and local levels as well, and if anything, the "new militancy" (Posey, 1968) of public employees in the 1960s was most important in urban settings. Again, the New York City teachers' strike of 1968 serves as an excellent example of a struggle for control over public policy. There, the central question was who should control the schools; elsewhere, similar matters arose with reference to almost the whole gamut of state and local functions. As protest, disruption, and dissent in the public service became more pronounced, some efforts were made by political authorities and public managers to stem its tide. However, under the doctrine of substantial interest, most of these activities were protected, and dismissal became highly impracticable *when an employee was willing to take his or her case to court.*

THE IDIOGRAPHIC APPROACH

Ultimately, the doctrine of substantial interest began to lose its forcefulness as a result of the lack of discipline that it encouraged in the public service and, equally important, the changing composition of the Supreme Court. Any comprehensive history of the Nixon presidency will have to concentrate not only on Watergate and foreign affairs, but also on his Supreme Court appointments and relationships with the bureaucracy. In some respects, the latter two were related. By the time he was inaugurated in 1969, Nixon had a well-established record of opposition to the federal bureaucracy. In the late 1940s and early 1950s, as a member of the House Un-American Activities Committee and a vice presidential candidate, he contributed to the notion that the bureaucracy was rife with subversives, and he vigorously sought to ferret out Communists. During the 1968 presidential campaign, when the Vietnam War and its domestic spillovers were a major issue, he stated that "the chief revolt in America is against 'an increasingly impersonal' Federal bureaucracy that saps individual initiative" (*New York*

Times, 1968:1). Once in office, he became convinced that elements in the bureaucracy were obstructing his policies and programs. A taped conversation (*New York Times,* 1974a:14) with George Schultz, Director of the Office of Management and Budget in 1971, is revealing in this regard:

> You've got to get us some discipline, George. You've got to get it, and the only way you get it, is when a bureaucrat thumbs his nose, we're going to get him. . . . They've got to know, that if they do it, something's going to happen to them. Where anything can happen. I know the Civil Service pressure. But, you can do a lot there, too. There are many unpleasant places where Civil Service people can be sent. We just don't have any discipline in government. That's our trouble.

While it is not unusual for presidents and other public executives to express such feelings, Nixon went further than any other president in trying to create an "administrative presidency" (Nathan, 1975) and to circumvent the merit system through various abuses (U.S. Civil Service Commission, 1976).

Nixon's desire for discipline in the bureaucracy was related to a desire for discipline throughout the society at large, and this became a theme in his nomination of justices to the Supreme Court. His basic desire was to create a court which was conservative in the sense of being passive and predisposed to uphold state power as against the rights and freedoms of individual citizens. For example, as he saw it, "the duty of a judge [is] to interpret the Constitution, and not to place himself above the Constitution or outside the Constitution. He should not twist or bend the Constitution in order to perpetuate his personal, political and social views" (Miller and Samuels, 1973:252). Despite his losing battles on behalf of his nominees Judges Clement Haynesworth and then George Carswell (whom the Senate refused to confirm), the evidence overwhelmingly suggests that Nixon was able to obtain the kind of court that he favored.

Once all four Nixon appointees were on the bench, the Court became clearly divided into three groups: the Nixon appointees, comprising Chief Justice Warren Burger and Justices William Rehnquist, Lewis Powell, and Harry Blackmun; the "liberal bloc," composed of Justices William Douglas, William Brennan, and Thurgood Marshall; and Justices Potter Stewart and Byron White, who provided the "swing votes." During the 1973-1974 term, the Nixon bloc was in the majority on all but one of the 103 occasions in which they voted together. This constituted about 75% of all cases decided. Justice White agreed with the Nixon bloc 85% of the time, and Justice Stewart agreed with them in 82% of the cases. On the other side, the liberal bloc voted together 74% of the time. However, 37% of their bloc votes were cast in minority opinions. Each member of the liberal bloc dissented about 50 times, as compared to a high of 25 dissents by Rehnquist among the remainder of the Court (*New York Times,* 1974b:10). These differences in judicial outlook had important consequences for the approaches governing the constitutional status of public employees.

The idiographic approach to the constitutional status of public employees is largely devoid of doctrinal, across-the-board judicial pronouncements, but rather consists of treating each case individually on its own merits. It was largely an outgrowth of the split between the liberal and Nixon appointees on the Court, which prevented achieving a high degree of agreement on approaches which would determine the direction of the law for years to come. Majorities could generally be formed only with reference to a specific set of facts and circumstances, and broad, sweeping generalizations were consequently precluded. In addition, such an approach was inherent in the doctrine of substantial interest, which required the courts to weigh the nature of the injury to the individual against the claimed benefit to the state in an effort to determine whether substantive constitutional rights had been violated and in order to decide what procedural safeguards, if any, had to be made available when adverse actions were taken against public employees. Thus, in several cases the Supreme Court and the rest of the federal judiciary required that, in applying general personnel regulations and principles to individual employees, public employers address the specific sets of facts and circumstances involved. For example, under this approach, it is possible to ban aliens from some, but not all public service positions; race might be used as a basis in making specific, but not general personnel assignments; and, apart from a regulation requiring a maternity leave very late in the term of a normal pregnancy, such a leave must be based on a medical determination of the individual employee's ability to continue in her job.

The idiographic approach had at least two major drawbacks. First and simplist, it led to an ever increasing amount of litigation in the already besieged courts. Since each case was to be treated separately on its own merits, rather than in accordance with broad, general doctrines, this was inevitable. Second, the approach placed public employers in a difficult position. There were few guidelines on how to resolve questions of constitutional law prior to judicial action. Neither the employer's nor the employee's rights were self-evident or adequately delineated by court decisions. Moreover, public personnel administrators found themselves being constantly second-guessed by the judiciary. In short, the approach led to an unstable and expensive situation which hampered effective public personnel administration. Fortunately, the idiographic approach appears to be transitory.

DECONSTITUTIONALIZATION

By 1975 the Nixon bloc had emerged as the dominant force on the Supreme Court. Their position was enhanced by the departure of Justice Douglas and the appointment of Justice John Paul Stevens. Under the leadership of Chief Justice Burger, the Court displayed greater awareness of its overcrowded dockets and sought means of reducing the number of cases that it had to decide and the number of opinions that it had to write. For example, Burger pointed out that the number of annually docketed cases had grown from 1,092 in 1935 to over 4,000 in 1975. In his view, consequently, "We must face up to the flinty reality that there is a necessity for a choice: If we wish to maintain the Court's historic function with a quality that will command public confidence, the demands we make on it must be reduced to what they were a generation ago" (*New York Times,* 1975:31). One approach for so doing has been to decide cases "summarily" without hearing arguments on them and without issuing written opinions—a trend much opposed by Justices Brennan and Marshall (*New York Times,* 1976:24). Another has been to deconstitutionalize areas of the law, such as that dealing with the constitutional status of public employees. Thus, for instance, in *Bishop* v. *Wood* (1976:693), a case dealing with the dismissal of a policeman, the Court per Justice Stevens reasoned that

> The federal court is not the appropriate forum in which to review the multitude of personnel decisions that are made daily by public agencies. We must accept the harsh fact that numerous individual mistakes are inevitable in the day-to-day administration of our affairs. The United States Constitution cannot feasibly be construed to require federal judicial review for every such error.

Politically, deconstitutionalization is beneficial to public employers in that it affords them greater freedom to deal with their employees. Before discussing the consequences of this approach, however, we should review some of the major remaining issues concerning the constitutional status of public employees at the present time.

PROCEDURAL DUE PROCESS

Under the doctrine of substantial interest, complex questions arose for the first time as to whether the due process clause of the Constitution requires that public employees be provided with the reasons for adverse personnel actions and an opportunity to rebut them. Diversity of opinion on these matters became rife as the idiographic approach emerged and subsequently the Supreme Court indicated its desire to deconstitutionalize this area of the law to an extent.

The Supreme Court addressed procedural due process requirements in *Board of Regents* v. *Roth* and *Perry* v. *Sindermann* in 1972. Roth involved an untenured professor at Wisconsin State University—Oshkosh whose one-year appointment was not renewed. The university's rules governing dismissals did not provide for a statement of reasons, review, or appeal. Professor Roth claimed that his procedural and substantive rights had been abridged by the university's failure to renew his appointment. In addressing the procedural issue, the Court, per Justice Stewart, attempted to limit the application of due process in this area by stating, "The requirements of procedural due process apply only to the deprivation of interests encompassed by the Fourteenth Amendment's protection of liberty and property. When protected interests are implicated, the right to some kind of prior hearing is paramount. But the range of interests protected by procedural due process is not infinite" (*Board of Regents* v. *Roth,* 1972:569-570). The majority found that Roth had neither a liberty nor a property interest in continued employment at the university and, therefore, that he had no constitutional right to a hearing.

In discussing the Roth facts, however, Justice Stewart identified several situations which would demand procedures consistent with the Constitution's requirement of due process. One was a situation in which the employee's reputation would be adversely affected by the public employer's action. Given that Roth was not dismissed for dishonesty, immorality, or "any charge . . . that might seriously damage his standing and associations in his community" (*Board of Regents* v. *Roth,* 1972:573), the Court reasoned that due process was not applicable. Yet, it went on to say, that "where a person's good name, reputation, honor or integrity is at stake because of what the government is doing to him, notice and an opportunity to be heard are essential" (*Board of Regents* v. *Roth,* 1972:573). In such a case, at a minimum, it would have been required that Roth have an opportunity to refute the charges against him before university officials.

Another situation, perhaps not fully distinguishable from the preceeding one, concerned a dismissal that diminished a public employee's future employability. Although Justice Stewart indicated that such a dismissal would create a deprivation of "liberty" within the context of the Fourteenth Amendment and would thus necessitate a prior hearing, he also asserted that "it stretches the concept too far to suggest that a person is deprived of 'liberty' when he is simply not rehired in one job but remains as free as before to seek another" (*Board of Regents* v. *Roth,* 1972:574-575), as was true of Professor Roth.

Third, the Court stated that if one had a "property" right or interest in a position, due process requirements would have to be met before employment could be terminated. The court, however, held that "to have a property interest in a benefit, a person clearly must have more than an abstract need or desire for it. He must have more than a unilateral expectation of it. He must, instead, have a legitimate claim of entitlement to it" (*Board of Regents* v. *Roth,* 1972:577). In the closely related *Sindermann* case (*Perry* v. *Sindermann,* 1972:602) the Court indicated that such a claim need not be based on a written contract or statutory protection, but might also be based on "an unwritten 'common law' . . . that certain employees shall have the equivalent of tenure." Finally, the Court indicated that due process was applicable where the dismissal was in retaliation for the exercise of such constitutionally protected rights as freedom of expression or association.

The liberal bloc dissented in *Roth* on the ground that greater procedural protection should be afforded to public employees in dismissals. In Justice

Marshall's words, "It can scarcely be argued that government would be crippled by a requirement that the reason be communicated to the person most directly affected by the government's action" (*Board of Regents* v. *Roth*, 1972:591).

It is evident that the Court's reasoning left about as many questions unanswered as resolved. This put the public personnel manager in a difficult position and also placed a difficult burden of proof on public employees *not* enjoying job protection. For example, how is one to know what constitutes a charge that might seriously and adversely affect an employee's reputation or chances of earning a livelihood in a chosen occupational area?

Such a question was inherently unanswerable under the doctrine of substantial interest and thereby contributed to the emergence of the idiographic approach because it is only after a specific set of facts has been ruled upon by the courts that the practitioner can receive sufficient guidance to deal constitutionally with several kinds of situations. However, by then it may be too late, as the number of cases reaching the courts in this area tends to confirm. Thus far, for instance, it has been held that removals or nonrenewals for fraud (*U.S.* v. *Rasmussen*, 1963), racism (*Birnbaum* v. *Trussell*, 1966), lack of veracity (*Hostrop* v. *Board*, 1972), and, at least in connection with high-level urban employment, absenteeism and gross insubordination (*Hunter* v. *Ann Arbor*, 1971) can violate the constitutional requirements of procedural due process in the absence of a hearing. It has also been strongly suggested that removal at an advanced age (*Olson* v. *Regents*, 1969), except as part of a general retirement system, tends to preclude subsequent employment and therefore requires the application of due process. On the other hand, a charge of being "antiestablishment" has been found not to create a sufficient impairment of reputation so as to afford constitutional protection (*Lipp* v. *Board of Education*, 1972).

One way out of this judicial morass was to deconstitutionalize some of the questions involved. The Supreme Court took an initial step in this direction in *Arnett* v. *Kennedy* (1974). Wayne Kennedy was an employee of the Chicago Regional Office of the Office of Economic Opportunity. He was dismissed by a supervisor whom he had allegedly accused in public of attempting to bribe a third party. Kennedy refused to reply to the charges against him and asserted that he had a constitutional right to a hearing prior to dismissal. He further argued that his allegations were protected by the First Amendment and therefore could not provide the basis for his discharge. In its holding on the procedural issue, a three-judge district court found that Kennedy's removal violated due process requirements in failing to provide a pretermination hearing. The Supreme Court, without majority opinion, reversed.

Justice Rehnquist announced the judgment of the Court in an opinion joined by Justice Stewart and the Chief Justice. He argued that Kennedy's dismissal was not constitutionally defective. In his view, since Congress had passed a statute, the Lloyd-LaFollette Act of 1912, protecting employees from removals except for such cause as would promote the efficiency of the federal service, Congress could also choose expressly to omit "the procedural guarantees which [Kennedy] insists are mandated by the Constitution" (*Arnett* v. *Kennedy*, 1974:152). This conclusion was tempered somewhat by the fact that Kennedy would be afforded considerable protections in a *post*termination hearing. Thus, these three justices expressed the view that, since Congress had provided federal employees with some protections against arbitrary removals, it could specify the procedures to be followed in dismissing them. From their perspectives, the Constitution was of little relevance in this regard, and consequently the matter should be deconstitutionalized.

Six justices disagreed, reasoning that, once Congress conveyed a property interest, the Constitution governed its deprivation. However, Justices Powell and Blackmun believed that the Constitution's requirements had been met in the case at hand; Justice White felt that they would have been satisfied if only Kennedy had been afforded the same process before an *impartial* official; and the liberal bloc argued that a full evidentiary hearing was necessary. Taken as a whole, then, *Arnett* v. *Kennedy* represents a limitation on the degree of due process which must be afforded employees prior to termination. Significantly, three justices endorsed a line of reasoning which would deconstitutionalize such dismissals almost entirely, and three others construed the constitutional requirements to be minimal.

By the end of the October 1975 term, this process had been carried considerably further by the Court's action in *Bishop* v. *Wood* (1976), as was mentioned earlier. By a 5 to 4 majority, it limited severely the circumstances that would create a property interest in a position and thereby afford constitutional protections in dismissals. Indeed, the Court's decision cast doubt upon the continued validity of the approach outlined in *Sindermann,* decided only a few years earlier. The case involved a policeman, who, after becoming a "permanent employee" by virtue of a city ordinance, was dismissed without a hearing or other procedural safeguard. The appropriate regulations provided that such an employee could be discharged if he or she failed to perform work up to the standard of classification or was negligent, inefficient, or unfit. Justice Stevens, speaking for the Court, however, reasoned that despite this ordinance, the employee held his position at the "will and pleasure" of the city because the regulation did not prohibit dismissals for other reasons. Hence, his "discharge did not deprive him of a property interest protected by the Fourteenth Amendment" (*Bishop* v. *Wood,* 1976:691).

This construction, although somewhat tortured and resting upon an equally divided lower court for authority, was all the more striking because the Court found itself compelled to "assume that his discharge was a mistake and based on incorrect information" (*Bishop* v. *Wood,* 1976:692). Moreover, the Court paid little heed to the employee's claim that his reputation and future employability had been damaged. It reasoned that since the basis for the action was communicated in private, no harm was done in this regard.

What accounts for such an outlook? Steven's closing paragraph suggests that the majority sought to take advantage of the case to engage in further deconstitutionalization of the legal status of public employees:

In the absence of any claim that the public employer was motivated by a desire to curtail or to penalize the exercise of an employee's constitutionally protected rights, we must presume that official action was regular and, if erroneous, can best be corrected in other ways. The Due Process Clause of the Fourteenth Amendment is not a guarantee against incorrect or ill-advised personnel decisions. [*Bishop* v. *Wood*, 1976:693]

Four justices expressed themselves in three dissenting opinions. Justice Brennan, joined by Marshall, argued that the Court's decision destroyed the "last vestige of protection of 'liberty' by holding that a State may tell an employee that he is being fired for some nonderogatory reason, and then turn around and inform prospective employers that the employee was in fact discharged for a stigmatizing reason that will effectively preclude future employment" (*Bishop* v. *Wood*, 1976:694). Justice White's dissent was joined by Justices Brennan Marshall, and Blackmun. He argued that the holding in this case was clearly contrary to the opinions expressed by a majority of justices in *Arnett* v. *Kennedy*. In his view, the city ordinance clearly conditioned dismissal on appropriate cause and thereby created a property interest, which could only be abridged consistent with the requirements of the due process clause.

On the basis of *Arnett* v. *Kennedy* and *Bishop* v. *Wood*, it appears that deconstitutionalization will be the order of the day in the future. Absent an explicit property interest, a majority of the Court does not consider the federal judiciary a proper vehicle for protecting public employees against arbitrary or improper dismissals. Moreover, as the opinions in *Arnett* suggest, even where such an interest exists, a majority of the Court now believes that the required constitutional protections are very minimal. As it stands today, only Justices Brennan and Marshall support elaborate procedural protections for public employees prior to dismissal. Hence, *Bishop* v. *Wood* is very significant because it signals an attempt to end an era of public personnel administration by law suit.

FREEDOM OF EXPRESSION, ASSOCIATION, AND RELATED MATTERS

Today, the courts, when dealing with public employees' rights in the area of freedom of expression, make a distinction between partisan and nonpartisan speech. Partisan speech is subject to far greater restriction. Federal regulations for political neutrality are typical of governmental efforts in the United States to depoliticize public administration by taking politics out of the civil service and the civil service out of politics. Although their origin can be traced back as far as the presidency of Thomas Jefferson, it was not until 1907 that political neutrality became an important feature of the federal service (Rosenbloom, 1971a:39-40, 94-118). In that year President Theodore Roosevelt changed the civil service rules to forbid employees in the competitive service from taking active part in political management or in political campaigns. At the same time, the rule explicitly allowed such employees to express privately their opinions on all political subjects.

In 1939, the First Hatch Act extended these restrictions to almost all federal employees, whether in the competitive service or not. It allowed federal employees to express their views on all political subjects, rather than only to express them privately. Nevertheless, it proclaimed that the act was intended to prohibit the same activities that the Civil Service Commission considered illegitimate under the earlier regulation, which allowed only private expression. The Second Hatch Act (1940) extended these regulations to positions in state employment having federal financing and allowed public employees to express their opinions on "candidates" as well as political subjects.

One of the major difficulties associated with these regulations has been the government's inability to define in advance exactly what they proscribe. In general, they prohibit being an officer in a political party, engaging in electioneering, and handling political contributions, but employees have also been disciplined for acts ranging from disparaging the President to stating "unsubstantiated facts about the ancestry of a candidate," and from failing to "discourage a spouse's political activity" to voicing "disapproval of treatment of veterans while acting as a Legion officer in a closed Legion meeting" (*National Association of Letter Carriers* v. *Civil Service Commission*, 1972:581).

The federal political neutrality program was first upheld by the Supreme Court in *United Public Workers* v. *Mitchell* (1947), a case decided under the then prevailing doctrine of privilege. The Court, divided 4 to 3, reasoned that "for the regulation of employees it is not necessary that the act regulated be anything more than an act reasonably deemed by Congress to interfere with efficiency of the public service" (*United Public Workers* v. *Mitchell*, 1947:101).

As the doctrine of privilege became less and less tenable, several courts concluded that the logic of the *Mitchell* case had been hopelessly undermined. (*Mancuso* v. *Taft*, 1972; *Hobbs* v. *Thompson*, 1971; *Bagley* v. *Washington Township Hospital*, 1966; *Fort* v. *Civil Service Commission*, 1964; *Minielly* v. *State*, 1966.) The Supreme Court readdressed the questions involved in *Civil Service Commission* v. *National Association of Letter Carriers* (1973). The postal union was seeking a declaratory judgment to the effect that the First Hatch Act, sustained in *Mitchell*, was unconstitutional. The lower court had reasoned that past decisions "coupled with changes in the size and complexity of public service, place *Mitchell*, among other decisions outmoded by passage of time" (*National Association of Letter Carriers* v. *Civil Service Commission*, 1972:585). The Supreme Court reversed, however, and in no uncertain terms:

> We unhesitatingly reaffirm the Mitchell holding that Congress had, and
> has, the power to prevent . . . [employees] from holding a party office,
> working at the polls, and acting as party paymaster for other party
> workers. An Act of Congress going no farther would in our view unques-
> tionably be valid. So would it be if, in plain and understandable language,
> the statute forbade activities such as organizing a political party or club;
> actively participating in fund-raising activities for a partisan candidate or
> political party; becoming a partisan candidate for, or campaigning for, an
> elective public office; actively managing the campaign of a partisan candi-
> date for public office; initiating or circulating a partisan nominating
> petition or soliciting votes for a partisan candidate for public office; or
> serving as a delegate, alternate or proxy to a political party convention.
> [*Civil Service Commission* v. *National Association of Letter Carriers*,
> 1973:556]

With reference to the challenge that the regulations were so vague and broad that
they would deter legitimate speech and activity along with that which was
prohibited, Justice White concluded that the restrictions were such that "the
ordinary person exercising ordinary common sense can sufficiently understand
and comply with [them], without sacrifice to the public interest" (*Civil Service
Commission* v. *National Association of Letter Carriers*, 1973:579).

Justice Douglas, joined by Justices Marshall and Brennan, dissented, contend-
ing that "Mitchell is of a different vintage from the present case" (*Civil Service
Commission* v. *National Association of Letter Carriers*, 1973:598) and that the
Hatch Act represented an unconstitutional interference with federal employees'
freedom of speech. In *Broadrick* v. *Oklahoma*, a companion case to the *Letter
Carriers* case," the Court upheld state regulations of a similar variety. While the
Court's action in these cases cannot be seen as deconstitutionalization, it does
represent a desire not to constitutionalize an area long left to the discretion of
governmental branches other than the judiciary.

Public employees can constitutionally be prohibited from engaging in partisan
speech and activity, but these limitations on their First Amendment rights have
been balanced by protections against dismissals for partisan reasons. Thus, in
Elrod v. *Burns* (1976) the Supreme Court held, without majority opinion, that
removals as part of a political patronage system created an unconstitutional
infringement on public employees' freedoms of belief and association. The case
arose out of practices in the Cook County (Illinois) Sheriff's Office. In the
plurality opinion, Justice Brennan reasoned that patronage dismissals were "not
the least restrictive means for fostering" legitimate governmental ends, such as
efficiency and effectiveness (*Elrod* v. *Burns*, 1976:565). His decision, however,
left open the question of such dismissals from policy making positions, although
these would presumably be constitutional. Justice Powell was joined by Justice
Rehnquist and Chief Justice Burger in a dissent arguing that "The judgment
today unnecessarily constitutionalizes another element of American life—an

element certainly not without its faults but one which generations have accepted on balance as having merit" (*Elrod* v. *Burns,* 1976:574). In his view, patronage dismissals could be justified on the basis that they served legitimate and important purposes of the state.

Public employees have been afforded considerably more freedom with reference to nonpartisan speech. In *Pickering* v. *Board of Education* (1968), the most significant case in this area, the Supreme Court held that the special duties and obligations of public employees cannot be ignored, but that the proper test is whether the government's interest in limiting public employees' "opportunities to contribute to public debate is . . . significantly greater than its interest in limiting a similar contribution by any member of the general public" (*Pickering* v. *Board of Education,* 1968:573). The Court identified six elements which would generally enable the government to abridge legitimately public employees' freedom of expression:

1. The need for maintaining discipline and harmony in the work force.
2. The need for confidentiality.
3. The possibility that an employee's position is such that his or her statements might be hard to counter due to his or her presumed greater access to factual information.
4. The situation in which an employee's statements impede the proper performance of work.
5. The instance where the statements are so without foundation that the individual's basic capability to perform his or her duties comes into question.
6. The jeopardizing of a close and personal loyalty or confidence.

In applying *Pickering,* the lower courts developed a fairly consistent pattern of relying on the idiographic approach. For instance, certain categories of speech have been held to be beyond the protection of the Constitution. These include public expression concerning matters not of public concern (*Clark* v. *Holmes,* 1972), "bickering and running disputes" with superiors (*Chitwood* v. *Feaster,* 1972), "intermeddling" and disruptive expression (*Rozman* v. *Elliott,* 1972), "extremely disrespectful and grossly offensive remarks" (*Duke* v. *North Texas State University,* 1973), and in-classroom statements made by a military teacher (*Goldwasser* v. *Brown,* 1969). Protected speech, on the other hand, has tended to be nondisruptive and to concern matters of general public interest. Thus the wearing of black armbands in protest of the war in Vietnam (*James* v. *Board of Education,* 1972; *Peale* v. *U.S.,* 1971), silent refusal to salute the flag and recite the Pledge of Allegiance (*Russo* v. *Central School District,* 1972; *Hanover* v. *Northrup,* 1970), and public criticism of an employer (*Donahue* v. *Staunton,* 1972) were held to be protected by the First Amendment.

Another aspect of the idiographic approach to public employees' freedom of expression has been related to the position occupied by the employee. Hence, it has been reasoned that "the less likely it is that the public will attach special importance to the statements made by someone in a particular position, the weaker is the argument that the state needs special restrictions on false or erroneous statements made by someone in that position to prevent substantial deception of the public" (*Donovan* v. *Reinbold,* 1970:743). Conversely, it was held elsewhere (*Fisher* v. *Walker,* 1972) that a fireman who was a union president could be suspended constitutionally for having published statements in a local union publication falsely criticizing an immediate superior and having thereby created a divisive effect within the fire department.

In 1976, the Supreme Court served notice that it might seek to deconstitutionalize some aspects of public employees' expression, at least where these are relatively trivial. Thus, in *Kelley* v. *Johnson* the court found no constitutional barrier to regulations placing limits on the length of hair and the nature of facial hair that could be worn by policemen. The regulations were found to be illegitimate by the Court of Appeals for the Second Circuit which reasoned that "choice of personal appearance is an ingredient of an individual's personal liberty" (*Kelley* v. *Johnson,* 1976:712) and is therefore protected by the Fourteenth Amendment. Consequently, in the absence of the state's ability to show a "genuine public need" for such standards, the regulations were deemed unconstitutional. The Supreme Court reversed in an opinion delivered by Justice Rehnquist, who gave short shrift to this approach. In his view, even if "the citizenry at large has some sort of 'liberty' interest within the Fourteenth Amendment in matters of personal appearance" it would be the burden of the employee challenging the regulations to "demonstrate that there is no rational connection between the regulation . . . and the promotion of safety of persons and property" (*Kelley* v. *Johnson,* 1976:714, 716). Thus, the burden falls upon the challenger of such a regulation to show that it "is so irrational that it may be branded 'arbitrary,' and therefore a deprivation of [the challenger's] 'liberty' interest in freedom to choose his own hair style" (*Kelley* v. *Johnson,* 1976:716). Given that several police forces throughout the nation had similar regulations and that one can find some rationale for them, Rehnquist found that the Fourteenth Amendment had not been violated.

Justices Marshall and Brennan dissented, finding no rational relation to some legitimate state interest. In their opinion, "An individual's personal appearance may reflect, sustain, and nourish his personality and may well be used as a means of expressing his attitude and lifestyle" (*Kelley* v. *Johnson,* 1976:717-718). Consequently, such expression could only be infringed where the government could demonstrate a rational interest in so doing.

Given the importance of freedom of expression, it appears improbable that wholesale deconstitutionalization will occur. Certainly, a return to the simplistic logic of the doctrine of privilege is unlikely. However, in a related area—that of freedom of association, broadly construed—the Court issued a decision which in a very real sense parallels the *Kelley* v. *Johnson* case. Public employees' right of freedom of association was broadly guaranteed by the Supreme Court in *Shelton* v. *Tucker* (1960). The case left open the questions of whether public employees could have membership in subversive organizations, organizations with illegal objectives, and unions. Their right to join the latter has been upheld. With regard to the former, however, it has been held that there can be no answer. Rather each case has to be judged on the basis of whether a public employee actually supports an organization's illegal aims, because, as the Supreme Court expressed it, "Those who join an organization but do not share its unlawful purposes and who do not participate in its unlawful activities surely pose no threat, either as citizens or as public employees" (*Elfbrandt* v. *Russell,* 1966:17). Consequently, it is incumbent upon public employers seeking to dismiss employees for membership in subversive organizations or those with illegal purposes to prove that the employees actually shared in their objectionable aims and activities. In regard to labor organizations, *McLaughlin* v. *Tilendis* (1968) and *American Federation of State, County, and Municipal Employees* v. *Woodward* (1969) held that public employees had a constitutional right to membership in unions, absent an intent to commit unlawful acts. In a later case, though, it was held that despite a right to union membership, public employees had no constitutional right to strike (*United Federation of Postal Clerks* v. *Blount,* 1971).

A measure of deconstitutionalization occurred in *McCarthy* v. *Philadelphia Civil Service Commission* (1976:369), a case involving residency requirements, which raised questions concerning the right to travel and freedom of expression and association. In the Court's view, the regulation, which applied to fireman, established a "bona fide continuing residence requirement" of constitutional acceptability. Given that the Court did not provide a comprehensive explanation of this position, it is unclear whether all such regulations are constitutional or whether there are limits on a public employer's right to require that its employees live within its jurisdictional boundaries. In view of the general trend toward deconstitutionalization, however, it appears unlikely that regulations of this nature will be successfully challenged in the future.

EQUAL PROTECTION

Perhaps no aspect of the constitutional status of public employees has been of greater importance in recent years than their right to equal protection of the law. This has been especially true in urban areas, where racial relationships often rank high among the political, social, and economic problems in need of resolution. Under the doctrine of privilege it was held that a public employer could exclude individuals from the public service on almost any grounds, including those of race and color (*Fursman* v. *Chicago,* 1917). Contemporary doctrine prohibits such practices and guarantees equal protection of the laws to all citizens seeking public employment. For example, in *Brooks* v. *School*

District (1959), it was stated that any practice of denying employment on the basis of race or color would be unconstitutional. The demise of the doctrine of privilege, however, has generated difficulties for the judiciary when addressing some facets of discrimination based on race, sex, and citizenship. In part this has been due to the fact that public employees were traditionally unprotected by the equal protection clause, but in part it also results from the complexity of the issues involved. In order to reduce the latter, the Supreme Court has recently deconstitutionalized some features of public employees' right to equal treatment by their employer.

The most acute equal protection issue facing public employers today concerns the constitutionality of merit examinations having discriminatory effects. In a typical case in this area, members of a minority group who have taken a public personnel examination and have not scored high enough to receive appointment challenge the constitutionality of the examination procedures, both on their own behalf and as a class action on behalf of all others similarly situated. They attempt to demonstrate that the exam has a disproportionate impact. The remedy generally sought consists of immediate employment in the position applied for, an injunction barring further use of the examination in question, and the introduction of an employment formula to redress the consequences of past discrimination.

It is evident that in such a case the method of determining whether discrimination exists is of crucial importance. Although public personnel administrators are wont to argue that the examinations that they administer are neutral and that differential performance on them between minorities and nonminorities is due to more general social, educational, and economic inequalities found in the nation as a whole, several judicial decisions have reasoned that *result* rather than purity of intent is of greater relevance. It has been held, for instance, that "When state action which is neutral on its face unintentionally disadvantages racial minorities in areas such as public employment, the state has the burden of demonstrating that the action in question has at least a substantial relation to a legitimate state interest" (*Vulcan Society* v. *Civil Service Commission*, 1973:1272). Thus, "harsh racial impact, even if unintended, amounts to an invidious *de facto* classification that cannot be ignored" (*Chance* v. *Board of Examiners*, 1972:1175), and the question becomes, therefore, what constitutes a harsh racial impact.

Here, two different, but related approaches have developed. For the most part, "Where the plaintiffs have established that the disparity between the hiring of whites and minorities is of sufficient magnitude, then there is a heavy burden on the defendant to establish that the examination creating the discrimination bears a demonstrable relationship to successful performance of jobs for which they were used" (*Bridgeport Guardians* v. *Bridgeport Civil Service Commission*, 1973:1337). In other cases, however, a less stringent test has been used:

The percentage of minority-group persons employed by [a public jurisdiction] is grossly and disproportionately less than the percentage of minority-group persons in the general population of the area. Thus, whatever may have been the good intentions of defendants, there is a prima facie case for predicating employment discrimination unless defendants justify the selection method. [*Western Addition Community Organization* v. *Alioto,* 1971:539]

Finally, in at least one case, both of these criteria have been used to establish a harsh racial impact (*Vulcan Society* v. *Civil Service Commission,* 1973).

It is possible, however, to defend successfully a discriminatory examination if it can be shown that job performance criteria are within an area of legitimate state interest. In other words, "Despite discriminatory impact, the law in this area clearly recognizes the possibility that examination procedures can be justified by the requirements of the job" (*King* v. *Civil Service Commission,* 1973:6248). Two judicial approaches for dealing with the validity of exams have developed. Given that, where matters of technical expertise are involved, the judiciary is often reluctant to substitute its own judgment for that of administrators, one approach places "primary emphasis . . . on the validity of the methods used in creating the examination not on the independent validity of the end product" (*Kirkland* v. *New York,* 1974:1373). However, in some cases, the courts have made an independent assessment of the "job relatedness" of the examination under challenge (*Bridgeport Guardians* v. *Bridgeport Civil Service Commission,* 1973; *Vulcan Society* v. *Civil Service Commission,* 1973). In either event, before the courts, public employers have had only very limited success in defending examinations having harsh racial impacts.

In terms of remedies for unconstitutionally discriminatory examinations, the courts, while agreeing that "hiring quotas are discriminatory since they deliberately favor minority groups on the basis of color," have nevertheless tended to sanction "hiring quotas to cure past discrimination" (*Bridgeport Guardians* v. *Bridgeport Civil Service Commission,* 1973:1340). Thus in several cases (*Armstead* v. *School District,* 1971; *Bridgeport Guardians* v. *Bridgeport Civil Service Commission,* 1973; *Officers for Justice* v. *Civil Service Commission,* 1973; *Carter* v. *Gallagher,* 1971; *NAACP* v. *Allen,* 1974), courts have sanctioned relief which required that the public employer "temporarily institute race as the final determination factor in their appointment of applicants" (*NAACP* v. *Allen,* 1974:618).

In *Washington* v. *Davis* (1976), the Supreme Court sought to deconstitutionalize much of this area of the law. The case involved the validity of a qualifying examination administered to applicants for positions as police officers in the District of Columbia Metropolitan Police Department. Those scoring high enough on this instrument were eligible for a 17-week training program; others were excluded. The examination was challenged on the basis that it had a harsh racial impact in disqualifying a disproportionately high number of black applicants. The court of Appeals for the District of Columbia treated the issue on a constitutional basis and found the examination to be in violation of equal protection. The Supreme Court reversed, arguing that the Court of Appeals had erroneously equated the standards under Title VII of the Civil Rights Act of 1964 with those of equal protection under the Constitution. In the Court's view, such an interpretation was highly undesirable:

> A rule that a statute designed to serve neutral ends is nevertheless invalid, absent compelling justification, if in practice it benefits or burdens one race more than another would be far reaching and would raise serious questions about, and perhaps invalidate, a whole range of tax, welfare, public service, regulatory, and licensing statutes that may be more burdensome to the poor and to the average black than to the more affluent white. [*Washington* v. *Davis,* 1976:612]

On the other hand, under Title VII, "Congress provided that when hiring and promotion practices disqualifying substantially disproportionate numbers of blacks are challenged, discriminatory purpose need not be proved, and that it is an insufficient response to demonstrate some rational basis for the challenged practices" (*Washington* v. *Davis,* 1976:611). In other words, since more rigorous standards and "a more probing judicial review" (*Washington* v. *Davis,* 1976:612) could be applied under Title VII, there was no need to treat the issue constitutionally.

In theory, therefore, deconstitutionalization in this area will strengthen public employees' legal right to equal treatment. However, it also affords legislators and other political authorities an opportunity to revise and weaken standards such as those contained in Title VII. Moreover, in applying Title VII in this case, the Court held that the examination in question was acceptably valid. This may signal the Court's desire to allow public employers greater leeway in developing and administering examinations. Indeed, Justices Brennan and Marshall dissented primarily on this ground. The Court did not deal with the issue of racially conscious relief.

Another equal protection issue concerning race involves the assignment of personnel. In *Baker* v. *City of St. Petersburg* (1968), 12 black policemen contested the constitutionality of a police department practice under which only blacks were assigned to an area that was predominantly black in population, and no blacks were assigned to other areas. The police department argued that blacks could better communicate with and identify other blacks, a rationale which was found acceptable in district court. On appeal, however, this decision was reversed on the ground that the assignment of blacks solely on the basis of race to a black enclave violated the equal protection clause. Yet the court went on to say, "We do not hold that the assignment of a Negro officer to a particular task because he is a Negro can never be justified" (*Baker* v. *St. Petersburg,*

1968:300-301). Consequently, the idiographic approach is to be followed in this area.

Turning to discrimination based on sex, equal protection cases have generally centered on mandatory pregnancy leave regulations. Typically, women were required to terminate their employment—often without provision for reinstatement—at a specified point in their pregnancy, often after the fourth or fifth month. After opposing conclusions had been reached in the lower courts, the Supreme Court addressed the issues involved in *Cleveland Board of Education* v. *LaFleur* and *Cohen* v. *Chesterfield County School Board* (1974). Although the cases had been argued largely on equal protection grounds, the majority of the Court, per Justice Stewart, decided the issue on the ground that the regulations mandating early, unpaid maternity leaves constituted a deprivation of "liberty" within the meaning of the Fourteenth Amendment. The Court reasoned that "by acting to penalize the pregnant teacher for deciding to bear a child, overly restrictive maternity leave regulations can constitute a heavy burden on the exercise of . . . protected freedoms" (*Cleveland Board of Education* v. *LaFleur*, 1974:640). Given that the regulations infringed upon protected liberties, while serving no valid state interest, they were deemed unconstitutional. The Court reasoned that an "individualized determination" of the employee's ability to continue in her job was necessary (*Cleveland Board of Education* v. *LaFleur*, 1974:644).

Elsewhere it has been held that a person may not be denied public employment, or the benefits of that employment, simply on the basis of sex. For example, in *Eslinger* v. *Thomas* (1973) it was held that an applicant could not be denied employment as a senate page in the South Carolina legislature solely because of her sex. The Fourth Circuit Court of Appeals concluded that the state had demonstrated no " 'fair and substantial relation' between the basis of the classification [appearances of impropriety] and the object of the classification [persons of the female sex]" (*Eslinger* v. *Thomas*, 1973:230-231). Thus, refusal to hire Ms. Eslinger was held to be in violation of the equal protection clause.

Similar lines of reasoning have been followed in cases involving aliens. Thus, in *Sugarman* v. *Dougall* (1973) and *Hampton* v. *Mow Sun Wong* (1976), the Supreme Court held general New York State and federal prohibitions on the employment of aliens unconstitutional on the grounds that they bore little relation to legitimate governmental interests. At the same time, however, the Court indicated that such a ban might be constitutional with reference to some positions.

Finally, in *Massachusetts Board of Retirement* v. *Murgia* (1976), another aspect of public personnel administration affecting equal protection was effectively deconstitutionalized. At issue here was a law, declared unconstitutional at the circuit court level, requiring all uniformed state police officers to retire upon reaching the age of 50, regardless of physical or mental fitness. The regulation was upheld by the Supreme Court in a per curiam opinion. It reasoned that because age 50 "marks a stage that each of us will reach if we live out our normal span," (*Massachusetts Board of Retirement v. Murgia,* 1976:525) the law did not discriminate against any specific category of individuals. Consequently, even though the mandatory retirement policy might be imperfect, it was nevertheless found to be constitutionally acceptable. Justice Marshall dissented on the grounds that mandatory retirements must be based on individual fitness if they are to meet the constitutional requirements of equal protection.

IMPLICATIONS FOR URBAN GOVERNMENT

For urban government the implications of the changing constitutional doctrines, and especially the development of the deconstitutionalization approach, are best understood with reference to collective bargaining and politics in general. As mentioned earlier, the 1960s were "the decade of the public employee" (see Moskow et al., 1970). For example, between 1962 and 1968, public employee organizational membership increased some 136%, as compared to a 5% increase over the same period in the private sector. The American Federation of State, County, and Municipal Employees (AFSCME), the dominant union in nonfederal public employment, doubled its membership during the 1960s, bringing the total up to 400,000 by 1969. Most of its members are city employees, with state and county employees being the next largest categories. Today, public employees constitute about a fifth of the work force, and about half of them are organized. In some sectors, unionization is continuing at an almost incredible pace. Thus, during the 1974-1975 membership year, the National Education Association gained over 4,000 new members per week. Whereas in the 1950s public employee unions were still fighting membership and recognition battles, in the late 1960s the struggle was not infrequently over matters of public policy. This "posed in starkest terms the collision of union and community interests—a collision that over the years is likely to put greater strains on the capacity for survival of all our big cities than the traditional 'battle for the buck' in contract negotiations" (Raskin, 1972:134). Indeed, both policywise and financially, it is not too much to say that the fate of America's cities is now on the bargaining table.

The doctrine of substantial interest and the idiographic approach obviously enhanced the position of organized labor. In fact, it was only after the emergence of the substantial interest doctrine that it became unconstitutional to bar employees from union membership. These approaches also afforded public employees vastly greater job protections—and on a constitutional basis. Moreover, they allowed and perhaps even encouraged public employees to speak out on the issues of the day, including those which brought them into conflicts with their employers. The idiographic approach, in particular, given its inability to afford general guidance to public employers and its reliance on the judicial process as a substitute for more orthodox public personnel administration, weakened the position of governments vis-à-vis their employees and made adverse actions difficult and costly, if not impossible. Yet, these approaches had important benefits. They protected the basic rights of a growing sector of society, and they enabled public employees to make more positive (as well as more negative) contributions to the political system as a whole.

From these prespectives, deconstitutionalization may be seen as an idea whose time has come. While it clearly weakens the employee vis-à-vis the public employer, it comes after a decade of rapid unionization and expanding employee power in the public sector. Whereas in the past employees needed constitutional protection if their rights and liberties were to be preserved, today many can rely on their unions for this purpose. Although matters such as political neutrality, mandatory retirements, residency, and grooming styles may fall outside the scope of collective bargaining in the public sector, as traditionally conceived, it is evident that they can nevertheless be included in negotiations, especially in view of the trend toward codetermination of policy matters (Nigro and Nigro, 1973:325-328). In addition, stronger unions are likely to carry more weight in their lobbying activities before legislative bodies and consequently can have a greater impact on policy matters in this context as well. Hence, deconstitutionalization may be seen as redressing the balance of power between public employer and employee by depriving the latter of the added advantage of being able to obtain in court what cannot be gotten at the bargaining table. This, of course, is not to suggest that the Supreme Court has returned the fate of the cities solely to urban governments, but it is rather to contend that deconstitutionalization gives the latter a greater potential to control their destinies. This is especially true in view of the Court's holding in *National League of Cities* v. *Usery* (1976), which voided the application of the wages and hours provisions of the Fair Labor Standards Act to state and local employees as a violation of the Tenth Amendment.

At the same time, however, two caveats must be mentioned. First, although the trend seems clear, deconstitutionalization is not likely to proceed to the point where the doctrine of privilege is for all intents and purposes resurrected. While the constitutional rights won by public employees in the 1950s and 1960s are not likely to be expanded by the Burger Court, nor are they likely to be obliterated. Deconstitutionalization has reached some fundamental questions, including the right to procedural due process, but public employees are likely to retain many of their First and Fourteenth Amendment rights. Second, the ramifications of *Elrod* v. *Burns* (1976) bear special attention. It may be that, in declaring patronage dismissals unconstitutional, the Court has dealt a deadly blow to the remaining vestiges of urban political machines, with all their integrative functions. If, for example, Chicago is a city that "works," will it work after *Elrod?*

Finally and more generally, as this question suggests, the current tendency toward deconstitutionalization of the legal status of public employees bears close scrutiny by all those concerned with urban affairs and public personnel administration. For if anything stands out in the history of public personnel activities in the United States, it is "the 'multiplier' importance of public service—great changes in a wide arena are instigated by small alterations in governmental personnel policy" (Krislov, 1967:5). Changes in the judicial doctrines affecting public employees are related to changes in the nature, mood, and tone of the public service in general, and it is only by keeping abreast of such developments, and even anticipating them, that we can hope to maximize our effectiveness as a society in dealing with expanding size, power, and complexity of the public sector.

REFERENCES

American Federation of State, County, and Municipal Employees v. Woodward (1969). 406 F2d 137.

Armstead v. School District (1971). 325 F. Supp. 560.

Arnett v. Kennedy (1974). 416 U.S. 134.

Bagley v. Washington Township Hospital (1966). 421 P2d 409.

Bailey v. Richardson (1950). 182 F2d 47; 341 U.S. 918.

Baker v. City of St. Petersburg (1968). 400 F2d 294.

Birnbaum v. Trussell (1966). 371 F2d 672.

Bishop v. Wood (1976). 48 L Ed 2d 684.

Board of Regents v. Roth (1972). 408 U.S. 564.

Bridgeport Guardians v. Bridgeport Civil Service Commission (1973). 482 F2d 1333.

Broadrick v. Oklahoma (1973). 413 U.S. 601.

Brooks v. School District (1959). 267 F2d 733.

Carter v. Gallagher (1971). 452 F2d 315.

Chance v. Board of Examiners (1972). 458 F2d 1167.

Chitwood v. Feaster (1972). 468 F2d 359.

Civil Service Commission v. National Association of Letter Carriers (1973). 413 U.S. 548.

Clark v. Holmes (1972). 474 F2d 928.

Cleveland Board of Education v. LaFleur (1974). 414 U.S. 632.

Cohen v. Chesterfield County School Board (1974). 414 U.S. 632.

Donahue v. Staunton (1972). 471 F2d 475.

Donovan v. Reinbold (1970). 433 F2d 735.

DOTSON, A. (1955). "The emerging doctrine of privilege in public employment." Public Administration Review, 15 (spring):77-88.

Duke v. North Texas State University (1973). 469 F2d 829.

Elfbrandt v. Russell (1966). 384 U.S. 11.

Elrod v. Burns (1976). 49 L Ed 2d 547.

Eslinger v. Thomas (1973). 476 F2d 225.

Fisher v. Walker (1972). 464 F2d 1147.

Fort v. Civil Service Commission (1964). 392 P2d 385.

Fursman v. Chicago (1917). 278 Ill. 318.

Gilmore v. James (1967). 274 F. Supp. 75.

Goldwasser v. Brown (1969). 417 F2d 1169.

Hampton v. Mow Sun Wong (1976). 48 L Ed 2d 495.

Hanover v. Northrup (1970). 325 F. Supp. 170.

HERSHEY, C. (1973). Protest in the public service. Lexington, Mass.: Lexington Books.

Hobbs v. Thompson (1971). 448 F2d 456.

Hostrop v. Board (1972). 417 F2d 488.

Hunter v. Ann Arbor (1971). 325 F. Supp. 847.

JAHODA, M. (1955). "Morale in the federal service." Annals of the American Academy of Political and Social Science, 300(July):110-113.

James v. Board of Education (1972). 461 F2d 566.

Kelley v. Johnson (1976). 47 L Ed 2d 708.

Kiiskila v. Nichols (1970). 433 F2d 745.

King v. Civil Service Commission (1973). 6 E.P.D. 6243.

Kirkland v. New York (1974). 374 F. Supp. 1361.

KRISLOV, S. (1967). The Negro in federal employment. Minneapolis: University of Minnesota Press.

Lipp v. Board of Education (1972). 470 F2d 802.

Mancuso v. Taft (1972). 341 F. Supp. 574.

Massachusetts Board of Retirement v. Murgia (1976). 49 L Ed 2d 520.

McAuliffe v. New Bedford (1892). 155 Mass 216.

McCarthy v. Philadelphia Civil Service Commission (1976). 47 L Ed 2d 366.

McLaughlin v. Tilendis (1968). 398 F2d 287.

MILLER, K., and SAMUELS, N. (eds., 1973). Power and the people. Pacific Palisades, Calif.: Goodyear.

Minielly v. State (1966). 411 P2d 69.

MOSKOW, M., et al. (1970). Collective bargaining in public employment. New York: Random House.

NATHAN, R. (1975). The plot that failed. New York: Wiley.

National Association for the Advancement of Colored People v. Allen (1974). 493 F2d 614.

National Association of Letter Carriers v. Civil Service Commission (1972). 346 F. Supp. 578.

National League of Cities v. Usery (1976). 49 L Ed 2d 245.

New York Times (1968). October 7, p. 1.

––– (1974a). July 20, p. 14.

––– (1974b). July 1, p. 10.

––– (1975). August 14, p. 31.

––– (1976). November 9, p. 24.

NIGRO, F., and NIGRO, L. (1973). Modern public administration (3rd ed.). New York: Harper and Row.

Officers for Justice v. Civil Service Commission (1973). 371 F. Supp. 1328.

Olson v. Regents (1969). 301 F. Supp. 1356.

Peale v. U.S. (1971). 325 F. Supp. 193.

Perry v. Sindermann (1972). 408 U.S. 593.

Pickering v. Board of Education (1968). 391 U.S. 563.

POSEY, R. (1968) "The new militancy of public employees." Public Administration Review, 28(March/April):111-117.

RASKIN, A.H. (1972). "Politics up-ends the bargaining table." Pp. 122-146 in S. Zagoria (ed.), Public workers and public unions. Englewood Cliffs, N.J.: Prentice-Hall.

ROSENBLOOM, D.H. (1971a). Federal service and the Constitution. Ithaca, N.Y.: Cornell University Press.

——— (1971b). "Some political implications of the drift toward a liberation of federal employees." Public Administration Review, 31(4):420-426.

ROSENBLOOM, D.H., and GILLE, J.A. (1975). "The current constitutional approach to public employment." University of Kansas Law Review, 23(2):249-275.

Rozman v. Elliott (1972). 467 F2d 1145.

Russo v. Central School District (1972). 469 F2d 623.

Shelton v. Tucker (1960). 364 U.S. 479.

STORING, H. (1964). "Political parties and the bureaucracy." Pp. 137-158 in R. Goldwin (ed.), Political parties, U.S.A. Chicago: Rand McNally.

Sugarman v. Dougall (1973). 413 U.S. 634.

United Federation of Postal Clerks v. Blount (1971). 325 F. Supp. 879.

United Public Workers v. Mitchell (1947). 330 U.S. 75.

U.S. Civil Service Commission (1976). "A self-inquiry into the merit system." Washington, D.C.: Author.

U.S. Senate, Judiciary Committee (1967). Protecting privacy and the rights of federal employees: Report 519. Washington, D.C.: Author.

U.S. v. Rasmussen (1963). 222 F. Supp. 430.

Vulcan Society v. Civil Service Commission (1973). 360 F. Supp. 1265.

Washington Post (1970). October 10, p. B1.

Washington v. Davis (1976). 48 L Ed 2d 597.

Western Addition Community Organization v. Alioto (1971). 330 F. Supp. 536.

5

Institutional Barriers to Equity in Local Government Employment: Implications for Federal Policy Derived from Attitudinal Data

FRANK J. THOMPSON

□ WHEN FEDERAL POLICYMAKERS CONTEMPLATE PROGRAMS to assist disadvantaged job hunters, these officials often turn to local government. To paraphrase Willie Sutton, this is partly because "that's where the jobs are." Excluding the military, local agencies contain well over half of all public jobs. Despite the recent recession and headlines proclaiming layoffs, employment in local government rose each year during the sixties and seventies. As of October 1975, local agencies employed the equivalent of 7,369,000 full-time personnel. This represents an 18% increase since 1970 and a 75% jump since 1960 (U.S. Bureau of the Census, 1975). Furthermore, the need for additional personnel to deliver important public services at the local level seems substantial. True, no one knows the precise point at which the marginal return on additional local employees will drop sharply. Nor would any serious student of unemployment argue that local units can productively absorb all those without jobs. There is nonetheless substantial agreement that city and county agencies can effectively utilize many more personnel (Harrison and Osterman, 1974; Levitan, 1975; Nathan, 1975; Sheppard, 1972).

Beyond sheer numbers, several other factors contribute to the attractiveness of local government as an employer. Local agencies generally require a less skilled work force than other levels of government. Public jobs in the cities also tend to be geographically accessible to the deprived, since most of the land allocated to "public administration" is in the central city. Manufacturing firms

AUTHOR'S NOTE: *The author wishes to thank Charles H. Levine and Benna Thompson for helpful comments on an earlier draft.*

by contrast tend to be located in the city's outskirts. Public jobs are, therefore, less likely to require the reverse commuting which only a small number of disadvantaged residents of the central city find profitable to undertake (Harrison and Osterman, 1974:314). Finally, local agencies are attractive employers because they operate in the primary labor market. This market features jobs that are stable, pay relatively well, and provide ample fringe benefits. Working for local government can, therefore, offer a meaningful exit from poverty.

Because of the position of local agencies in the labor market, federal programs often attempt to use them as instruments of manpower policy. One subset of these programs has sought to encourage local units to reduce institutional barriers to employment of the deprived. Efforts to promote equal employment opportunity are a conspicuous example in this regard, but the federal government's foray against institutional barriers goes beyond this. It includes such programs as Project PACE Maker (Public Agency Career Employment Maker), which called on the National Civil Service League to provide technical assistance to state and local agencies interested in removing legal, administrative, and other barriers to hiring the disadvantaged. It also includes those sections of the Emergency Employment Act of 1971 and the Comprehensive Training and Development Act of 1973, which urged local officials to reduce hiring barriers.

At the heart of the attack on obstacles to employment is a challenge to the human capital approach to manpower policy. This approach, which dominated the thinking of federal officials during much of the sixties, attributed the difficulties that the deprived face in finding jobs to their relative lack of education, experience, and skills. Proponents of the institutional view have played down this emphasis (e.g., Harrison and Osterman, 1974; Connolly, 1976). As the National Civil Service League (1973:3) put it, "The old and tired assumption that the people did not fit the jobs, and therefore could not be employed, was discarded in favor of an assumption that the system was not working right."

The success of federal efforts to reduce institutional barriers to employment is far from clear. On balance, however, federal officials have not placed a high priority on encouraging local governments to remove these obstacles. Although the Emergency Employment Act and the Comprehensive Employment and Training Act led local officials to bend civil service rules, these acts precipitated little effort to eliminate artificial hiring and promotion barriers (Fechter, 1975:135; Hamilton and Roessner, 1972; Harrison and Osterman, 1974: 330-331). Nor have problems in the equal employment arena vanished. Minorities continue to be "underrepresented" in many types of positions (U.S. Equal Employment Opportunity Commission, 1974). Furthermore, the Equal Employment Opportunity Commission appears to be so starved for resources and so riven with internal conflict that it cannot vigorously enforce equal employment requirements in local government. As of early 1976, the commission had a backlog of between 120,000 and 150,000 cases (*Federal Times*, 1976;

see also the U.S. Commission on Civil Rights, 1975). Problems involving institutional barriers do, then, persist.

PURPOSE, HYPOTHESES, METHOD

This essay focuses on the attitudes which local personnel officials, or manpower specialists, hold concerning institutional barriers to the employment of the disadvantaged.[1] In this regard, it explores:

- The fundamental hiring values of these officials, such as their commitment to the merit principle.
- Their orientations toward specific selection criteria, such as written tests, education, and experience.
- Their attitudes toward affirmative action for racial minorities.
- Their orientations toward job creation and training programs for the deprived.

An analysis of these values and beliefs will help elucidate the degree of attitudinal support present for breaking down institutional barriers. Such analysis has, among other things, implications for those federal officials who must grapple with issues of manpower policy.

For purposes of this essay, "institutional barrier" will refer to any behavior by officials within a public agency (including a conscious decision to do nothing) which impedes the opportunities of the disadvantaged to obtain jobs. In focusing on barriers, the author will make no attempt to distinguish between those that are artificial and those that are not. This is partly because practices that are artificial in one light (e.g., from the perspective of hiring the most adroit) are often quite rational from another vantage point (e.g., in terms of reducing the economic costs of recruitment). Moreover, the meaning of "artificial" is ambiguous even if one accepts "hiring the most competent" as the sole value to be considered in assessing recruitment processes. It is hard enough to define and measure competence, let alone firmly label certain selection criteria as truly predictive of performance. Given these uncertainties, calling some barriers artificial and others not is to risk entrapment in intellectual quicksand.

PROFESSIONALISM AND SOCIOPOLITICAL BELIEFS: GENERIC SOURCES?

In addition to describing attitudes related to institutional obstacles and suggesting some of their policy implications, this essay explores two potential sources of these attitudes: the professionalism of manpower specialists, and their sociopolitical views. Professionalism has often been touted as a major determinant of how administrators will perform. In this regard, some observers have suggested that professionalism tends to retard responsiveness to the disadvantaged (e.g., Fainstein and Fainstein, 1974; Palumbo and Styskal, 1974). Despite these claims, the author expected that personnel officials who were more professionalized would favor a lowering of institutional barriers. This seemed likely because many centers of professional thought (e.g., personnel associations, universities) have stressed the need for administrators to be sensitive to the problems of minorities and other disadvantaged groups.

In addition to professionalism, the author expected that certain sociopolitical beliefs (namely, localism, liberalism, and awareness of inequality) would contribute to an understanding of attitudes pertaining to institutional barriers. Administrators with localistic orientations are primarily interested in community as opposed to national or international affairs and pay highest respect to individuals with community reputations rather than national ones (Dye, 1963). In the case of institutional obstacles to employing the deprived, it seems likely that administrators who are more localistic will be less inclined to reduce these obstacles. Much of the support for removing barriers comes, after all, from such national institutions as the federal court system, the Congress, and such federal bureaucracies as the Equal Employment Opportunity Commission. Political leaders and other elites at the local level are seldom in the vanguard of those stressing the need to modify hiring practices in order to assist the disadvantaged.

In addition to suggesting the importance of localism, existing evidence indicates that normative political perspectives do much to shape attitudes toward work related issues (e.g., Mechanic, 1974:69-87). In this vein, political liberalism may well correlate with attitudinal support for measures favorable to deprived job seekers.

Perceptions of inequality comprise a third kind of sociopolitical outlook likely to be important. Irving Kristol (1968:110) observed that in contemporary political thought, "Every inequality is on the defensive [and] must prove itself against the imputation of injustice and unnaturalness." To be aware of inequality is in many instances to favor programs designed to mitigate it. Such a reaction seems especially probable in the case of racial inequality, since few can easily justify the special handicaps that minorities have faced in their efforts to compete for the economic prizes of this society. Hence the author hypothesized that those who are more aware of racial inequality will be more likely to hold attitudes conducive to reducing institutional barriers.

The extent to which the data support the hypotheses concerning professionalism and sociopolitical beliefs has implications for federal policy. If measures of professionalism correlate with attitudes conducive to reducing institutional barriers, one implication is to lend support to those who see the "low quality" of local personnel as a major obstacle to the effective implementation of federal programs (Pressman, 1975:9). It would imply that efforts by the federal government to encourage the professionalization of personnel staffs might well pay dividends in the attainment of other manpower policy objectives.[2] By contrast, if sociopolitical views are more powerful predictors of attitudes concerning institutional barriers to hiring, the problem for federal policy makers becomes much more complex. Who knows a feasible means for transforming the basic normative views (e.g., liberalism) of officials? The difficulties involved in modifying sociopolitical views which are essentially cognitive or perceptual (e.g., awareness of inequality) are probably less but are still formidable.

THE SAMPLE

In order to test the hypotheses concerning professionalism and sociopolitical views and to accomplish the more basic purposes of this study, the author sampled the attitudes of a group of personnel administrators. Among those officials influencing recruitment to government, these administrators were among the most important. An analysis completed by the National Civil Service League (1972, 1973) testified to their significance. After studying obstacles to employing the disadvantaged in local governments, the league concluded that almost all necessary reforms could be implemented by personnel directors through administrative interpretations. Modifications of statutes or other formal changes in the rules were not necessary. There is no question that personnel officials tend to exert influence over decisions which affect the *degree* to which institutional barriers will be reduced, the *rate* at which such a reduction will occur, and the *political costs* which those outside local government must pay in an effort to bring about change (e.g., Thompson, 1975:112-139).

The importance of these manpower specialists prompted the author to mail questionnaires to 980 members of the International Personnel Management Association in 1975.[3] This sample was randomly drawn from a total of 1,351 individuals identified in the association's membership directory as working for city hall, county government, or some local "independent district" (e.g., school district). 630 or 64% of the sample returned the questionnaires.[4] Of those responding, 68% worked for city government, 19% for county units, and 13% for special districts. Nearly half (47%) worked in large cities of 100,000 or more; 20% were employed in medium-sized cities of from 50,000 to 99,999; 33% worked in small communities of less than 50,000. The respondents came from all regions of the country. As might be expected, they were overwhelmingly white (95%) and male (86%).

It is impossible to say precisely how representative of all personnel administrators in the country these respondents were.[5] Suffice it to note that the subjects occupied influential positions. Nearly all personnel directors in cities of 100,000 or more belonged to the professional association at the time of this survey. Many directors of county personnel departments were also members. It should come as no surprise, therefore, that 46% of the subjects indicated that they were personnel directors or chief personnel officers. Many others reported that they held jobs of comparable status such as assistant city manager in charge of personnel. Moreover, respondents certainly perceived themselves as influential in hiring processes. Only 12% agreed with the statement: "I don't think top officials care much what people in my position think about hiring." Only 14%

concurred that "People like me don't have much say about government recruitment practices." Finally, a mere 10% thought that they had "little decision-making power" on their jobs.

The fact that respondents appear to have considerable say over recruitment heightens the behavioral significance of their attitudes. Attitudes are, after all, more likely to affect behavior if an official has the resources with which to translate these sentiments into deeds.[6] The survey data in the following section should, then, provide important clues concerning the propensities of local officials to reduce institutional barriers to the hiring of the underprivileged.

ATTITUDES CONCERNING INSTITUTIONAL BARRIERS: A SUPPORTIVE MILIEU?

A milieu supportive of reductions in institutional barriers to employment would be characterized first by hiring values which strongly emphasize the social responsibility of government to recruit the deprived and which deemphasize the traditional merit principle (i.e., "hire the most competent applicant available"). An administrator who adheres to these views is apt to listen to arguments on behalf of the disadvantaged. One such argument is that hiring deprived applicants can at times do more to enhance organizational effectiveness than recruiting the most competent individuals. The typical example here is the local agency that somehow manages to recruit the most adroit but fails to be effective because its work force is racially "unrepresentative" and engenders hostility in the minority community. Another argument often presented is that recruitment processes should embody a concern not only with organizational effectiveness but with social equity as well (e.g., Kranz, 1974). Proponents of these arguments often see virtue in "compensatory hiring" or "hiring the qualified" rather than the most skilled.

Certain attitudes concerning recruitment techniques would also be part of a supportive mileu. In this respect, the milieu would be characterized by uncertainty on the part of administrators as to how well they achieve merit objectives and the best means for promoting these objectives. Uncertainty is often the mother of flexibility. Officials who feel less confident about how to improve their scores in the merit game or who doubt that they know what the score is may be more open to modifying selection processes. In particular, they may be more receptive to arguments that certain changes likely to benefit disadvantaged job hunters will also be consistent with merit objectives.[7] A supportive milieu would also feature the endorsement of those selection strategies least likely to damage the job prospects of the disadvantaged. In addition, officials would indicate a willingness to reduce the use of those criteria which work against the interests of the deprived.

Besides featuring certain beliefs concerning recruitment procedures, a supportive milieu would be characterized by favorable views toward affirmative action. Officials would be willing to endorse such strategies as special hiring campaigns in the minority community and the use of employment targets and timetables.

A supportive milieu would also be filled with officials who are enthusiastic about special job and training programs sponsored by the federal government. They would approve of on-the-job training programs for the disadvantaged, welcome federal subsidies for the creation of new jobs, and favor redesigning jobs so that they do not call for as much education and experience.

To the extent that administrators vary in their attitudes, a supportive milieu would be one in which those working in large cities (100,000 or more) hold attitudes that are the most conducive to reducing institutional barriers. Problems of deprivation, unemployment, and general social pathology are probably most acute in large cities. Help for the disadvantaged may well be more crucial in such settings.

These, then, are some of the attitudinal characteristics of a milieu which would tend to facilitate reductions in institutional barriers. The data should provide clues concerning the degree to which such a milieu exists.

OVERARCHING HIRING VALUES

In order to explore basic values toward hiring, the questionnaire included three Likert items pertaining to merit and one concerning social responsibility. Analysis of the responses to these questions indicates that a supportive milieu was partly present (see Table 1). A substantial majority of administrators did agree that government has a social responsibility to recruit the disadvantaged. Moreover, officials in large cities were more likely to endorse this view than administrators in cities of less than 50,000. By the same token, however, an even larger proportion of officials endorsed the traditional merit precept. Regardless of the size of the cities in which they worked, at least nine out of 10 administrators adhered to the view that government ought to hire the most competent people available.

What explains the capacity of officials to endorse both merit and social responsibility? Perhaps, they had never explicitly considered the possible trade-offs between the two values. Or perhaps these administrators had thought about the issue but had concluded that there are no serious trade-offs between the two objectives. Or it may be that officials saw both merit and responsibility goals as praiseworthy but ranked one or the other higher on their list of priorities depending on the circumstances of a given case. Whatever the explanation, the respondents' overwhelming endorsement of the traditional merit principle could spell trouble for efforts to reduce institutional barriers to employment of the deprived.

TABLE 1
ATTITUDES TOWARD MERIT AND RESPONSIBILITY
TO THE DISADVANTAGED BY CITY SIZE[a]

Statement	100,000+		50,000-99,999		Less than 50,000	
	Agree	Disagree	Agree	Disagree	Agree	Disagree
Merit						
"Governments should try to hire the most competent people available."	97% (N = 295)	2%	99% (N = 122)	1%	98% (N = 210)	0%
"Governments should try to hire the most competent people available even if others oppose such efforts."	91 (N = 295)	4	93 (N = 122)	3	92 (N = 209)	3
"It's impractical for public agencies to try to hire the most competent people available."	6 (N = 293)	92	6 (N = 122)	92	8 (N = 210)	89
Social Responsibility						
"Government has a social responsibility to recruit members of socially disadvantaged groups."[b]	74 (N = 293)	15	70 (N = 122)	20	57 (N = 208)	30

a. The percentage of those who said they did not know or had no opinion is not presented. A similar practice is followed in subsequent tables.

b. Chi square indicates that differences among the three groups are significant at the .01 level.

MEANS

If a supportive milieu is only partially present in the case of general values toward hiring, it may nonetheless exist with respect to orientations toward hiring techniques. For all their commitment to merit hiring, administrators might be uncertain about how to accomplish it; they might also favor selection criteria likely to benefit deprived job seekers.

Uncertainty

Whatever the advantages of uncertainty, most of the respondents revealed little of it in responding to two items on the questionnaire (see Table 2). Roughly half of the subjects affirmed that they could tell whether their departments were hiring the most competent applicants available. So, too, more than half rejected the statement that it was hard to tell what the best methods for

TABLE 2
UNCERTAINTY ABOUT MERIT BY CITY SIZE

Statement	100,000+		50,000-99,999		Less than 50,000	
	Agree	Disagree	Agree	Disagree	Agree	Disagree
"Most of the time government officials can tell whether their departments hire the most competent applicants available."	48%	37%	62%	31%	55%	32%
		(N = 293)		(N = 121)		(N = 208)
"It's hard to tell what the best methods for recruiting skilled employees are."[a]	36	60	39	56	45	43
		(N = 291)		(N = 120)		(N = 207)

a. Chi square differences among the three groups are significant at the .01 level.

recruiting skilled employees are. There was some, though not marked, variation in responses by city size. Respondents working in large cities were the least likely to claim that they knew whether their agencies hired the most competent applicants; but these same respondents were also the least likely to acknowledge uncertainties concerning the best methods for recruiting skilled employees.

Overall, the dominant impression conveyed by the data is that many officials felt little uncertainty about how to promote merit objectives. This absence of uncertainty might well make it more difficult to reduce institutional barriers to the disadvantaged.

Screening Criteria

Aside from the issue of uncertainty, a supportive milieu would be more in evidence if officials deemphasized selection criteria which tend to work against the interests of disadvantaged job hunters. Among the criteria that would be played down are written tests, formal education, extensive job experience, and background checks of moral character. Written tests have gained notoriety for screening out minorities disproportionately (e.g., Jencks et al., 1972:81-82; C. Moore et al., 1969; Sadacca, 1971). Educational requirements have often kept the disadvantaged out of jobs and appear to be set at needlessly high levels in many instances (Berg, 1971). Then, too, job experience requirements have worked against the interests of the disadvantaged. The deprived often spend time shifting from one poorly paying job to the next in the secondary labor market. For that reason their work histories are too checkered to suit the tastes of those personnel officials who attach great importance to previous experience. In addition, an emphasis on experience poses a particularly formidable barrier to minority youths in core cities who tend to lack work histories and who suffer from extremely high unemployment rates. Background checks into "moral character" have also taken a particularly heavy toll on the disadvantaged. Often these checks have been used capriciously to screen out those from "different" class and ethnic backgrounds.

While a supportive milieu would play down certain criteria, it would also put stronger emphasis on the oral examination or interview. Interviews have in the past provided officials with a handy means of discriminating against minorities. More recently, however, the oral examination has won recognition as one of those criteria which do the least damage to disadvantaged job hunters (National Civil Service League, 1972:225). Among other things, it gives more discretion to officials to take into account strengths that deprived applicants may have.

In an effort to probe attitudes toward selection, a number of standard screening criteria appeared on the questionnaire. The instructions requested the respondents to check all those criteria that they considered "very important" in selecting, first, white-collar or professional employees, and, second, unskilled or subprofessional applicants.

Responses indicate that a supportive milieu was only partially present (see Table 3). Interview performance did rank relatively high on the list of important screening criteria and most officials did not attach great importance to background checks of moral character, written tests, or education (especially for lower level posts). But there is another, less Panglossian side of the picture. Previous job experience attracted overwhelming endorsement as a very impor-

TABLE 3
SELECTION CRITERIA CONSIDERED VERY IMPORTANT BY CITY SIZE

Criteria	100,000+ % Labeling as Very Important	50,000-99,999 % Labeling as Very Important	Less than 50,000 % Labeling as Very Important
Professional-White Collar			
Previous job experience[a]	87	96	91
Interview performance	61	66	71
Education	56	52	61
Written tests[b]	34	19	18
Recommendations	23	30	33
Background check of moral character	17	16	22
Subprofessional-Unskilled			
Previous job experience	85	85	87
Interview performance[a]	45	49	56
Recommendations	26	30	33
Background check of moral character	18	13	21
Written tests	17	15	15
Education	10	11	9

a. Chi square indicates differences significant at the .05 level.
b. Chi square indicates differences significant at the .01 level.

tant criterion. Well over 80% of all respondents believed that experience is important for both professional and subprofessional positions. Nor did larger cities sustain a more supportive milieu for reducing institutional barriers. In fact, manpower specialists in large cities were more than twice as likely as those in communities of less than 50,000 to consider written tests a very important device in selecting people for higher level posts. In general, however, variation among officials working in cities of different sizes was small.

Change?

The question of which criteria administrators considered very important is related to but is not precisely the same as the question of which hiring practices they were disposed to change. A supportive milieu would be one in which administrators directly expressed a willingness to reduce institutional barriers to employment. In order to probe this orientation toward change, the questionnaire included three items concerning educational requirements, demands for previous experience, and the use of written tests. The responses suggest that there was considerable support for the status quo (see Table 4). About half of the administrators believed that educational requirements for most public jobs were about what they should be while a quarter thought that these requirements

TABLE 4

BELIEFS ABOUT CHANGING SCREENING CRITERIA BY CITY SIZE

Statement	100,000+	50,000-99,999	Less than 50,000
Educational requirements for most government jobs:[a]			
should be increased	11%	14%	19%
are about right as they are	50	47	46
should be reduced	30	23	16
don't know	10	16	19
	(N = 291)	(N = 121)	(N = 208)
Job experience requirements for most government positions:			
should be increased	16%	23%	18%
are about right as they are	49	48	52
should be reduced	22	19	14
don't know	13	10	16
	(N = 293)	(N = 122)	(N = 209)
Believe that:			
written tests should be discontinued in most cases.	8%	8%	8%
in one form or another, written tests should be continued.	89	91	89
other or don't know.	3	1	3
	(N = 291)	(N = 120)	(N = 207)

a. Chi square indicates differences are significant at the .01 level.

should be reduced. Furthermore, about half thought that current requirements for previous experience were satisfactory while less than a quarter favored their reduction. Nearly nine out of 10 administrators favored the continuation of some form of pencil-and-paper tests. In the cases of experience and education, specialists in large cities were more inclined to reduce requirements. But half of them were still committed to the status quo.

THE SPECIAL CASE OF AFFIRMATIVE ACTION FOR MINORITIES

A supportive milieu for reducing institutional barriers would also find personnel specialists favoring affirmative action for minorities. Responses to four questions concerning affirmative action indicate that officials were willing to endorse some practices but not others (see Table 5). The great bulk of administrators believed that government should make a special effort to advertise job vacancies in the minority community. Most respondents also thought that a minority job hunter should be given preference over a white one where underrepresentation exists and the minority and white applicants are of comparable ability. But there were severe limits as to what officials felt it was proper to do

TABLE 5
ATTITUDES TOWARD AFFIRMATIVE ACTION MEASURES BY CITY SIZE

Proposal	100,000+		50,000-99,999		Less than 50,000	
	Agree	Disagree	Agree	Disagree	Agree	Disagree
Make a special effort to advertise job vacancies in the minority community.[a]	92% (N = 292)	4%	89% (N = 119)	7%	81% (N = 210)	11%
Have public agencies establish hiring targets for minorities as well as timetables for obtaining them.[a]	43 (N = 293)	46	40 (N = 122)	45	31 (N = 208)	54
Where minority and white applicants are of about equal ability and minorities are underrepresented on the department's work force, give preference to minority applicants.	75 (N = 293)	17	71 (N = 122)	19	68 (N = 209)	20
Government recruiters should *not* hire a minority applicant if there is a more competent applicant who wants the job.	63 (N = 293)	20	67 (N = 122)	20	75 (N = 207)	14

a. Chi square indicates that differences among cities of different sizes are significant at the .01 level.

on behalf of nonwhite job seekers. A plurality of respondents opposed the use of hiring targets and timetables for nonwhites. Nearly two-thirds of them believed that government should not hire a minority applicant if there is a more competent individual who wants the job.

The findings concerning affirmative action are consistent with the data presented earlier on commitment to merit. Those proposals which were seen as involving the least threat to merit hiring evoked the widest support. Thus, officials were most amenable to altering the way in which they advertised for employees, since that involves fewer obvious risks to competence levels than do modifications in selection processes (Thompson, 1975:131). Similar logic undergirds the "other things being equal, hire a minority" rule. Use of such a rule permits administrators to have their cake and eat it too. They presumably sacrifice nothing in competence but still "prefer" minorities. By contrast, targets and timetables have often been portrayed as a threat to recruitment based on merit. The U.S. Civil Service Commission has, for example, long resisted adoption of these devices on grounds that they erode open competition for jobs and threaten merit hiring (Rosenbloom, 1973). The reluctance to sacrifice skill is even more manifest in the way that administrators reject the idea of hiring a minority person if it means passing over a more competent applicant.

Since officials are loath to sacrifice skill in order to hire minorities, the future of affirmative action in part hinges on the belief that minority employees will increase the competence of the civil service. When questioned in this regard, however, officials were not especially optimistic (see Table 6). Less than one in 20 believed that if the government rapidly hires more minorities, competence will increase. Roughly half suggested that such hiring would make no difference one way or the other. Nearly half were of the opinion that there would be some

TABLE 6
BELIEFS ABOUT THE IMPACT OF MINORITY HIRING ON COMPETENCE
BY CITY SIZE

Belief	100,000+	50,000-99,999	Less than 50,000
What do you think is most likely to happen if government greatly increases the proportion of minority employees over the next few years?			
Government employees will be *more competent* than ever.	4%	3%	3%
There will be *no impact* on the overall competence of government employees.	50	50	45
There will be *some* decline in the overall competence of government employees.	47 (N = 288)	48 (N = 117)	52 (N = 207)

decline in the competence of the civil service if more nonwhites rapidly joined the payrolls.

JOB CREATION AND TRAINING

A supportive milieu for reducing institutional barriers to employment would also be one in which officials endorse a variety of job creation and training programs. If personnel administrators are averse to such programs, they may fail to seize opportunities to establish manpower projects; or they may offer half-hearted support when the time comes to implement the projects.

Responses to three questions indicate that officials were the most favorably disposed toward on-the-job training programs (see Table 7). Regardless of city size, close to two-thirds of the manpower specialists supported increases in subsidies for such training. Support of this magnitude evaporated when officials expressed their views concerning the desirability of subsidies to create public jobs for the unemployed. Slight pluralities in cities over 50,000 favored such a step, while the reverse held in smaller cities. Finally, the restructuring of jobs precipitated the most opposition.[8] In cities of all sizes, at least a plurality of officials opposed redesigning jobs so that they would not call for as much education and experience.

In terms of immediate work-related considerations, the limited endorsement that job programs receive may stem primarily from notions about the burdens that these programs impose on personnel staffs. In particular, redesigning jobs

TABLE 7

ATTITUDES TOWARD ON-THE-JOB TRAINING, JOB RESTRUCTURING, AND JOB CREATION BY CITY SIZE

Statement	100,000+		50,000-99,999		Less than 50,000	
	Agree	Disagree	Agree	Disagree	Agree	Disagree
"Allocate more funds for extensive on-the-job training of disadvantaged applicants in public agencies."	65% (N = 292)	23%	67% (N = 121)	26%	61% (N = 206)	25%
"Redesign jobs so that they do not call for as much education and experience."[a]	42 (N = 293)	49	34 (N = 121)	59	25 (N = 209)	63
"Greatly increase federal subsidies for the creation of public jobs to help the unemployed."	43 (N = 292)	40	48 (N = 121)	40	39 (N = 208)	44

a. Chi square indicates differences among the three city sizes are significant at the .01 level.

calls for much skill and effort in task analysis (Fine, 1972).[9] Accepting federal subsidies to establish positions also creates costs for local personnel specialists. True, most studies suggest that local units have a number of needs that can readily be filled by such employees. Bringing new employees on board to fill these needs is, however, a complex and time-consuming task. Since federal funding commitments are often of short duration, many officials may feel that job programs are not worth the bother. Beyond this, participation in such programs may bring manpower specialists under fire from their local constituencies. The hackles of union leaders may be raised, for example, if personnel officials attempt to circumvent civil service rules in order to fill subsidized positions. Complaints from the unions may become particularly bitter if employees who have been laid off cannot apply for federally funded jobs. Whatever their precise rationale, personnel administrators as a group are not very enthusiastic about special job programs for the disadvantaged.

In sum, then, an examination of attitudinal data suggests that a supportive milieu for reducing institutional barriers to hiring the disadvantaged exists to a limited extent only. To be sure, officials do endorse the abstract notion that government has a responsibility to recruit the deprived. But when it comes to specific practices which would aid the underprivileged, support is much less in evidence. This finding is consistent with the findings of other studies which also point to a conflict or tension between abstract, general beliefs and specific attitudes (e.g., Westie, 1964:585-586).

THE ROLE OF PROFESSIONALISM AND SOCIOPOLITICAL BELIEFS

Before examining some of the policy implications of the data presented in the previous section, it is important to examine the role played by professionalism and sociopolitical beliefs in shaping attitudes concerning institutional barriers. Earlier in the paper, it was hypothesized that professionalism, liberalism, and awareness of inequality would correlate with attitudes which are more favorable to disadvantaged job hunters. Localism, by contrast, was hypothesized to be a source of more negative attitudes. Do the data support these hypotheses?

PROFESSIONALISM: BOON OR BANE?

Adopting Wilbert Moore's definition, officials may be seen as more professionalized to the extent that they practice a full-time occupation, reveal commitment to a calling, identify with their professional peers (often through formalized associations), possess skills based on specialized training of great duration, exhibit a service orientation, and have autonomy on the job (W. Moore, 1970:5-6; see also Mosher, 1968; Ritzer and Trice, 1969). Limited space on the questionnaire inhibited attempts to operationalize all dimensions of profession-

TABLE 8
PROFESSIONALISM AND SELECTED ATTITUDES CONCERNING
INSTITUTIONAL BARRIERS TO THE EMPLOYMENT OF THE DISADVANTAGED

Attitudes Concerning Institutional Barriers	Correlation r with Professional Characteristics						
	Greater Education	Graduate Specialization in Personnel	Belief in University	Full-time in Personnel	Greater Participation at Professional Meetings	More Contacts with Other Specialists	Belief in Reading Journals
Overarching Values							
Commitment to merit[a]	.06	-.06	.08c	-.06	.04	.04	.08c
Social responsibility	.09b	.05	.10b	.09b	.07c	.13b	.04
Selected Screening Criteria							
Very important criteria for unskilled workers:							
Job experience	-.06	.00	.04	.00	-.02	-.03	-.01
Written tests	.01	-.01	.04	.01	-.03	-.04	-.07c
Education	-.04	.01	.05	.01	-.09	.04	-.01
Interview	.02	-.03	.07c	-.04	.07c	-.02	-.02
Increased education requirements	-.03	-.05	.09b	-.22b	-.09b	-.07c	.13b
Increased experience requirements	-.11b	-.01	-.06	-.10b	-.09b	-.06	.02
Abolish written tests	-.02	.01	.00	-.02	.03	.06	-.05
Jobs and Training							
Job restructuring	.06	.00	-.04	.23b	.11b	.05	.03
Federal subsidies for jobs	.05	.02	.01	.06	-.01	.05	.06
Belief in training	.09b	-.10c	-.01	.07c	.01	.04	.09b
Affirmative Action for Minorities							
Advertise in minority community	.09b	.10c	.03	.21b	.07c	.12b	.07c
Belief in targets and timetables	.04	.04	.02	.13b	-.02	-.01	.04
Would prefer a minority person, other things being equal	.03	.15b	.05	.13b	.01	.05	.08c
Would *not* hire a minority person over a more competent white applicant	-.07c	-.14b	-.04	-.10b	-.08c	-.04	-.05

a. The three questions form a merit scale with a standardized Alpha coefficient of .6.

b. Significant at the .01 level.

c. Significant at the .05 level.

alism. A number of measures were, however, included. Three of them focused on the educational element. The author collected data on the level of education that the officials had attained, the degree to which they specialized in personnel administration during their graduate careers, and the extent to which they believe that the principles of administration taught in universities are relevant to their work.[10] Four other measures of professionalism dealt with the occupational involvement of manpower specialists and, as such, are rough measures of commitment to their calling and their identification with professional peers. These measures indicate whether officials work full-time in personnel, regularly attend professional conferences, hold frequent meetings with other personnel specialists to discuss common problems, and believe that reading professional journals is essential for effective performance.

Do these measures of professionalism correlate with attitudes which are favorable to disadvantaged job seekers? The data suggest that the relationship is tenuous at best (see Table 8).[11] Out of a total of 112 correlations, only 37, or about one-third, are significant at the .05 level or better. Of all the attitudes previously discussed, those relating to affirmative action are most closely linked to professionalism. Half of the coefficients in the minority hiring cluster are significant at the .05 level. Yet even in this cluster nearly all the significant correlations are very low. In fact, only three coefficients out of all those examined are greater than .20. Overall, the level of professionalism attained by administrative officials appears to have only a miniscule effect on their orientations concerning institutional barriers.[12]

Why does professionalism prove to be such an anemic predictor of attitudes concerning hiring impediments? One possibility is that the author failed to extract the theoretically relevant dimensions of professionalism and to measure them adequately. Another (and, the author believes, more compelling) possibility is that the universities and professional associations which socialize personnel officials have not in the past and do not at present speak with one voice on occupational matters. True, various centers of professional thought have recently emphasized the need for manpower specialists to do something for the disadvantaged. But there is probably no firm consensus (or community of values) concerning what that something is.

If this explanation is off target and such a consensus does exist, then other factors could account for the findings concerning professionalism's limited impact. Perhaps rank-and-file personnel officials do not accurately perceive what the consensus of opinion is in centers of professional thought. Or perhaps they do know what the consensus is, but take it with a grain of salt and look to other reference groups in formulating their views.

SOCIOPOLITICAL PERSPECTIVES

If professional characteristics possess such limited explanatory power, do sociopolitical perspectives fare much better? In order to gauge the role of localism in orienting officials toward hiring barriers, a measure developed by Thomas Dye (1963) was used. A typical question in this scale reads: "I have greater respect for a man who is well established in his local community than a man who is widely known in his field but has no local roots." In an attempt to tap political liberalism on domestic economic issues, the author used a scale developed by Herbert McClosky (1964). A typical item in this scale is: "Every person should have a good home even if the government has to build it for him." Awareness of inequality was measured by asking officials to compare the current

TABLE 9

LOCALISM, LIBERALISM, AWARENESS OF INEQUALITY, AND
ATTITUDES CONCERNING INSTITUTIONAL BARRIERS TO EMPLOYMENT

Attitudes Concerning Barriers	Localism Second Order[a]		Liberalism Second Order		Optimism about Black Condition Second Order	
	r	Partial r	r	Partial r	r	Partial r
Overarching Values						
Commitment to merit	.00	−.05	−.21[b]	−.20[b]	.08[c]	.06
Social responsibility	−.23[b]	−.15[b]	.34[b]	.30[b]	−.28[b]	−.22[b]
Selected Screening Criteria						
Very important criteria for unskilled workers:						
Job experience	.09[b]	.08[c]	−.01	.02	.08[c]	.06
Written tests	.04	.03	−.07[c]	−.06	.03	.01
Education	.02	.01	.01	.03	.08[c]	.09[b]
Interview	−.03	−.04	.00	.00	.05	.06
Increase education requirements	.12[b]	.07[c]	−.09[b]	−.05	.21[b]	.18[b]
Increase experience requirements	.15[b]	.11[b]	−.11[b]	−.07[c]	.23[b]	.20[b]
Abolish written tests	−.03	−.02	.01	.00	−.05	−.05
Affirmative Action for Minorities						
Advertise in minority community	−.21[b]	−.16[b]	.12[b]	.06	−.27[b]	−.22[b]
Belief in targets and timetables	−.15[b]	−.08[c]	.21[b]	.17[b]	−.29[b]	−.24[b]
Prefer a minority person other things being equal	−.19[b]	−.14[b]	.09[b]	.04	−.22[b]	−.18[b]
Would *not* hire a minority person over a more competent applicant	.17[b]	.12[b]	−.16[b]	−.11[b]	.21[b]	.17[b]
Jobs and Training						
Job restructuring	−.09[b]	−.03	.24[b]	.21[b]	−.22[b]	−.18[b]
Subsidies for job creation	−.12[b]	−.04	.52[b]	.51[b]	−.12[b]	−.03
Belief in training	−.15[b]	−.07[c]	.36[b]	.33[b]	−.23[b]	−.18[b]

a. In each case the second order partial represents the association between the independent and dependent variable controlling for the two other sociopolitical beliefs.

b. Significant at the .01 level.

c. Significant at the .05 level.

socioeconomic status of blacks with that attained by other social groups (Rossi et al., 1974). Possible responses ranged from a statement that blacks were treated better than any other segment of the population, to an assertion that they were treated worse.

The data provide some support for the hypotheses concerning sociopolitical beliefs (see Table 9). Out of a total of 48 correlations, 36, or 75%, are significant at at least the .05 level and are in the expected direction. The strength of the correlations is less impressive. Nonetheless, nearly six times as many sociopolitical coefficients (17) attain a level of .20 or greater than was the case with professionalism.

The three sociopolitical beliefs differ in the degree to which they are linked with attitudes concerning institutional obstacles. Optimism about the black condition emerged as the most consistent predictor of attitudes toward hiring barriers, achieving a statistically significant relationship with 13 of the 16 attitudes. Nine of these coefficients are greater than .20. Political liberalism also ranks high, attaining a significant relationship with 12 of the attitudes concerning institutional barriers. Six of these significant relationships are above .20, and three are greater than .30. Localism by contrast is the least impressive predictor. This scale achieves a significant association with 11 of the 16 dependent variables, but only two correlations are greater than .20.

Knowing the relationship between each sociopolitical variable and selected attitudes concerning institutional barriers does not, of course, tell us very much about the independent impact of each sociopolitical belief. It is possible, for example, that the apparent effects of localism derive from a tendency for locals to be more conservative politically or more optimistic about the black condition. If, however, one examines the partial correlations between each of the sociopolitical beliefs and attitudes concerning hiring barriers while controlling for the other two sociopolitical variables (see Table 9), the results suggest that localism, liberalism, and awareness of inequality each contribute to the explanation of attitudes concerning hiring (albeit to a limited degree).

Whatever the precise role of each sociopolitical variable, it should be emphasized that the explanation generated by sociopolitical beliefs varies from one cluster of attitudes concerning institutional barriers to the next. The strongest and most consistently significant relationships were evident with respect to affirmative action and job programs. Perceptions of racial inequality comprised the most powerful predictor of attitudes toward affirmative action. Those aware of racial inequality apparently viewed this social condition as unjust and saw affirmative action as an appropiate response to it. By contrast, political liberalism was more strongly related to support for job and training programs. This is not surprising, since support for job programs has almost become synonymous with liberalism itself.[13] Sociopolitical beliefs are also consistently related to overarching hiring values. As with job and training programs, political liberalism achieves the strongest associations with these hiring values. Officials were apparently able to see the implications of their political ideology in this area.

The relationships between sociopolitical beliefs and specific screening criteria were by far the least impressive of those examined. A likely explanation is that specialists do not bother to relate their more basic social perceptions and ideologies to the nuts and bolts of their administrative lives. Alternatively, they may try to forge a mental link between sociopolitical convictions and attitudes towards recruitment, and yet reach different conclusions concerning the implications of these convictions for hiring practices. Or even if officials agree on the hiring implications of their sociopolitical convictions, they may remain committed to certain selection criteria out of deference to other values.

In sum, sociopolitical beliefs (especially liberalism and awareness of inequality) are more powerful predictors of attitudes concerning institutional barriers to hiring than the professional characteristics of officials. While sociopolitical beliefs do not provide a definitive explanation,[14] the data presented in this and the previous section are sufficiently cogent to consider some of their ramifications for policy.

POLICY IMPLICATIONS

A discussion of policy inferences requires clarification of underlying assumptions about the labor market. Borrowing from the work of economist Lester Thurow (1975), we may see the labor market less as a market for existing skills than as a market where training slots must be allocated to different applicants. The task of the employer is typically to select that applicant who will attain the most acceptable ratio of performance to on-the-job training costs within a certain period of time. Translated into merit terminology, the most competent applicant becomes the one predicted to have the most "acceptable ratio." Employers, of course, use various screening criteria to predict the competence and training costs associated with a job seeker. On the basis of these criteria they determine which job seekers to include in the labor queue for a position and how to rank them.

Given this conception of the labor market, this author can consider three approaches for federal policy makers concerned with reducing institutional barriers to the disadvantaged. One approach emphasizes a fundamental value transformation. It asks local officials to abandon or suspend their commitment to hiring the candidate perceived as having the "best" competence/training ratio. Another approach is essentially technical. It accepts the merit orientation of officials as a given but strives to alter the perceptions of officials concerning the most suitable means for attaining merit objectives. More specifically, it involves changing perceptions in ways that will lead more disadvantaged workers to get into labor queues and to attain a higher ranking in these queues. A third approach is basically expansionist. It does not challenge basic hiring values or perceptions of appropriate screening criteria but instead seeks to increase the willingness of officials to create new jobs for which the disadvantaged can compete. These approaches are not, of course, mutually exclusive. They are isolated for analytic purposes only.

VALUE TRANSFORMATION: THINKING THE UNFEASIBLE

Some students of public administration have contemplated a fundamental value transformation within bureaucracies. Articles assessing the administrative implications of John Rawls' egalitarian theories have appeared in this connection (e.g., Hart, 1974). Moreover, a number of students of personnel administration have essentially dismissed the traditional merit precept as passé (e.g., Stanley, 1974). The data collected for this study, however, offer little encouragement to those who would help disadvantaged job hunters through a fundamental transformation in hiring values. True, most of our manpower specialists believed that government has a social responsibility to recruit the deprived. But even larger majorities endorsed the traditional merit precept. Moreover, less than one in every four administrators would hire a minority applicant if a more competent applicant were seeking the job. The fact that the values of meritocracy have persisted during a period of challenge and turbulence in personnel administration provides ample testimony to their staying power. Further compounding the difficulties involved in modifying these hiring values is the fact that one of their stronger correlates is political conservatism rather than professionalism. If hiring values have roots in more fundamental political ideology, they may be especially resistant to change.[15] Thus proposals which depend on officials reducing their concern with attaining the most "appropriate" levels of performance in relation to training costs are in for rough sledding.

A TECHNICAL APPROACH: CREATE UNCERTAINTY AND FOCUS VALIDATION

If approaches rooted in a basic value transformation are hardly feasible, a technical approach seems more promising. Such an approach aims at altering the ways in which public agencies advertise job vacancies and at modifying notions about the appropriate criteria to be used in including and ranking applicants in labor queues. In this regard, certain attitudes held by administrators at least have the potential for being translated into behavior beneficial to the deprived. Thus the great majority of officials believed that public agencies should go out of their way to inform minority job hunters of vacancies. Most officials did not think that education or written tests are very important screening criteria. Greater percentages believed that the interview is an important selection mechanism. Then, too, when other criteria placed minority and white applicants in roughly the same position in the labor queue, most manpower specialists favored breaking the tie in a way that benefited nonwhites. The fact that convictions about appropriate screening criteria are not closely related to more generic sociopolitical views also provides a basis for optimism. It suggests that attitudes toward specific selection methods might undergo change without altering or threatening more fundamental political identities.

On the other hand, most respondents opposed the abolition of written tests. Officials showed great respect for work experience as a predictor of competence. Most of them were not inclined to reduce education or experience requirements for jobs. Many of them appeared to have drawn firm conclusions as to what the best methods for recruiting skilled employees are. Then, too, those devices— targets and timetables—which place more pressure on officials to include minorities in labor queues and rank them favorably failed to win the endorsement of most specialists. The fact that nearly half of the respondents believed that rapidly hiring more minorities would damage civil service competence suggests that, without the pressure induced by such devices as targets and timetables, gains in minority employment are likely to come slowly.

Proponents of affirmative action might conceivably take some consolation in the coefficients which suggest that mere recognition of inequality (rather than localism or liberalism) engenders support for recruiting nonwhites. This relationship suggests that campaigns to make administrators more aware of the facts concerning inequality may be helpful. But getting officials to recognize social disparities is no simple matter. The fact that abundant evidence points to the existence of inequality does not mean that recognition of it will sink into the consciousness of officials. Although there is much evidence to the contrary,[16] for example, almost half of the respondents believed that blacks are better off than whites, treated equally, or are free from discrimination if one controls for social class.

If greater awareness of inequality is difficult to promote, is there anything that federal officials can do to encourage changes in attitudes toward recruitment practices? Recognizing that nearly all feasible alternatives are likely to have a limited impact, at least two strategies deserve consideration.

The first strategy would consist of greater effort by the federal bureaucracy to promulgate consistent and explicit statements concerning the research imperatives of strict, criterion validation.[17] In so doing, the epistemological limitations of other, less rigorous validation methods (which are more feasible) would be emphasized. More uniform federal adoption of the validation standards endorsed by the Equal Employment Opportunity Commission would serve as a useful starting point in this regard. Largely because of resistance by the U.S. Civil Service Commission, EEOC standards have not been accepted by all the federal agencies which impinge upon local personnel systems. In fact, as of May 1975, the Bureau of Intergovernmental Personnel Programs of the U.S. Civil Service Commission had issued no guidelines at all concerning validation (U.S. Commission on Civil Rights, 1975:31-59, 179-181). EEOC validation guidelines fall short of the strictest standards of selection validation. Nonetheless, they are on firmer methodological ground than most of their alternatives. Accepting EEOC standards implies more than endorsing and promulgating them. The Bureau of Intergovernmental Personnel Programs should also insist that the validation studies that it funds through its grant program come very close to meeting these standards. Subsidizing validation studies which fall short of rigorous methodological requirements may actually inhibit officials from attempting more penetrating reassessments of their recruitment practices.

The point of emphasizing more stringent standards of selection validation is not that all local governments can reasonably be expected to conduct research which will conform to these standards. Nor is there good reason to believe that the courts will uniformly adopt these guidelines in adjudicating cases involving recruitment to local government. Instead, the main point is to help undermine the psychological closure than many officials appear to have achieved concerning the "best methods" for recruiting skilled employees. With officials more sensitive to the fundamental uncertainties involved in hiring workers, disadvantaged job hunters may be more likely to receive the benefit of a doubt. Individuals previously barred from labor queues may be permitted to join them. Selection criteria which place the deprived at the end of the queue may be deemphasized. Moreover, complaints about reverse discrimination may become less persuasive to officials if they acknowledge the fundamental complexities involved in hiring (Kranz, 1974).

Stimulating uncertainty about the predictive capacity of selection processes may serve the deprived as well as or better than actually conducting validation studies. Completed studies might, after all, buttress a case for using certain criteria which are detrimental to the interests of disadvantaged job hunters (Rosenbloom, 1972). Nonetheless, in an effort to promote merit hiring, federal officials may quite understandably wish to continue their subsidies of validation research. In doing so, they can pursue a second strategy of potential help to the deprived: they can focus research on widely used selection criteria which do disadvantaged job seekers much harm.[18] Consider job experience. This study indicates that local officials place great faith in this criterion; other studies have drawn a similar conclusion. A survey conducted by the National Civil Service League found, for example, that 60% of all counties and cities require applicants for unskilled positions to have job experience. By contrast 34% required a written test and 21% a high school diploma (Rutstein, 1971:8). Given the emphasis on job experience, validation research could usefully address the question: To what degree, if any, do varying amounts of different types of work experience actually predict future performance?

RECEPTIVITY TO EXPANSION

This discussion of validation should not, however, divert attention from a more fundamental issue. Reduction in institutional barriers to employment of the deprived without an increase in the overall number of job opportunities may well function to create a new set of disadvantaged individuals. One study indicates, for example, that, without any compensating increase in the stock of jobs in the economy, the elimination of occupational discrimination by race would reduce the job perquisites of white males with limited educations (Harrison, 1972:44). Whatever the advantages of fostering greater equality between blacks and whites, this kind of success will do little to address the problems of disadvantaged job hunters more generally.

Subsidies to local governments to create more jobs are obviously one means of making a reduction in institutional barriers to employment less of a zero sum game.[19] But the data of this study indicate that such programs are not likely to receive an enthusiastic response from manpower specialists. Less than half of the respondents favored substantial increases in federal subsidies to create public jobs. Moreover, these sentiments were clearly a function of how liberal the officials were politically. Embedded in basic ideology, the attitudes may be difficult to modify.

Still, it is likely that certain alterations in federal job programs would bolster support for them. Longer term commitments to funding may well be one means of building more support within local bureaucracies. Short-term commitments make for harried officials who feel that they lack the time to use federal dollars efficiently. As Mayor Coleman Young of Detroit noted before a U.S. House of Representatives subcommittee, the limitation of Comprehensive Employment and Training Act funding to a year "almost eliminated any lead time for thoughtful development of public service employment projects" (U.S. House of Representatives, 1976:14). Another problem with ephemeral employment programs is that they resemble the job opportunities present in the secondary labor market. This market consists of poorly paid, unstable jobs which yield a worker few fringe benefits. Confronted with employment of this nature, workers have little incentive to take their jobs seriously and are probably less productive than they would otherwise be (Harrison and Osterman, 1974:326). If federal funding were guaranteed over a longer time period (probably a minimum of three years), manpower specialists in local bureaucracies might become more convinced that such programs are worth their time and effort.

A LINK BETWEEN TECHNICAL AND EXPANSIONIST APPROACHES

An expansionist approach built on long-term commitments to funding could pave the way for successful implementation of a technical approach. In this regard, federal subsidies for jobs can be tied to actions by local governments to reduce institutional barriers to hiring the deprived.[20] Such a policy would provide incentives for local officials to analyze their recruitment procedures critically and to modify these procedures. Moreover, linking funding to analysis and change could give local manpower specialists a needed resource with which to overcome local opponents to reform. Thus personnel specialists could use the excuse that "our hands are tied" as they pursued various alternatives to the status quo.

In combining technical and expansionist approaches, one problem would be to build in an adequate payment-for-performance mechanism. Simply having localities submit some plan for studying and reducing barriers to employment as a prerequisite for funding would not be enough to stimulate reform. Instead, planning must be tied to implementation (Pressman and Wildavsky, 1973). At a minimum, yearly renewal of funding should be contingent on a demonstration by the agency that it has made progress in scrutinizing and reducing institutional barriers.

Efforts to unite technical and expansionist approaches would, of course, entail many problems beyond those involved in linking payment to performance. Despite the potential hazards, the union of those approaches is one of the more promising options for reducing institutional obstacles to employment of the deprived.

NOTES

1. For purposes of this essay, the "disadvantaged" are those who have suffered from long-term unemployment, are on welfare, and/or have incomes below the poverty level.

2. Ironically, of course, professionalization would tend to reduce opportunities for the disadvantaged to win jobs on personnel staffs.

3. The International Personnel Management Association is a major national association of public personnel officials. The survey of local administrators was part of a more extensive study of administrators from all levels of government.

4. Over one-third of the sample failed to return the questionnaire. What bias does this create? While the author lacks definitive proof, it could be that individuals with less sympathy for reducing institutional barriers to the employment of the disadvantaged are more reluctant to share their views. The data may, therefore, overestimate the degree to which officials are predisposed to reduce such barriers.

5. There is, however, no obvious reason for believing that the attitudes held by respondents are highly unrepresentative of those held by influential personnel officials in local government.

6. In this vein, one study of personnel directors working for private firms in core cities found that the attitudes of these officials contributed significantly to an explanation of minority hiring rates (Rossi et al., 1974:262-326; see also Thompson, 1975:126-127).

7. It is possible that those officials who are more certain about how to recruit the skilled are also the most likely to deemphasize those selection criteria that are major barriers to employment of the deprived. In this case, uncertainty would not be as beneficial to the disadvantaged. Analysis indicates, however, that, among respondents, certainty is not strongly correlated with attitudes toward selection criteria.

8. The New Careers Program was one of the major federal efforts to encourage job restructuring. This program sought to persuade local officials to do careful task analysis, carve the more menial tasks out of professional roles, and bundle these tasks into new positions which those with less education and experience could perform (Institute for Local Self Government, 1969).

9. The tendency of officials in large cities to manifest greater support for job restruc-

turing may reflect the fact that they supervise larger personnel staffs which possess more expertise in dealing with the task of redesigning jobs.

10. Belief in the practical applications of university training was measured with a three-item index. A typical item in the index reads: "Administrators should expect that their university training will be of little practical value in their departments." Responses to all three questions are strongly correlated with one another.

11. Some analysts emphasize that Pearson correlations should only be employed with interval data and that Likert scale items are ordinal. A countervailing view (one adopted here) is that by assuming the data to be interval one can use more powerful statistical techniques at minimal risk of misinterpretation (Tufte, 1969). To guard against error, however, the author also computed Tau correlations for the relationships among the ostensibly ordinal measures. A comparison of the Tau coefficients and the Pearson coefficients indicates that there are minimal differences between them. This finding further strengthens the case for using the Pearson r.

In exploring and presenting correlations, those attitudes concerning institutional barriers to employment which probably have the most direct implications for disadvantaged job hunters receive primary attention. For example, correlations are presented with respect to screening criteria for subprofessional positions.

12. This is not to argue that professionalism makes no difference whatsoever in shaping performance. Highly professionalized officials may adopt certain role orientations (Haga et al., 1974) or may be more skilled in performing various tasks.

13. Endorsement of federal subsidies for job creation may be one dimension of political liberalism. Whether this is in fact the case does not, for purposes of this study, matter very much. Rather, the analysis simply seeks to determine whether support for job creation is somehow intertwined with the more fundamental political ideologies of officials.

14. The unimpressive magnitude of many of the correlations between sociopolitical beliefs and attitudes concerning hiring impediments suggests the need for additional research. The reader should be cautioned against the assumption that an analysis of the demographic characteristics of personnel specialists—e.g., their age, race, or sex—would greatly increase the variance explained. The author examined the correlations between these demographic variables and attitudes toward barriers to hiring and generally found them to be smaller than those generated by sociopolitical beliefs.

15. Since modifying more basic values requires a redefinition of the self, these values are often difficult to change. Specific attitudes which are intertwined with these basic values are also difficult to modify. See Katz (1963).

16. For an overview, see Levitan et al. (1975).

17. At a minimum, the method requires that the relationship between selection scores and subsequent performance on the job be assessed. For a further description of the methodological imperatives of strict criterion validation, see Thompson (1975:176-180).

18. One could, of course, take the opposite approach and focus on validating those criteria which seem likely to do the least harm to disadvantaged job hunters. The author's hunch is, however, that any single criterion will possess marginal predictive power. It will, in sum, be easier to show that some criterion forecasts relatively little than to show that it is a powerful predictor.

19. A host of other policies could, of course, also reduce the zero sum nature of the employment game. Adjustments in monetary policy involving the expansion or contraction of funds by the Federal Reserve Board could play a role. So, too, could fiscal policy which emphasizes the manipulation of a large variety of tax and expenditure instruments (holding the monetary supply constant). See Hall (1975).

20. Up to this point, federal programs have not made a serious attempt to forge a link between technical and expansionist approaches. See Mangum and Walsh (1973:137-138).

REFERENCES

BERG, I. (1971). Education and jobs: The great training robbery. Boston: Beacon.

CONNOLLY, H.X. (1976). "The economic status of black America: A critical restrospective." Pp. 61-80 in Proceedings of a Conference on Employment Problems of Low Income Groups: A special report of the National Commission on Manpower Policy. Washington, D.C.: U.S. Government Printing Office.

DYE, T. (1963). "The local-cosmopolitan dimension and the study of urban politics." Social Forces, 41(March):239-246.

FAINSTEIN, N.I., and FAINSTEIN, S.S. (1974). Urban political movements. Englewood-Cliffs, N.J.: Prentice-Hall.

FECHTER, A.E. (1975). "Public service employment: Boon or boondoggle?" Pp. 123-146 in Proceedings of a Conference on Public Service Employment: A special report of the National Commission for Manpower Policy. Washington, D.C.: U.S. Government Printing Office.

Federal Times (1976). "Perry admits EEOCC abuses." 12(May 24):6.

FINE, S.A. (1972). "Job task analysis." Pp. 297-301 in H.L. Sheppard et al. (eds.), The political economy of public service employment. Lexington, Mass.: D.C. Heath.

HAGA, W.J., GRAEN, G., and DANSEREAU, F., JR. (1974). "Professionalism and role making in a service organization: A longitudinal investigation." American Sociological Review, 39(February):122-133.

HALL, R.E. (1975). "The role of public service employment in federal unemployment policy." Pp. 79-97 in Proceedings of a Conference on Public Service Employment: A special report of the National Commission for Manpower Policy. Washington, D.C.: U.S. Government Printing Office.

HAMILTON, G.S., and ROESSNER, J.D. (1972). "How employers screen disadvantaged job applicants." Monthly Labor Review, 95(September):14-21.

HARRISON, B. (1972). "Public employment and the theory of the dual economy." Pp. 41-75 in H.L. Sheppard et al. (eds.), The political economy of public service employment. Lexington, Mass.: D.C. Heath.

HARRISON, B., and OSTERMAN, P. (1974). "Public employment and urban poverty." Urban Affairs Quarterly, 9(March):303-336.

HART, D.K. (1974). "Social equity, justice and the equitable administrator." Public Administration Review, 34(January/February):3-11.

Institute for Local Self Government (1969). Local government new career implementation tactics. Berkeley, Calif.: Author.

JENCKS, C., et al. (1972). Inequality. New York: Basic Books.

KATZ, D. (1963). "The functional approach to the study of attitudes." Pp. 340-350 in E.P. Hollander and R.G. Hunt (eds.), Current perspectives in social psychology. New York: Oxford University Press.

KRANZ, H. (1974). "Are merit and equity compatible?" Public Administration Review, 34(September/October):434-439.

KRISTOL, I. (1968). "Equality as an ideal." In D.L. Sills (eds.), International encyclopedia of the social sciences. New York: Macmillan and Free Press.

LEVITAN, S.A. (1975). "Creation of jobs for the unemployed." Pp. 148-170 in Proceedings of a Conference on Public Service Employment: A special report of the National Commission for Manpower Policy. Washington, D.C.: U.S. Government Printing Office.

LEVITAN, S.A., JOHNSTON, W.B., and TAGGART, R. (1975). Still a dream. Cambridge, Mass.: Harvard University Press.

MANGUM, G.L., and WALSH, J. (1973). A decade of manpower development and training. Salt Lake City: Olympus.

MCCLOSKY, H. (1964). "Consensus and ideology in American politics." American Political Science Review, 58(June):361-382.

MECHANIC, D. (1974). Politics, medicine and social science. New York: John Wiley.

MOORE, C.L., JR., MACNAUGHTON, J.F., and OSBURN, H.G. (1969). "Ethnic differences within an industrial selection battery." Personnel Psychology, 22(winter):473-482.

MOORE, W. (1970). The professions: Roles and rules. New York: Russell Sage Foundation.

MOSHER, F. (1968). Democracy and the public service. New York: Oxford University Press.

NATHAN, R.P. (1975). "Public service employment-'Compared to what?' " Pp. 99-121 in Proceedings of a Conference on Public Service Employment: A Special Report of the National Commission for Manpower Policy. Washington, D.C.: U.S. Government Printing Office.

National Civil Service League (1972). "Overcoming civil service barriers to employment in the public sector." Pp. 215-229 in H.L. Sheppard et al. (eds.), The political economy of public service employment. Lexington, Mass.: D.C. Heath.

――― (1973). "Pacemaker: A handbook for civil service change." Good Government, 90(spring):1-14.

PALUMBO, D.J., and STYSKAL, R.A. (1974). "Professionalism and receptivity to change." American Journal of Political Science, 18(May):385-394.

PRESSMAN, J. (1975). Federal programs and city politics. Berkeley: University of California Press.

PRESSMAN, J., and WILDAVSKY, A.B. (1973). Implementation. Berkeley: University of California Press.

RITZER, G., and TRICE, H.M. (1969). An occupation in conflict. Ithaca, N.Y.: State School of Industrial and Labor Relations.

ROSENBLOOM, D.H. (1972). "Equal employment opportunity: Another strategy." Personnel Administration/Public Personnel Review, (July/August):38-41.

――― (1973). "The Civil Service Commission's decision to authorize the use of goals and timetables in the federal equal employment opportunity program." Western Political Quarterly, 26(June):231-251.

ROSSI, P.H., BERK, R.A., and EIDSON, B.K. (1974). The roots of urban discontent. New York: John Wiley.

RUTSTEIN, J.J. (1971). "Survey of current personnel systems in state and local governments." Good Government, 87(spring):1-28.

SADACCA, R. (1971). The validity and discriminatory impact of the Federal Service Entrance Examination. Washington, D.C.: Urban Institute.

SHEPPARD, H.L. (1972). "The nature of the job problem and the role of new public service employment." Pp. 13-39 in H.L. Sheppard et al. (eds.), The political economy of public service employment. Lexington, Mass.: D.C. Heath.

STANLEY, D.T. (ed., 1974). "The merit principle today." Public Administration Review, 34(September/October):425-452.

THOMPSON, F.J. (1975). Personnel policy in the city: The politics of jobs in Oakland. Berkeley: University of California Press.

THUROW, L.C. (1975). Generating inequality: Mechanisms of distribution in the U.S. economy. New York: Basic Books.

TUFTE, E. (1969). "Improving data analysis in political science." World Politics, 21(July):641-654.

U.S. Bureau of the Census (1975). Public employment in 1975. Washington, D.C.: U.S. Government Printing Office.

U.S. Commission on Civil Rights (1975). The federal civil rights enforcement effort, 1974: To eliminate employment discrimination. Washington, D.C.: U.S. Government Printing Office.

U.S. Equal Employment Opportunity Commission (1974). Minorities and women in state and local government, 1973. Washington, D.C.: Author.

U.S. House of Representatives, Committee on Education and Labor (1976). Authorization of appropriations for fiscal year 1976 for carrying out Title VI of the Comprehensive Employment and Training Act of 1973. Washington, D.C.: U.S. Government Printing Office.

WESTIE, F.R. (1964). "Race and ethnic relations." Pp. 576-618 in R.E.L. Faris (ed.), Handbook of modern sociology. Chicago: Rand McNally.

6

Technology, Professionalization, Affirmative Action, and the Merit System

ROBERT W. BACKOFF
HAL G. RAINEY

☐ A NUMBER OF POLITICAL, social, technological, economic, and legal changes in American society in the past decade have had a significant impact on the management of human resources in the public sector. The changes have produced demands and supports for adaptive personnel policies and practices at all levels of government; these policies in turn have generated changes in the operating personnel management systems. The focus of our study is on three of these societal changes: the demand for equal employment opportunity for all citizens, the increasing professionalization of our society, and the introduction of new technologies. More specifically, we are interested in exploring the interrelationships among technological change and innovation, professionalization, the pursuit of equal employment opportunity through affirmative action programs, the traditional and updated merit principles, and the management of the major public personnel/human resource functions or processes in urban governments, e.g. recruitment, selection, etc. Each of these subjects has been given a great deal of attention in its own rights, but there have been few attempts to develop a systematic analysis of the interactions between all five.

There has been extensive debate as to the impact of affirmative action programs on the traditional principles of the merit system of personnel management (Glazer, 1975; Kranz, 1974; Nigro, 1974). The debate tends to highlight apparent conflicts in the application of norms derived from both policies. As

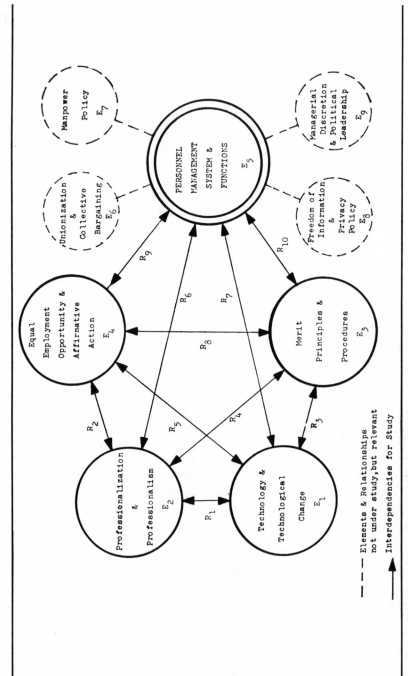

Figure 1: ELEMENTS AND RELATIONSHIPS OF RESEARCH PROBLEM

well, some have suggested that the increasing professionalization of the public service is undermining elements of the traditional merit system (Mosher, 1968; Henry, 1975). A relatively unexplored question concerns the potential conflict between affirmative action practices and the pressure for higher professional standards in personnel decisions, for upgrading job descriptions, and for higher credentials. Most discussions of technological change concern the enhancement of the role of specialists, technicians, and professionals, and the loss of jobs for routine workers. Unexplored is the impact of various types of technological change on the efforts toward affirmative action, on the merit system, and more generally on the personnel/human resource management system.

The basic components of the research problem (E_1-E_5) are portrayed in Figure 1, with the ten potential interdependencies (R_1-R_{10}). The remainder of this essay provides an analysis of these components and relationships. Obviously, we have omitted important sources of change, such as unionization and collective bargaining, federal public service employment programs (Sheppard et al., 1972), freedom of information and privacy legislation, and others (see E_6-E_{10} in Figure 1).

The paper is divided into five sections. The first defines the urban personnel management system and provides a model of that system (E_5). Next we discuss the nature of technology and technological change (E_1) in the context of urban personnel management (R_1, R_3, R_5, R_7). The third section examines professionalization, equal employment opportunity and affirmative action, and merit principles of public personnel management; the focus is on defining the central norms or prescribed rules of conduct associated with each (E_2, E_3, E_4). Derived from this analysis is a discussion, in the following section, of the interdependencies (R_2, R_4, R_8) between norms. The concluding section briefly considers the utility of the present approach and the major probable outcomes of the interaction of the norms discussed in the previous section.

THE URBAN GOVERNMENT PERSONNEL MANAGEMENT SYSTEM

Since we will be referring to impacts on and within the personnel management system, conceptualization of such a "system" is required. Figure 2 presents a highly simplified version of our model of an Urban Government Personnel Policy and Management System (UGPPMS).[1] Models for personnel systems of private corporations (Glueck, 1974; Miner and Miner, 1973; Blakeney et al., 1974) and the federal government (Levine and Nigro, 1975), but ours represents the system for a relatively large American city.[2]

By personnel management we mean activities of persons in government to achieve their goals through effective and efficient use of human resources. The activities are those associated with procurement, development, utilization, and maintenance of human resources (French, 1974; Miner and Miner, 1973; Nigro

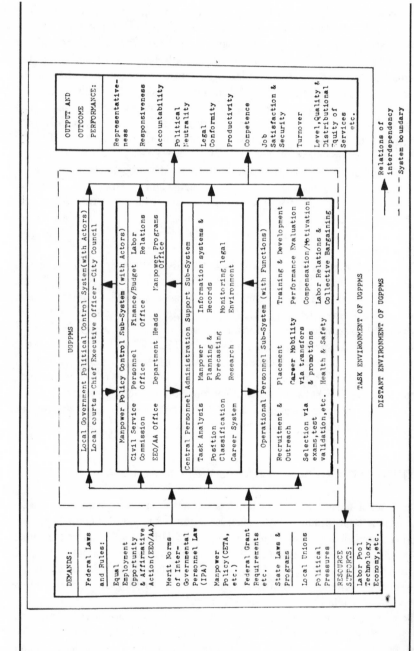

Figure 2: URBAN GOVENMENT PERSONNEL AND MANAGEMENT SYSTEM (UCPPMS) MODEL

and Nigro, 1976; Beatty and Schneier, 1977), and the UGPPMS includes all elements and relationships in urban government that are related to such activities, and that affect the relationship of such activities to performance (Backoff, 1974a; Levine et al., 1975).

In the model, the UGPPMS has a hierarchy of four subsystems: (1) a political control subsystem (mayor or city manager, city council, and local courts) which produces general policy for the city; (2) a manpower policy control subsystem (civil service commission, personnel officer, finance department, affirmative action/EEO officer, heads of operating departments, and others) which produces human resources policy, subject to general guidance from the political control subsystem; (3) a personnel administration support subsystem (usually the central personnel office staff) which performs such specialized personnel activities as task analysis, position classification, human resource planning and research, information collection and storage, etc.; (4) an operational personnel subsystem, including the central personnel staff and many others who are involved in a number of operational functions—recruiting, selection, placement, training, evaluation, compensation and motivation, labor relations, etc.

Obviously, the UGPPMS and its components are subject to such influences from the task environment as Supreme Court decisions, federal legislation and programs, and labor market conditions and will show adaptations to these impacts; the system may, in turn, influence the task environment. There will also be continuing mutual influences and adjustments among UGPPMS components, often as a result of task environment changes or inputs.

In discussing, below, these influences and adaptations, two terms from Haas and Drabek (1973) are useful. *Stress* will refer to discrepancy between demands made on the UGPPMS (and its components) and its capacity to sustain performance. *Strain* refers to discrepancies or incompatibilities between components or elements within components. Here we address one type of strain, *inconsistent norms* for decision and action.

TECHNOLOGICAL CHANGE AND
URBAN GOVERNMENT PERSONNEL MANAGEMENT

The rapid increase in demands on local government for new or increased, services, along with tighter fiscal capacity, have generated stress for urban policymakers and managers. One response is an attempt to find new ways to enhance capacity through introduction of new technologies. The new technologies have and will have an impact on the management of human resources in local government; the exact nature and extent of this impact is not clear as yet. In this section of the paper we shall briefly explore this question by reviewing some of the empirical evidence and raising some conjectures as to the general impact of technological innovation and change on the personnel management functions, on the professionalization of local government occupational groups, and on the implementation of affirmative action programs.

CHARACTERISTICS OF TECHNOLOGICAL CHANGE AND INNOVATION

There is a great deal of disagreement over the concept "technology." Although the traditional use of the term applied to machines and tools, we will use the broader contemporary conception of technology as actions performed by individuals on some object, with or without the aid of tools or mechanical devices, to transform the object in some way (Perrow,1967). The transformation involves the use of knowledge (including that embodied in machines) via a set of techniques with defined sets of action and decision rules for specified circumstances (Nelson et al., 1967). The objects to be transformed or the raw materials worked on may be human beings, symbols, or inanimate/physical objects.

In considering technological change and innovation, the key attribute we shall consider is the degree of *routinization* involved in the transformation or conversion process associated with a particular technology (March and Simon, 1958; Perrow, 1967; Hage and Aiken, 1969; Mohr, 1971; Van de Ven and Delbecq, 1974; Hunt, 1970; Lynch, 1974). Routine conversion processes or technologies are uniform, simple (as against complex), easily understood, predictable, lacking in variability, and require little search for appropriate processing procedures or rules. The reverse holds for nonroutine technologies.

For the purposes of this essay, technological change will involve any variation in technologies (relative to some social, physical, and temporal context), and technological innovation refers to technological change that involves the invention, adoption, and utilization of *new* or novel technologies (Backoff, 1974b). The variations in technologies may involve their structural attributes as well as their functional applications. Technological change would involve processes such as adding more machinery or doing more information processing or decision making in the same ways. Technological innovation in our conception implies not greater or lesser use of existing technologies for the same purposes or functions, but rather the use of new technologies with new structures and/or functional uses. Most conceptions of innovation in organizational contexts define "new" relative to the inventing, adopting, implementing organization or government.

Thus, the question we have posed in this study concerns the nature of technological innovation and change in local government. What types of technologies are being used? What new technologies have been or are likely to be adopted? At what rate? To serve what governmental and managerial functions, and with what consequence for personnel management, the merit system, professionalism, and affirmative action? No extensive research has been completed to date which answers all these questions; in the next section we review the existent evidence and provide some conjectures on the impact of technological change and innovation on the local government personnel management system.

TRENDS OF TECHNOLOGICAL CHANGE AND INNOVATION IN URBAN GOVERNMENTS

In Table 1, we have summarized the empirical findings on the adoption of technological innovations in terms of computer technologies (Kraemer et al., 1976; Watlington, 1971), operations or material technologies (Feller et al., 1976; Yin et al., 1976), and managerial technologies (Fukuhara, 1977). The types of technologies are related to areas of application and to extent of use; clearly more

TABLE 1
TYPES OF TECHNOLOGICAL INNOVATION IN LOCAL GOVERNMENT

1. Computers [a]	*Adoption rate:* all cities over 50,000 population have adopted in past 20 years; gradual pattern of adoption following "S" curve; manageable degree of stress placed on local government system.
	Resource allocation: for cities over 100,000 population, 1% of operating budgets and 1% of personnel allocated for computer related expenditures.
	Uses: for cities over 50,000 population, the main uses with percentage of applications are: record keeping (42%); calculating and printing (35%); record search (8%); record restructuring (6%); sophisticated analytics (5%); and process control (3%).
	Functional areas of use (in order of use): accounting, treasury/collection, police protection, budgeting and management, utilities, personnel, purchasing inventory, data processing, assessment, central garage/motor pool, fire protection, planning and zoning, and sanitation.
	Personnel management uses:[b] (1) payroll records and reports (87% of all respondents use computers for these functions), (2) employee benefit records and reports (66%), (3) personnel budget and accounting (63%), (4) pay statistics and reports (62%), (5) personnel statistics and reports (58%), (6) position statistics and reports (46%), (7) medical statistics and reports (29%), (8) test scoring (16%), (9) performance evaluation (16%), (10) recruiting lists and analysis (12%), (11) collective bargaining reports (12%), (12) manpower planning reports (8%), (13) skills inventory (4%), and (14) work sampling (4%).
2. Operational and managerial	*Line agency operational or material technologies:* no comprehensive descriptive statistics available; samples of technologies in four functional areas studies—air pollution control, fire fighting, traffic control, and solid waste.[c]
	Managerial technologies: over 50% of cities studies report use of: (1) program, zero-base, or performance budgeting; (2) cost-accounting; (3) management information systems; (4) program evaluation; and (5) productivity improvement programs. Less frequently used managerial technologies: (1) management by objectives (46% of cities reporting), (2) project management using PERT or CPM (35%), (3) performance auditing (35%), (4) management incentive programs (15%), and (5) productivity bargaining (8%). All data for cities over 50,000.[d]

a. Data reported from Kraemer et al. (1976).
b. Data from survey of 24 of the largest 50 U.S. cities, from Tomeski and Lazarus (1974).
c. Data detailed in Feller et al. (1976).
d. Data reported U.S. cities by Fukuhara (1977).

research is needed to precisely document the extent and rate of innovation across all types of technologies.

In the sudies noted above, the general impact of introducing new technologies on jobs and skill requirements were frequently addressed. The major conclusions to be drawn are these: (1) the introduction of computers did not lead to staff reductions, although the rate of increase of clerical staff was reduced; (2) no real reductions in costs were achieved from introducing the computers as they generated additional information processing demands; (3) there has been an increase of jobs for clerks to enter data and for technically trained personnel, both in the computing unit and in user departments; (4) new job classifications have been developed for computer supervisors, senior systems analysts, programmers, equipment operators, and keypunch operators; and (5) new jobs have been created for professionals, specialists, and technicians to manage and apply the new operations and managerial technologies (e.g, policy analysts, or, in air pollution control, stack sampling teams).

At this point we wish to offer our conjectures as to the likely impact of technological innovation on the personnel management system in the near future. With regard to the local government employee profile, we would expect that the results noted above will continue. Where routine technologies are adopted or more extensively applied, the rate of increase in clerical staff will be modest. Where nonroutine technologies are adopted (computer, operations, or managerial), we expect to see increased employment of professionals, scientists, and technicians and the generation of new job classifications. A second-order effect of this shift in employment pattern will be further increases in the adoption of technological innovations spurred by the new professionals.

How will the current merit system personnel management practices be affected by the introduction of technologies for routine and nonroutine tasks? For routine tasks, the new technologies should be largely compatible with the norms and practices of the merit system. If objective procedures can be applied to easily analyzable, repetitive, uniform, and predictable tasks and jobs, then objective merit-based tests, training technologies, job performance evaluations, pay standardization and equalization can be more easily applied. On the other hand, if technologies are rapidly introduced which generate nonroutine tasks, we would expect a great deal of difficulty in developing valid tests, job performance assessment techniques, position classification techniques, and reliable training methods.

The impact of new technologies on efforts to implement affirmative action is rather indirect; it may be that introducing routine technologies at the operational level of the local government will provide more minorities with employment opportunities in the bottom grades. This would not assure minorities of upward mobility, however. In the case of nonroutine technologies, it is likely that new jobs created will not advantage minority members. The pool of professionally and technically trained minorities is very limited, so in the short run no significant increases in representativeness would be observed in professional and technical jobs. In the next section we will turn to more detailed discussion of professionals, the merit system, and affirmative action.

OTHER SOURCES OF STRESS AND STRAIN FOR PERSONNEL MANAGEMENT

We now turn to definition of the other three sources of impact on the UGPPMS—increasing professionalization in local government, the defining principles of merit systems, and the new policy directives for affirmative action. Later we will examine strains between these factors, particularly the normative inconsistencies or conflicts.

PROFESSIONALIZATION AND PROFESSIONALISM

For at least half a century, "professionals" and "professionalization" have been the subjects of voluminous discussion by sociologists and others, yet it is still frequently noted that no definitive conceptualization of these phenomena has emerged (Vollmer and Mills, 1966; Mosher, 1973; Harries-Jenkins, 1970). In spite of the complexities, however, the topic is of such importance that academic and popular interest continues. Recently the importance of professionals in public service has been emphasized, and certain unique considerations have been pointed out (Mosher, 1968; Bowen, 1973).

Also, in spite of the difficulties, the sociological and organizational literature offers some useful clarifications (Bobbitt et al., 1974). A *profession* is an occupation which possesses a high level of the attributes mentioned in the next paragraph. Thus, "profession" can be regarded as an ideal construct, with individual occupations varying in their degree of *professionalization,* that is, in their state of development in the process of acquiring the set of attributes (or the levels of those dimensions). A member of a professionalized occupation is a *professional,* but professionals vary in their *"professionalism,"* according to the degree to which they possess a multidimensional set of attributes—attitudes, usually—which correspond to the attributes of the profession. Occupations, then, can vary along a continuum from unprofessionalized to highly professionalized, and individuals can vary in their professionalism, within the same profession or between professions.

Yet while these distinctions are of some help to us in conceiving of what, in the abstract, we mean by professional, professionalization, etc., there is immense difficulty in specifying the appropriate set of attributes or dimensions for use in classifying individuals and occupations. Concerning the professionalization of occupations, there is considerable variation in the sets of defining attributes specified by different authors. One way of dealing with this difficulty is to review definitions and compile or distill a set of the most frequently mentioned attributes. Hickson and Thomas (1969) report Millerson's compilation, from a review of some 20 definitions, of the following list of most often cited "elements" of the definition of a profession: (1) skill based on theoretical knowledge, (2) required education and training, (3) testing of competence (via exams, etc.), (4) organization (into a professional association), (5) adherence to a code of conduct, and (6) altruistic service.

Concerning individuals, the attributes which are suggested as characteristic of a "professional" are usually attitudinal (Filley et al., 1976; Hall, 1972). Recent compilations of most commonly cited characteristics have included the following: (1) belief in the need to possess *expertise* or training in a body of abstract knowledge; (2) belief that he and fellow professionals should have *autonomy* in work activities and decision making; (3) *identification* with the profession and fellow professionals; (4) *commitment* to the calling, to the work of the profession; (5) a feeling of *ethical obligation* to render service to the public without self-interest and with emotional neutrality; (6) a belief in *self-regulation* or *collegial maintenance* of standards, i.e., that fellow professionals are best qualified to judge and police each other. These individual professional attributes have been the focus of discussion and empirical research, with a major topic of this research being the potential and actual conflict between the professional attitudes just noted and the characteristics of bureaucratic organizations. For example, a highly bureaucratized organization would tend to stress hierarchical authority and formal rules and procedures, which would conflict with a professional's concern for autonomy.

Mosher (1968, 1973) has argued that, while professional-bureaucratic conflict is an important concern for students of government, an equally important problem in government is the domination of an agency or organization by a particular profession, to the detriment of responsiveness and political control. In government there are occupations whose members have virtually no alternative means of practicing the occupation other than employment in a particular agency; military officers and foreign service officers are examples. The drive for control over their own personnel practices, by these "public service professionals," may be of greater concern than professional-bureaucratic conflict because of the threat to public accountability inherent in their desire for autonomy and self-regulation in these matters.

Three attributes of professionalization and professionals will be examined in detail in the next section of this paper: the norm requiring use of theoretical or systematic knowledge, the norm requiring education and training (usually at universities), and the norm supporting self-control and policing of professionals by other professionals, and the concomitant emphasis on autonomy in decision making. These norms appear to be the most likely sources of strain for the personnel management system, given the normative pressures of the existing merit principles and the new stresses created by equal employment opportunity laws, and particularly affirmative action mandates.

MERIT NORMS OR PRINCIPLES

Any discussion of new pressures on public personnel administration must attempt to clarify the nature of the merit system and merit principles as they are currently understood and practiced. Historically, there has been elaboration and differentiation of new norms, expectations, and decision rules to guide the conduct of those managing human resources in the public service (Van Riper, 1958; Mosher, 1968; Gallas, 1976). As these norms developed in response to new demands, stresses were placed on incumbents in the public service and a number of normative strains have been felt.

For this essay we are going to rely upon the norms reaffirmed and slightly elaborated in the Intergovernmental Personnel Act of 1970; this legislation and the consequent implementing guidelines and regulations place emphasis on six principles which are widely referenced today and likely to guide personnel managers in the near future. The six IPA principles are the following (U.S. Civil Service Commission, 1974):

(1) recruiting, selecting, and advancing employees on the basis of their relative ability, knowledge, and skills, including open consideration of qualified applicants for initial appointment;

(2) providing equitable and adequate compensation;

(3) training employees, as needed, to assure high-quality performance;

(4) retaining employees on the basis of the adequacy of their performance, correcting inadequate performance, and separating employees whose inadequate performance cannot be corrected;

(5) assuring fair treatment of applicants and employees in all aspects of personnel administration without regard to political affiliation, race, color, national origin, sex, or religious creed and with proper regard for their privacy and constitutional rights as citizens;

(6) assuring that employees are protected against coercion for partisan political purposes and are prohibited from using their official authority for the purpose of interfering with or affecting the result of an election or a nomination for office.

The following prescribed norms of conduct for public personnel managers represent a synthesis of the more traditional norms (and terms) and the IPA list (and language) and will be used in our subsequent analysis: (1) norm of competence via appointment and promotion on basis of merit (ranked), (2) open competition via civil service exams (ranking on lists), (3) objectivity, impartiality, and political neutrality, (4) equal opportunity (without regard to race, color, sex, etc.), (5) training to assure high performance, (6) retention of employees on basis of adequacy of their performance, correcting inadequacy, and separation if not corrected, and (7) equitable and adequate compensation or "equal pay for equal work." One norm, objectivity, is particularly important to highlight for our analysis. It is derived from the general value placed on the scientific method and derivatively on the use of practical procedures or tests. Under these principles, for example, to treat people alike, one is enjoined to standardize procedures, to assess relative merit, and to measure individual differences in valid and reliable ways using standardized tests. Objective, standardized descriptions of tasks and jobs should be the basis for the tests.

One caveat is in order here. In our analysis we are treating the merit norms as prescribed rules of conduct, not necessarily assuming that they are in fact fully practiced by contemporary personnel managers (Savas and Ginsburg, 1973; Kranz, 1974); this is an empirical question that has not been adequately researched to draw such a conclusion (Boesel, 1974; Rutstein, 1971). We turn now to the norms associated with the implementation of the equal employment opportunity and affirmative action (E_4 in Figure 1).

EQUAL EMPLOYMENT OPPORTUNITY AND AFFIRMATIVE ACTION

While technological change, professionalization, and shifting merit norms apparently generate modest stress and strain for contemporary personnel management, it seems that pressures for greater equal employment opportunity and for affirmative action create greater normative conflicts and pressures. One might expect, then, that local government personnel management systems will exhibit coping behaviors aimed at enhancing capacity to apply these norms, or at avoiding or decreasing such demands. But for consideration of those pressures and the responses to them, one needs to specify the nature of the norms implied by the demands and requirements. That task is made difficult by the considerable controversy over what affirmative action is and what it requires, but in briefly reviewing some of the points of controversy, we can move to conclusions as to the norms under affirmative action.

Equal employment opportunity connotes norms under which employers base personnel decisions solely on individual merit and fitness. Merit should be related to specific jobs, and judged without regard to race, sex, political affiliation, physical disability, and other nonmerit factors. This conception, of course, is the same as the fifth IPA principle concerning fair treatment of all people. Yet affirmative action tends to prescribe rules and actions for *achievement* of equal employment opportunity and more representative public organizations at all levels of government. With increasing pressure for such numerical representation of women and minorities, to the deemphasis of norms stressing selection of the best qualified, one might expect increasing strain for a system under pressure to adhere to certain of the merit principles.

Yet, this suggestion of a normative inconsistency between certain merit principles and affirmative action would be disputed by proponents of affirmative action. Many would argue that there is no inconsistency, that fair and equal treatment is now one of the IPA principles, and is simply a matter of enlightened, systematic personnel procedure. The Civil Service Commission has taken action against the use of "quotas" by local governments (Nigro and Nigro, 1976), and the federal agencies with primary responsibility for affirmative action policy have issued a joint memorandum aimed at distinguishing the "goals and timetables," which are required under affirmative action, from quotas. Goals and timetables are said to be more reasonable and flexible, useful primarily as management tools for implementation of affirmative action. So these sources might argue that affirmative action does not conflict with norms requiring selection of the best qualified. One also hears assertions that merit principles dictating open selection of the best qualified have never been adequately implemented, and that invalid tests and preferences have systematically excluded certain groups. Affirmative action would actually contribute to establishment of truly merit-based procedures by removing such obstacles (Kranz, 1974). These arguments are often accompanied by arguments concerning the desirability of representativeness in government.

Critics of affirmative action, on the other hand, would argue that it nevertheless involves pressure against hiring of the best qualified, and other merit norms. They point out that there is vagueness in affirmative action guidelines, to an extent that leads Sowell (1976) to comment that "there is no way to determine the meaning of affirmative action." A number of laws, executive orders, and other policy actions have given a number of agencies, as well as the courts, responsibilities for affirmative action policy. In addition to this source of complexity, there are technical difficulties in actually carrying out affirmative action guidelines—for example, in determining the levels of representativeness to be achieved, what alterations of personnel procedures are to be made, and how.[3] Faced with such uncertainties, it is argued, employers try to meet affirmative action requirements by hiring on the basis of race or sex, whether it is formally required or not (Seligman, 1974).

On balance, then, it is reasonable to conclude that affirmative action involves pressure to achieve numerical representativeness by race and sex, although the extent and nature of this pressure is a matter of dispute. This is one of the primary norms under affirmative action to be stressed in the discussion to follow, but there are related requirements, of a more procedural nature, which receive emphasis under affirmative action and which are among the major normative pressures.

As noted, there is uncertainty as to the exact requirements under affirmative action, but guidelines have been established which suggest the nature of those requirements. For example, one might look to the Civil Service Commission's *Guidelines for the Development of an Affirmative Action Plan* (1975) for indications as to what is actually required of employers in local governments. Although these guidelines are presented only as an outline of sound practices, they imply that an affirmative action plan for an organization or a jurisdiction should involve a number of potentially far-reaching actions or changes. These include preparation and publication of a formal affirmative action plan, assignment and training of staff for affirmative action responsibilities, and extensive record keeping and reporting to affirmative action authorities. An analysis of the external labor market and internal workforce are to be undertaken, as basis for identifying "underutilized" groups, and for designing remedies for such under-utilization. Goals and timetables are to be set for the implementation of such steps and the achievement of adequate representativeness. The remedial steps might include analysis of, and modification of, procedures for recruitment, selection, promotion, and placement, as well as possible alterations in provisions for career planning, job structuring, and training. It seems, then, that in addition to general pressures for numerical representativeness, affirmative action involves a number of major procedural requirements.

On the basis of the foregoing discussion, it is reasonable to underscore the following three norms and normative pressures as characterizing affirmative action:

(1) Special efforts of analysis, validation, reporting and justification of personnel procedures, aimed at removing obstacles to equal employment opportunity (EEO).

(2) Pressures for alterations of personnel procedures and structure:
    (a) for lowering or removal of certain standards, requirements, or practices (such as certain written tests or high school diploma requirements) that are not job-related.
    (b) for alteration of selection and certification procedures, and other employment procedures (sometimes, for "selection certification," for "hire, then train" procedures, for special pre-examination tutoring, for restructuring of tasks to allow for upward mobility), to remove obstacles to EEO.

(3) Pressure for numerical representativeness—for a degree of preference for minorities/females in personnel procedures, where failure to prefer them cannot be substantiated on the basis of clearly valid or job-related criteria, and where their presence in the organization is clearly lower than their presence in the local population.

Given these general pressures under affirmative action, one would expect a number of complicated interrelationships with other factors influencing local government personnel management systems.

CONTEMPORARY STRAINS WITHIN THE LOCAL GOVERNMENT PERSONNEL SYSTEM

Having briefly identified the central attributes and norm-sets of each of the three elements (E_2 E_3 E_4 of Figure 1), we are able to turn to the nature of the normative inconsistencies (or conflicts) within and between the norm-sets. As we noted earlier in introducing our model of stress and strain for the personnel system, we shall use the term strain to refer to normative incompatibilities that exist in the system. One type of incompatibility is emphasized here, normative inconsistency, or the presence of inconsistent or conflicting norms for decision and action within the local government personnel system. By norm, we mean a prescribed rule of conduct for the behavior of individuals, work units, or organizations in the personnel management system (Morris, 1956; Gibbs, 1965; Mitnick, 1974; Jackson, 1975). Norms as rules of conduct may prescribe or proscribe behaviors; they may or may not involve sanctions. The degree of acceptance of the norms as standards for conduct may vary from widely shared to narrowly shared.[4] Our concern here is with the extent to which conformity to or application of one rule of conduct is perceived and/or experienced directly as inconsistent or conflicting with the pursuit or another rule of conduct.

In addressing the nature of potential inconsistencies a further distinction is necessary. On the one hand we can examine two norms a priori and speculate on their potential inconsistency (conflict in principle); on the other hand, we might observe directly the application of the norms to specific situations and note actual inconsistency (operational conflict). In the discussion that follows, we shall frequently note whether the conflict is at the general level of analysis ("in principle") or arises in specific applications. In the latter case, we frequently note that a particular procedure or sub-norm instrumental to the more general norm may conflict with it (e.g., the possibility that written exams actually conflict with the merit norm of appointment on the basis of competence and open competitive exams—the tests may not be truly open and may not be valid indications of ability to do the job).

Our assessment is based on a review of current norms and practices of public personnel management in local government; of course, many of the conflicts appear to be generalizable to other levels of government. To assist the reader we have developed a matrix in Table 2 of the potential interdependencies between the norms noted previously for the merit system, affirmative action, and professionals (professions). Each cell in the table has been numbered so that our discussion of conflicts can be easily referenced in the table; the cells have "+," "-," "0," or "+ or -" symbols contained within them. The signs refer respectively to a consistent inter-norm relationship, an inconsistent or conflicting relation-

TABLE 2
MATRIX OF INTERDEPENDENCIES BETWEEN CENTRAL PUBLIC PERSONNEL MANAGEMENT NORMS

LEGEND

M — Merit norm
A — Affirmative action norm
P — Norms of professionals and professions
+ — Supportive (reinforcing)
0 — No impact
− — Non-supportive (conflict)

	M-1 Competence via appointment & promotion on merit (ranked)	M-2 Open competition via Civil Service exams (lists ranked)	M-3 Objectivity, impartiality, and political neutrality	M-4 Equal opportunity (w/o regard to race, sex, etc.)	M-5 Training to assure high performance	M-6 Retention, correcting, and separating on performance	M-7 Equitable & adequate compensation; equal pay for equal work	A-1 Analyze/justify personnel procedures	A-2 Eliminate non-job related criteria	A-3 Modify personnel procedures to remove obstacles to EEO	A-4 Achieve numerical representativeness	P-1 Rely on systematic knowledge	P-2 Require extensive education and training	P-3 Self regulation by professional group
M-1 Norm of competence via appointment & promotion on merit (ranked)		1 + or −	2 +	3 +	4 +	5 +	6 0	7 +	8 + or −	9 + or −	10 −	11 +	12 +	13 +
M-2 Open competition via Civil Service exams (ranking on lists)			14 + or −	15 +	16 +	17 + or −	18 0	19 0	20 −	21 −	22 −	23 0	24 −	25 −
M-3 Objectivity, impartiality, and political neutrality				26 +	27 0	28 +	29 +	30 + or −	31 +	32 + or −	33 −	34 +	35 0	36 + or −
M-4 Equal opportunity (without regard to race, color, sex, etc.)					37 +	38 +	39 +	40 +	41 +	42 +	43 + or −	44 +	45 + or −	46 + or −
M-5 Training to assure high performance						47 +	48 0	49 0	50 0	51 +	52 0	53 +	54 +	55 0
M-6 Retention of employees on basis of adequacy of their performance, correcting inadequacy, and separation if not corrected							56 0	57 +	58 +	59 +	60 −	61 +	62 0	63 + or −
M-7 Equitable and adequate compensation; "equal pay for equal work"								64 0	65 0	66 0	67 0	68 0	69 0	70 0
A-1 Analyze and justify personnel procedures to remove obstacles to EEO									71 +	72 +	73 −	74 0	75 0	76 −
A-2 Eliminate non-job related criteria										77 +	78 −	79 + or −	80 −	81 −
A-3 Modify personnel procedures to remove obstacles to EEO											82 −	83 0	84 −	85 −
A-4 Achievement of numerical representativeness												86 0	87 −	88 −
P-1 Rely on systematic knowledge													89 +	90 +
P-2 Require extensive education and training														91 +
P-3 Self-regulation by professional group														

ship, no apparent inter-norm relationship, and a relationship with both consistency and inconsistency depending on the level of analysis—in principle or in specific application. In the discussion that follows we shall briefly note the nature of the clearly inconsistent cases or the mixed cases. The table shows more cases of consistency. This cannot be taken to imply that the consistencies "outweigh" the inconsistencies, since the norms are not necessarily equally weighted—some might be the foci of greater pressure than others and sources of greater influence. The discussion highlights inconsistencies as a means to pinpointing problem areas. However, the consistencies should not be overlooked, as indications of frequent congruence among the major factors and the associated norms.

INCONSISTENT MERIT NORMS

Three pairs of merit norms may be inconsistent at the operational level (cells 1, 15, and 17—Table 2 cell numbers). In our statement of norms we have subdivided principle number one of the earlier Intergovernmental Personnel Act into two parts, one dealing with the norm of appointing and promoting the most competent and the other with the use of open competition through exams (with a ranked list as a result). There is no inconsistency in principle between these two norms (M_1 and M_2), but an inconsistency is often noted at the operational level, where it is pointed out that the exam scores and other ranking criteria are of questionable validity as indications of true merit and qualifications. It is pointed out, for example, that ranking by exam scores bases personnel decisions on exam-taking ability rather than truly valid qualifications (Beaumont, 1974). It is also frequently argued that the exams cannot possibly measure "merit" so accurately as to justify ranking on the basis of a difference of a few decimal points in test score.

The second pair of inconsistent norms (cell 15) involves open competition via exams and equal opportunity; again there appears to be *general* consistency, but in application, exams have not necessarily been open, have not necessarily been valid, and have tended to screen out minorities. They have not been open in the sense of providing minorities with adequate awareness of their occurrence.

The third pair of norms that may be inconsistent in applications involves again the practice of open, competitive exams (M_2), on the one hand, and, on the other, retention, correction, and separation based on job performance (M_6). In practice, the use of exams in making such decisions would conflict if the tests are not valid predictors or reflectors of performance.

INCONSISTENT AFFIRMATIVE ACTION NORMS

There are three pairs of inconsistent affirmative action (AA) norms (cells 73, 78, and 82). All three inconsistencies involve the norm prescribing use of numerical representativeness practices (considering race and sex in personnel decisions). In principle, analysis and justification of personnel procedures to attain true equal employment opportunity (EEO) would conflict with the additional norm of achieving numerical representativeness by giving special consideration to race and sex. This same conflict characterizes the norm pair (cell 82) involving changing personnel procedures to remove obstacles to EEO. In the case of cell 78, numerical hiring goals, and other efforts to achieve numerical representativeness, involve nonjob related criteria, except in a few cases where a minority member may be most qualified (dealing with minority clients, for example). Numerical representativeness will involve consideration of race and sex in personnel decisions, and these are nonjob related in all but a few cases.

INCONSISTENCY BETWEEN AFFIRMATIVE ACTION NORMS AND MERIT NORMS

The most widely noted conflicts between norms are those between affirmative action/EEO, and the merit principle of competence. There has not been much empirical evidence accumulated on the extent and nature of the conflict, at the local government level, either in principle or in practice (Thompson, 1976; E. Levine, 1976).[5] We wish to note three inconsistencies between the norm of competence and the affirmative action norms (cells 8, 9, and 10). First, at the general level, elimination of nonjob related criteria is consistent with basing appointment and promotion on qualifications, merit, and competence—such practices would actually be a step toward valid criteria of competence and merit. Inconsistency is at the level of operationalization, where many of the currently used indicators of merit—exam scores, educational attainment, veteran preferences, no arrest record—are seen by affirmative action proponents, with considerable justification, as being nonjob related. Conflict may be manifested in resistance or resentment where there are demands that certain criteria be relaxed or removed, or where there is criticism or court action over failure to remove them.

Regarding M-1 and A-3 (cell 9), many of the guidelines or suggested steps for altering personnel procedures to remove obstacles to EEO might clear the way for decisions and procedures which are based on more valid criteria of merit, and indeed make for a more truly merit-based system. However, some of the guidelines for change will conflict with existing operationalizations of merit and competence. For example, outreach to find qualified blacks and women is a special effort and expense which is clearly justifiable under a norm of representativeness, but is not necessarily justifiable under a norm of competence and merit; the same would hold for special training and special career ladders to facilitate the advancement of minorities. M-1 and A-4 are even more clearly in conflict, both at the general and operational level. In general M-1 (competence norm) would dictate selection of the best, regardless of race, creed, or color. At a more specific level, merit practices often put minorities and women at a disadvantage and fail to provide numerical representativeness (A-4). The inconsistency is quite visible when federal government agencies exert pressure for use of hiring goals, or when courts impose remedial quotas.

The second merit principle, the prescription for open competition for jobs via exams, also conflicts with three affirmative action norms (A-2, A-3, and A-4). Concerning A-2, there is no necessary conflict in principle, but there is a manifest conflict—courts have thrown out exams, and affirmative action proponents have harshly criticized them, because of their failure to be job-related. Affirmative action requirements involve pressure to validate exams and change ranking procedures, because they often are obstacles to provision of equal opportunity. Thus, M-2 and A-3 (cell 21) are also inconsistent in practice; the conflict over appropriate testing procedures has been dealt with by the courts (Washington versus Davis and on back to Griggs versus Duke Power). Similarly, emphasis on numerical representativeness (A-4) is a major impetus to opposition to exams. Here there is conflict in principle as well as practice, since open competition should not consider race, color, or sex. In practice, pressures for numerical representativeness are often directed at changing the ranking procedures.

The third merit norm concerns objectivity, impartiality, and political neutrality; three conflicts are noted (with A-1, A-3, and A-4). Regarding A-1, a thorough analysis and justification of personnel procedures, as called for under AA guidelines (CSC, EEOC, etc.) is supported by AA proponents and officials as a means of actually establishing objective, neutral, impartial personnel practices. However, since the aim of the analysis, some people argue, is achievement of AA goals, and numerical representation, it could be viewed as a special effort on behalf of women and minorities and, in fact, is not neutral and impartial. This same argument holds for the potential inconsistency with the norm modifying personnel procedures to remove obstacles to EEO. The major point of conflict concerns the norm of numerical representativeness; the pursuit of this norm violates the norm of objectivity and impartiality and, in principle, builds preference for certain groups. The only way to reconcile the two in principle is to assume, as do many AA proponents, that if there were truly objective and impartial procedures then numerical representativeness would occur naturally—an assumption which may not be warranted, in view of various other forms of discrimination which make females and minorities less able to compete, in many cases. In practice, one can point to the many criticisms of AA as violating the norm of objective impartial decision making, since it involves special provisions and preferences for some.

The fourth merit principle, emphasizing equal opportunity, may conflict with numerical representativeness (A-4, cell 43). Here, in what seems to be one of the major points of inconsistency, AA calls for a certain degree of preference for women and minorities as a means to EEO, yet such preference violates EEO, in principle. Critics insist that no matter how they try to devise euphemisms for quotas (goals, etc.) AA proponents are still in the position of calling for special preferences, rather than EEO. AA proponents argue that there is consonance in principle, since pressure for numerical representativeness is just temporary or remedial, is due to historical patterns of discrimination, and is aimed to EEO in the long run.

No conflicts between merit principle five—on training—and AA norms is proposed here; also consonant are the merit norm concerning equal pay for equal work, and AA norms. However, one conflict is noted between the merit norm of retaining employees on the basis of adequacy of performance, and the AA norm of numerical representativeness (cell 60). In principle the AA norm could override the job performance norm. However, the main conflict is operational, where some public administrators apparently balk at firing and correcting minority members out of fear of an affirmative action suit, even though affirmative action officers may tell them they need have no such hesitation, where an effective case can be made. There have been reverse discrimination cases by whites claiming they were punished more than blacks; there is probably a tendency on the part of some whites to experience and express feelings of inequity in this regard (Sowell, 1976; Seligman, 1974).

INCONSISTENCY BETWEEN AFFIRMATIVE ACTION
NORMS AND PROFESSIONAL NORMS

The last sets of norm relationships to be considered involve the relations of professional norms to merit and to affirmative action norms. Turning first to the AA norms, A-1, analysis and justification of procedures for removal of obstacles to EEO, may conflict with professionals' desires for autonomy or discretion (P-3 on self-policing and discretion). The requirement for formal systematic analysis and justification will be regarded as encroachment, and an imposition of red tape (Sowell, 1976). Professional personnel procedures involve judgment and discretion, and cannot be so readily submitted to quantifiable, clearly delimited criteria or processes, particularly with regard to nonroutine decisions. Members of professions in the public service which are becoming more professionalized (police, for example) will resent being told what criteria to use in selecting applicants.

The second AA norm is also inconsistent with the three professional norms (P-1, P-2, and P-3). Attempts to remove non-job related criteria in hiring, etc., may in practice mean deemphasizing the norm of applying ever more systematic knowledge to the job; this may be acutely felt in paraprofessional jobs. There are more significant conflicts with the norm emphasizing education and training (P-2), over whether educational and training requirements are actually job-related. Another inconsistency involves the autonomy of the professionals; as noted above, efforts to remove or lower criteria can be seen as encroachments on professional autonomy or self-regulation.

Norms prescribing modification of personnel procedures to remove obstacles to EEO create inconsistencies for professional norms by (1) suggesting removal of educational requirements—no high school degree for police and firemen, etc., and (2) by transferring decision-making authority from professionals in terms of who decides how procedures will be modified and supervised (dictated by AA officials or decided by the professionals themselves? (Cell 85).

The norm emphasizing numerical representativeness generates inconsistencies with the professional norm of education and training; AA goals cannot be met due to the paucity of females and minorities with the requisite training. A-4 also is inconsistent with the desire for professional autonomy and self-regulation (P-3); professions do not emphasize numerical representativeness, but rather the supervision of the competent application of the subject matter or set of techniques. AA pressures may be viewed, then, as interference to the detriment of professional standards.

INCONSISTENCY BETWEEN MERIT NORMS AND PROFESSIONAL NORMS

Several conflicts or inconsistencies are suggested below between the merit norms and the professional norms discussed above. First, the merit norm emphasizing open competition via exams may be inconsistent with the professional norm of requiring education and training (cell 24). Educational requirements for professionals and groups moving in the direction of professionalization tend to replace the exam as a hiring mechanism, and the selection process is delegated to educational institutions (Mosher, 1968). The training requirements also reduce the openness of the process—one must have the training, even to take an exam (in the case of licensing exams); laymen cannot just come in and compete. Thus merit system coverage is attenuated for professional positions. The second merit norm may also be inconsistent with the professional's norm of autonomy; professionals tend to prefer an "unassembled review" of candidates (Mosher, 1968). In the case of the use of exams, the professionals would want to devise the exams, rather than leave it to the civil service commission staff—e.g., data processors wish to define the needs for recruiting professionals in their area.

The merit norm relying on objectivity, impartiality, and political neutrality is inconsistent to some degree with the professional norm of autonomy, depending on the perspective taken. Professionals, of course, are supposed to regulate themselves according to professional criteria and trained judgment, which should be objective and impartial. However, "professional socialization" and professional self-interest are among the factors which could reduce impartiality, and certain professions might screen females and minorities (and in some cases show more acceptance, also). Definitely, decisions will be less objective or impartial in the sense of being systematic and formalized for purposes of accountability.

Another merit norm where we note inconsistency with professional norms concerns emphasis on equal employment opportunity (M-4). First, professional requirements for extensive training (P-2) are an obstacle to EEO in practice, in view of the difficulties for women and minorities in obtaining and completing training. However, they might be consonant with EEO in that once a person has the training, he or she has access as a professional and a member of the professional group; AA proponents might actually argue for the sanctity of such educational credentials, once obtained. Second, the EEO norm may conflict with professional desire for self-regulation and discretion, if the professionals build barriers to entry into the profession and resist accountability for their decisions, as noted earlier.

Finally, the retention of employees on the basis of job performance (M-6) may be inconsistent, in certain cases, with the desire of professionals for autonomy. Where centralized, standardized performance evaluation procedures are proposed, professionals may display resentment or resistance because they prefer self-regulation.

CONCLUSIONS

We have suggested a number of interrelationships among factors which are of major importance to contemporary personnel administration. The impacts are not yet entirely clear, but a multi-level model or schema as proposed here, together with the stress-strain perspective, represents a useful initial step toward a unified analysis of such impacts. While important early research has been acknowledged here, there is clearly a need for further inquiry into "stresses and strains," as well as resultant coping or adaptive behaviors. To that end, we might draw some brief conclusions as to major trends and apparent outcomes or responses. These suggestions are informed by the results of such research as there has been, and our own judgments and observations, and they might well be pursued in further research.

The impact of technological change and innovation appears to be gradual in its impact on local government personnel systems. To the extent that "routine" technologies are established, opportunities for implementation of affirmative action should be enhanced through greater availability of new types of jobs which do not require advanced training. Increasing instances of "nonroutine" or relatively complex technology should be accompanied by requirements for highly trained or educated applicants, thus diminishing prospects for affirmative action.

Technological advancements should, of course, show a relationship to professionalization. There seem to be relatively few highly professional categories (doctors, scientists, etc.) in local governments, but increasing numbers of high professionals should represent an obstacle for affirmative action and for the extension of certain of the merit principles, due to advanced education requirements and emphasis on professional "judgment" or autonomy in personnel decisions. The emergence of newer specializations or incipient professions (data processing, policy analysis) may result in similar conflicts or resistances, concerning application of affirmative action and certain merit procedures (such as the design of competitive exams).

Yet, as noted, technology and professionalization are apparently of less immediate impact and concern than affirmative action. One can observe, and should expect, concerns among local administrators over technical and procedural requirements under affirmative action: What is required? How can we do it? How are we vulnerable? There are inevitable strains and conflicts over required changes to purportedly merit-based procedures, such as changes in the "rule of three," removal or alteration of hiring tests, job classification requirements and procedures. Noteworthy, also, are examples of apparent congruence between affirmative action and certain merit principles, in that affirmative action requirements support efforts of central personnel administrators to implement systematic personnel procedures. Another type of conflict—over resources—can arise due to the need for increased staff and funding for carrying out the affirmative action guidelines (test validation, job classification, etc.). Still another, more frequently discussed, form of strain which may accompany affirmative action is the potential for resentment and hostility on the part of individuals, and between individuals and groups. These problems may arise not just where there are preexisting prejudices, but as a result of perceived conflicts already noted—feelings that "merit" is being deemphasized on behalf of certain groups or, from the other viewpoint, that invalid operationalizations of "merit" are being used as tactics of resistance to fair and equal treatment.

These and numerous other possibilities should be pursued in further research. Greater understanding of these and related phenomena can certainly be of use in meeting the challenge to provide a public service that is equitable as well as competent and effective in the face of continuing social and technological change.

NOTES

1. A more elaborate description of this UGPPMS model, with more attention to definitions and assumptions, is available from the authors. The model draws on recent extensions of general systems research (Sagasti, 1970; Laszlo, 1975), and proposes a multi-level, multi-goal, multi-function adaptive system representation of public personnel systems (Baumgartner et al., 1976; Whyte et al., 1969).

We regard the terms "model" and "conceptual scheme" as interchangeable. They refer to a qualitative representation of a concrete system—the urban human resource management system. The model is for heuristic purposes; it suggests a simplified, abstracted set of elements and relations selected for this research problem. It is not a theory; it does not provide a direct basis to explain and predict behavior. Further research is needed to refine it for these other scientific goals.

2. For discussion of the range of possible differences between public and private organizations, which might also be applicable to public and private "systems," see Rainey, Backoff, and Levine (1976).

3. See Berwitz (1975) and Newgarden (1976) on technical difficulties of implementing the AA program. To avoid such technical difficulties, to reduce the uncertainty of guidelines, and to show proof of affirmative action, employers may try to fill quotas (Seligman, 1974).

4. In this essay we cannot do justice to many issues related to norm-governed behavior. The study of normative change in public personnel systems, given the stress and strain

associated with affirmative action, should be a fruitful area of research. Research concerning the variation in conflict, behavior, and performance for public employees, as a function of norm attributes such as severity of sanctions, modes of enforcement, consistency in enforcement, latitude in application, degree of conformity, etc., would be worth pursuing.

5. Within the public sector literature, few studies of this conflict have been published to date; no doubt many will be available soon. Rainey (1977) has found, in studying a small sample of managers in a federal agency, that 71% believed there was a conflict between affirmative action and the norm of hiring on the basis of qualifications and merit. At the state level, Milward (1976) collected data from state personnel officers of 32 states; only six states reported no conflict in implementing affirmative action in the context of their merit systems. Thompson (1976) surveyed personnel officers in local government concerning their attitudes toward affirmative action and the merit principle of competence (see his essay in this volume). Although he did not pose the question in terms of perceived conflict, his data suggests such a perception exists, given the respondents' extensive opposition to hiring targets (48% opposed, only 40% favored) and to hiring less qualified minorities in preference to more qualified whites (67% opposed, 19% favored).

REFERENCES

BACKOFF, R.W. (1974a). "Operationalizing administrative reform for improved governmental performance." Administration and Society, 6(May):73-106.
––– (1974b). Organizational innovation. Dissertation, Indiana University.
BAUMGARTNER, T., BURNS, T., DAVID, L., MEEKER, L., and WILD, B. (1976). "Open systems and multi-level processes: Implications for social research." International Journal of General Systems, 3:25-42.
BEATTY, R.W., and SCHNEIER, C.E. (eds., 1977). Personnel administration: An experiential/skill-building approach. Reading, Mass.: Addison-Wesley.
BEAUMONT, E. (1974). "A pivotal point for the merit concept." Public Administration Review, 34(September/October):426-430.
BERWITZ, C. (1975). The job analysis approach to affirmative action. New York: John Wiley.
BINGHAM, R.D. (1975). The adoption of innovation by local government. Milwaukee, Wis.: Marquette University Office of Urban Research.
BLAKENEY, R.N., MATTESON, M.T., and HUFF, J. (1974). "The personnel function: A systemic view." Public Personnel Management, (January/February).
BOBBITT, H.R., BREINHOLT, R.H., DOKTOR, R.H., and McNAUL, J.P. (1974). Organizational behavior: Understanding and prediction. Englewood Cliffs, N.J.: Prentice-Hall.
BOESEL, A.W. (1974). "Local personnel management: Organizational problems and operating practices." Municipal Year Book.
BOWEN, D.L. (ed., 1973). "Public service professional associations and the public interest." Philadelphia: American Academy of Political and Social Science, monograph 15.
CARR, F.J. (1971). "Industry and the city: The problem of technological transfer." Municipal Year Book 38.
DANZINGER, J.N. (1975). "EDP's diverse impacts on local governments." Nation's Cities, (October):24-27.
ETZIONI, A. (1964). Modern organizations. Englewood Cliffs, N.J.: Prentice-Hall.
FELLER, I., MENZEL, D.C., and KOZAK, L.E. (1976). Diffusion of innovations in municipal governments. University Park, Pa.: Pennsylvania State University Institute for Research on Human Resources, Center for the Study of Science Policy.
FILLEY, A.C., HOUSE, R.J., and KERR, S. (1976). Managerial process and organizational behavior. Glenview, Ill.: Scott, Foresman.

FRENCH, W. (1974). The personnel management process. Boston: Houghton Mifflin.
FUKUHARA, R.S. (1977). "Productivity improvement in cities." Municipal Year Book 44:193-205.
GALLAS, N.M. (1976). "The selection process." In W.W. Crouch (ed.), Local government personnel administration. Washington, D.C.: International City Management Association.
GIBBS, J.P. (1965). "Norms: The problem of definition and classification." American Journal of Sociology, 70(March):586-594.
GILLESPIE, D.F., and MILETI, D.S. (1977). "Technology and the study of organizations: An overview and appraisal." Academy of Management Review, 2(January):7-16.
GLAZER, N. (1975). Affirmative discrimination: Ethnic inequality and public policy. New York: Basic Books.
GLUECK, W.F. (1974). Personnel: A diagnostic approach. Dallas, Texas: Business Publications.
HAAS, J.E., and DRABEK, T.E. (1973). Complex organizations: A sociological perspective. New York: Macmillan.
HAGE, J., and AIKEN, M. (1969). "Routine technology, social structure, and organizational goals." Administrative Science Quarterly, 14(September):366-376.
HALL, R.H. (1972). Organizations: Structure and process. Englewood Cliffs, N.J.: Prentice-Hall.
HARRIES-JENKINS, G. (1970). "Professionals in organizations." In J.A. Jackson (ed.), Professions and professionalization. Cambridge: Cambridge University Press.
HAYES, F.O.R. (1972). "Innovation in state and local government." In F.O.R. Hayes and J. Rasmussen (eds.), Centers for innovation in the cities and states. San Francisco: San Francisco Press.
HENRY, N. (1975). Public administration and public affairs. Englewood Cliffs, N.J.: Prentice-Hall.
HICKSON, D.J., and THOMAS, M.W. (1969). "Professionalization in Britain: A preliminary measurement." Sociology, 3(June):38-53.
HUNT, R.G. (1970). "Technology and organization." Academy of Management Journal, 13(September):235-252.
JACKSON, J.M. (1975). "Normative power and conflict potential." Sociological Methods and Research, 4(November):237-263.
KAST, F.E., and ROSENZWEIG, J.E. (eds., 1973). Contingency views of organization and management. Chicago: Science Research Associates.
KRAEMER, K.L., DANZINGER, J.N., and LESLIE, J.L. (1976). "Information technology and urban management in the United States." Monograph of the URBIS Group, Public Policy Research Organization. Irvine, Calif.: University of California.
KRANZ, H. (1974). "Are merit and equity compatible?" Public Administration Review, 34(September/October):434-440.
LASZLO, E. (1975). "The meaning and significance of general system theory." Behavioral Science, 20(January):9-23.
LEVINE, C.H., BACKOFF, R.W., CAHOON, A.R., and SIFFIN, W.J. (1975). "Organizational design: A post Minnowbrook perspective for the 'new' public administration." Public Administration Review, 35(July/August):425-435.
LEVINE, C.H., and NIGRO, L.G. (1975). "The public personnel system: Can juridical administration and manpower management coexist?" Public Administration Review, 35(January/February):98-106.
LEVINE, E.L. (1976). "The institutional impact of equal opportunity laws on state-local personnel systems: A preliminary inquiry." Paper presented at the Region V (Southeast) Convention of the American Society for Public Administration.
LYNCH, B.P. (1974). "An empirical assessment of Perrow's technology construct." Administrative Science Quarterly, 19(September):338-356.
MARCH, J.G., and SIMON, H.A. (1958). Organizations. New York: John Wiley.

MILWARD, H.B. (1976). Unpublished research data. University of Kansas.

MINER, J.B., and MINER, M.G. (1973). Personnel and industrial relations: A managerial approach. New York: Macmillan.

MITNICK, B.M. (1974). "The theory of agency: The concept of fiduciary rationality and some consequences." Ph.D. dissertation. University of Pennsylvania.

MOHR, L. (1971). "Organizational technology and organizational structure." Administrative Science Quarterly, 16:444-456.

MORRIS, R.T. (1956). "A typology of norms." American Sociological Review, 21:610-613.

MOSHER, F. (1968). Democracy and the public service. New York: Oxford University Press.

――― (1973). "Commentary on the Gallas-Smith paper." In D.L. Bowen (ed.), "Public service professional associations and the public interest," monograph 15. Philadelphia: American Academy of Political and Social Science.

NELSON, R., PECK, M., and KALACHECK, E. (1967). Technology, economic growth, and public policy. Washington, D.C.: Brookings Institution.

NEWGARDEN, P. (1976). "Establishing affirmative action goals for women." Public Administration Review, 36(July/August):369-374.

NIGRO, F.A., and NIGRO, L.G. (1976). The new public personnel administration. Itasca, Ill.: F.E. Peacock.

NIGRO, L.G. (ed., 1974). "Minisymposium on affirmative action in public administration." Public Administration Review, 34(May/June):234-246.

PALMER, J.D. (1976). "Recruitment and staffing." In W.W. Crouch (ed.), Local government personnel administration. Washington, D.C.: International City Management Association.

PALUMBO, D.J., and STYSKAL, R.A. (1974). "Professionalism and receptivity to change." American Journal of Political Science, 18(May):385-394.

PERROW, C. (1967). "A framework for the comparative analysis of organizations." American Sociological Review, 32(April):194-208.

PRICE, J.L. (1968) Organizational effectiveness. Homewood, Ill.: Irwin.

Public Technology, Inc. (1973). The urban consortium for technology initiatives. Washington, D.C.: Author.

RAINEY, H.G. (1977). Unpublished data for use in Ph.D. dissertation. Ohio State University.

RAINEY, H.G., BACKOFF, R.W., and LEVINE, C.H. (1976). "Comparing public and private organizations." Public Administration Review, 36(March/April):233-244.

RUTSTEIN, J.J. (1971). "Survey of current personnel systems in state and local governments." Good Government, 87(spring):1-24.

SAGASTI, F. (1970). "A conceptual and taxonomic framework for the analysis of adaptive behavior." General Systems Yearbook 15:151-160.

SAVAS, E.S., and GINSBURG, S.G. (1973). "The civil service: A meritless system?" Public Interest, 32(summer):72.

SELIGMAN, D. (1974). "How 'equal opportunity' turned into employment quotas." Pp. 404-409 in W.C. Hamner and F.L. Schmidt (eds.), Contemporary problems in personnel: Readings for the seventies. Chicago: St. Clair Press.

SHEPPARD, H.L., HARRISON, B., and SPRING, W.J. (1972). The political economy of public service employment. Lexington, Mass.: Lexington Books.

SIMON, H.A. (1965). The shape of automation for men and management. New York: Harper and Row.

――― (1977). "What computers mean for man and society." Science, 195(March):1186-1191.

SOWELL, T. (1976). " 'Affirmative action' reconsidered." Public Interest, 42(winter):47-65.

STANFIELD, G.L. (1976). "Technology and organization structure as theoretical categories." Administrative Science Quarterly, 21(September):489-493.

THOMPSON, F.J. (1976). "Meritocracy, equality and employment: Commitment to minority hiring among public officials." Paper presented at the 1976 annual meeting of the American Political Science Association.

THOMPSON, J.D. (1967). Organizations in action. New York: McGraw-Hill.

TOMESKI, E.A., and LAZARUS, H. (1974). "Computerized information systems in personnel—A comparative analysis of the state of the art in government and business." Academy of Management Journal, 17(March):168-172.

U.S. Civil Service Commission (1974). "On the move: Personnel systems improvements in state and local governments." Washington, D.C.: Author.

——— (1975). "Guidelines for the development of an affirmative action plan." Washington, D.C.: Author.

VAN de VEN, A., and DELBECQ, A. (1974). "A task-contingent model of work unit structure." Administrative Science Quarterly, 19:183-197.

VAN RIPER, P. (1958). History of the United States civil service. New York: Harper and Row.

VOLLMER, H.M., and MILLS, D.L. (1966). Professionalization. Englewood Cliffs, N.J.: Prentice-Hall.

WATLINGTON, M. (1971). "Municipal use of automated data processing." Municipal Year Book 38.

WHISLER, T.L. (1970). The impact of computers on management. New York: Praeger.

WHYTE, L., WILSON, A., and WILSON, D. (eds., 1969). Hierarchical structures. New York: American Elsevier.

YIN, R.K., HEARL, K.A., VOGEL, M.E., FLEISCHAUER, P.D., and VLADECK, B.C. (1976). A review of case studies of technological innovations in state and local services. Santa Monica, Calif.: Rand.

7

Public Sector Unionism

FELIX A. NIGRO
LLOYD G. NIGRO

□ IN THIS CHAPTER, we are concerned with how increased unionism has affected urban management: the problems it has brought and the opportunities it offers.

CHANGING RELATIONSHIPS

Sometimes neglected in the concern over the rising costs of government and funding problems is the impact of collective bargaining on the mental sets of public officials, the employees, and the public. Speaking first of public officials, we note that many of them have great difficulty in comprehending and accepting collective bargaining. The first reaction is that the union wants to displace management. It is startling for officials who in the past have listened to employee suggestions—but always with the security that they need do so only as long as they chose—to find that they now *must listen until a contract is negotiated.*

The pattern of wanting to deal on employment matters with individuals, not with a committee the total membership of which is designated by an exclusive bargaining agent, sometimes continues because management finds the principle of "exclusivity" repugnant. That principle is, of course, a sina qua non of collective bargaining: like it or not, once legislation requiring collective bargain-

[141]

ing has been passed, management commits an unfair labor practice if it insists on negotiating conditions of employment with individual employees or minority unions. The majority union is the negotiating voice of *all* workers, whether or not they are members of that union or of any union; this is hard for some management officials to swallow, accustomed as they are to trying to keep each "member of the family" happy. Nor should one be harsh in commenting on the adjustment difficulties of these officials. As it has developed in this country, collective bargaining has certain distinctive features like exclusivity which are now virtually unquestioned in the private sector but represent modes of relationships with employees which have been alien inside government. (Nigro, 1970)

The collective bargaining agreement, particularly if it provides for binding grievance arbitration, makes officials recoil before employee proposals that they otherwise would not find worrisome. If an employee representative makes a request which the official believes improper because it would negate management rights, the only damage is to the official's sensibilities. In a collective bargaining negotiation, the situation is very different: these are demands some of which management may have to accept if a contract is to be signed. Once it grants these demands—such as class-size limits in the schools or maximum patient loads in the hospitals—its chances of obtaining the elimination of the particular provision in future contracts are not good. Because of their long experience with collective bargaining, private sector management officials are inured to the periodic battle during the bargaining sessions to maintain management rights. In government, collective bargaining is still too new for many officials to accept it as private companies have after a solid 40 years of experience since passage of the Wagner Act in 1935.

Under collective bargaining, the employer must adjust not only to the bilateral decision making process with the union but also to proving its case before third parties. The third party may be the state public employment relations board or commission which is investigating union charges of unfair labor practices by management. It may be an arbitrator who is hearing a grievance case or a fact finder who is studying the facts in a bargaining impasse in order to make recommendations for settling the dispute—a report which cannot be ignored since it is intended to serve as the basis for inducing the parties to resolve their differences. Fact finders may have subpoena powers for obtaining financial and other data that the employer would prefer not to make available; they may swear in witnesses who are required to tell the whole truth and nothing but the truth just as in a court. Management is forced to justify its actions and proposals; it cannot simply do as it sees fit.

These third parties are not infallible, so an all-out effort must be made to persuade them to make "correct" decisions. In the private sector this justification process, although regarded as onerous in many respects, has existed so long that management takes it for granted. The expense of arbitration and other third party services dismays economy-conscious public officials; the labor relations program means substantially increased costs just when the city must make economies.

Many public officials are adapting to collective bargaining and increased worker participation; note this statement by a former First Deputy Commissioner of the New York City Police Department:

> I believe that governing with the consent of the governed, while more troublesome, is in the long run more effective. The authoritarian model was fine for the Gestapo and may work from time to time in a democratic state, but it does not fit into my notions of how to run a police department in America in 1974. [International Association of Chiefs of Police, 1974:74]

A basic mental adjustment problem for public employees and their leaders has been how to use their new power. There has been much exaggeration, but certainly the successes at the bargaining table of so many public employee unions during the late 1960s and early 1970s led some of them to flex their muscles and bring relentless pressures. The present situation is one in which the unions, while not in full retreat, confront managements which insist that they have little additional to give to the workers and indeed want to take back some of the benefits that they granted in previous contracts. The financial position of governments has worsened to the point that money is not available to finance the size settlements to which many unions had become accustomed.

The teachers strike in New York City in September 1975 is a good example. The teachers were hardly victorious; indeed, their loss in fines (two days pay for every day on strike) was considerable. In a revealing interview after the strike, Albert Shanker, president of the United Federation of Teachers (UFT), said that he had opposed the walkout but was "almost alone" among UFT leaders in realizing that because of the city's grave financial difficulties "it was obvious that not much was to be gained."

> "There was really no time to discuss what had changed in the city and the world and the negotiating picture," Mr. Shanker said. "People left in June and the old world was still there. So most of the people came back with the usual expectations that everything would be settled at the wire with more money."

> "People became conditioned to using certain techniques and weapons and believing that they will always work .... I had to spend many hours convincing, including our own executive board and negotiation committee, after the final settlement." [New York Times, September 30, 1975]

Shanker chose his words well. Living in the "old world" and not perceiving the new one seems to be a weakness of many people. As previously indicated, it obviously afflicts employers, as well as employees—and also the general public.

It took much of the public several years to get used to collective bargaining and militant activity by government workers; many persons still cannot reconcile themselves to such a marked change on the social scene. The biggest problem for the public (to be discussed more fully later in this article), is to understand what is taking place currently in labor relations in government and to develop means of taking effective action to protect its interests in contract settlements. As in so many other areas, the public needs to educate itself and exert its influence opportunely.

MUNICIPAL UNIONISM AS A REVOLUTION OF PROFESSIONALS

What gives unionism in government its distinctive character is the much higher proportion of organized workers in the professional and white-collar ranks than in the private sector. U.S. Bureau of the Census statistics (1974:1) have shown that, as of October 1972, nearly 70% of all full-time teachers were organized (teachers account for almost one-third of all full-time employees in state and local governments). Increasingly, formerly independent associations of predominantly white-collar workers are affiliating with the American Federation of State, County, and Municipal Employees (AFSCME), which is an AFL-CIO union. Professional organizations like the American Nurses Association (ANA) and the National Education Association (NEA), while they reject affiliation with the outside labor movement, are outspoken advocates of collective bargaining and of militant tactics including strikes if necessary (U.S. House of Representatives, 1972). Organizations of social workers have been active from the earliest days of the new militancy; recently, attorneys (legal aid) and medical doctors (interns and residents) have employed the strike. Collective bargaining is making inroads, too, among librarians (Guyton, 1975).

The evidence is clear that, in government, professionalism and unionism, far from being antithetical, are mutually reinforcing: the union helps to achieve professional goals. Many professionals choose government employment in order to be able to contribute public service; loyalty to profession is combined with zeal to help the community. They identify with the mission of their organizations with an intensity seldom found in professionals working for private entities. And this is why the management rights concept is threatened by collective bargaining much more in government than in the private sector. For the most part, unionized private workers (most are in the blue-collar ranks) are not concerned with the firm's product or service. Their transgression on management rights is not in this sphere—whereas in government it definitely is.

To unionized professional workers in government, "management" is very different from "management" in a private firm. The objective of a public agency is service to the public, not profit. The ultimate management is the electorate, and employees at all levels of the organization have a responsibility for serving the public. Obviously, in any organization someone must have a final decision making responsibility, so "administrators" are necessary, but it is entirely appropriate in government for these administrators to share this responsibility with staff members whose special training makes it possible for them to contribute a great deal in matters of agency mission. This is why collective bargaining in government is so congenial to so many professional workers: it provides a means of both making concrete and enforcing the collegial principle.

In a prepared statement submitted to the Senate Committee on Labor and Public Welfare, which was holding hearings on a proposed National Public Employment Relations Act, the NEA stated that the scope of bargaining in the public service would have to be broader than in the private sector where it involves pay, hours, or other conditions of employment. The proper scope in government was "terms and conditions of employment and *other matters of mutual concern relating thereto*" (italics added). Noting that collective bargaining in government had been marked by a high degree of involvement of salaried professionals, the NEA stressed that teachers because of their training felt a "special identification with the standards of the 'practice' and the quality of the service provided to the 'clientele.' " (U.S. Senate, 1974:130-132)

In practice, teachers regard all phases of education as potentially appropriate for bargaining. Typical matters that teacher organizations want to bargain over besides class size include procedures for faculty access to supplies, equipment, and material; discipline in the classroom and corporal punishment of pupils; teachers' responsibilities for supervising pupils outside the classroom during the regular duty day; requirements for keeping records and reports; availability of outside telephones for teachers' use; teaching loads; preparation periods; instructional assignments; classroom interruptions; and responsibility for collecting lunch and milk money or donations in various drives.

Administrators in an organization made up predominantly of professionals understandably feel menaced by collective bargaining. It is not just a question of having to yield on some of their views as to how the taxpayers' money should be spent. It is also a matter of resisting attempts to relieve them of large amounts of discretion in noneconomic areas. Private executives may feel that they have lost their shirts in a negotiation with a union; public officials may conclude that they have practically lost their jobs.

Of course, it is the point of view that determines the sensation of victory or defeat: the union could argue that the best administrator in an organization of professionals is one who accepts shared decision making. But this ignores some of the realities facing the administration. Let us illustrate with the budget. Unions generally do not attempt to make the entire budget negotiable because there is no feasible way of doing this under long-established procedures in government. Yet some unions would like to bargain as much of the budget as possible, and they are not deterred by the fact that some of the needed funds are provided by higher levels of government. It has been argued, for example, that city welfare administrators should agree to promise in the collective agreements to try to persuade state and federal authorities to fund increased welfare allowances.

Under the present hierarchical arrangements in government, agency administrators are responsible to legislators and the public for the program results achieved with the public's funds. Unions may be excoriated for restraints that they impose on administrators at the bargaining table, but they are not considered part of management by the legislators or the citizenry. In any case, there is no practical way in which administrators can evade this responsibility and pin it on the unions.

IMPACT OF THE INCREASED UNIONISM ON THE POLITICAL PROCESS

There are two indisputable aspects of collective bargaining in government which require detailed comment. First, political factors affect the bargaining and the settlements much more than in private sector labor relations. Second, closed bargaining sessions and the process of reaching decisions as a whole hamper greatly the public's ability to influence the outcomes.

EVALUATION OF POLITICAL FACTORS

Public employees and members of their families are voters. Elective officials depend upon voting support, so they prefer that the employees be reasonably well satisfied with the contract terms that they are offered. Those running for public office need campaign contributions and the support of volunteers in various clerical and other election chores. The picture is often drawn of the typical elective official as ready to grant almost any union request in return for this help. While in our opinion this is a gross exaggeration, there is no denying that, as stated by Lee C. Shaw and R. Theodore Clark (1975:3-13), public officials must take into account the possibility of direct retribution by the employees. They can punish the officials by not voting for them, whereas in private industry workers disgruntled with a contract settlement usually cannot do much damage to the careers of the company negotiators. Collective bargaining is often referred to as an adversary relationship. What kind of adversaries can public officials be if they must worry about losing the workers' support?

The further point is made that elective officials cannot afford to take strikes because the public, while often enraged with the strikers, is mostly concerned with its own convenience and with getting public services restored as quickly as possible. To support this interpretation, we may note that most work stoppages in the public service have lasted several days only; long, drawn-out strikes of which there are plenty of examples in the private sector have been very rare. There is much evidence to support this analysis, although events in 1975 disproved the assumption that public attitudes would *never* permit public management to dig in for possible long walkouts. Political situations change, and at certain times public opinion can form solidly against a public service strike. But this is no comfort to those concerned about what they view as dangerous political coziness between public management and public employees.

Since the political power of public employees has been much discussed, it is appropriate to refer to research findings on the subject. The most complete study is still the *UCLA Law Review's* "Project: Collective Bargaining and Politics in Public Employment" (1972), carried out with the financial assistance of the Labor Management Relations Service (LMRS) of the National League of Cities, the U.S. Conference of Mayors, and the National Association of Counties. This report stated that, except for a few places like New York City, "the problem of union power seems more an emergent possibility than a present reality" and that it was "unlikely that many of the smaller unions would ever possess excessive political power." *(UCLA Law Review,* 1972:887-1051) Similar conclusions were reached by Paul E. Gerhart (1974) in an analysis based on interview data from Brookings Institution research in 1968 in 40 cities and counties.

The assumption that the unions have too much political power has influenced much of the discussion of labor relations in government, including that in scholarly works such as the book by Harry H. Wellington and Ralph K. Winter (1971). They stressed that market constraints which moderate union demands in the private sector do not exist in government. In industry, the unions know that if they push their wage and other demands too far, the effect may be to drive the employer out of business. They also are aware that the firm has low-cost, nonunionized competitors and that there is a limit to the increased wage costs that the employer can pass on to the consumer in the form of higher prices. Public employers do not have to watch costs in the same way because they usually are the only suppliers of the particular service. They cannot move to other locations or close down. If, in addition, the employer confronts tremendous union political power, the case is complete for modifying the collective bargaining process in government in such ways as to keep the unions under control. One could conclude from this analysis that collective bargaining in government is at best a necessary evil and that certainly the right to strike should be denied or very strictly controlled.

If, in fact, unions generally do not have excessive political power, the conclusions will be very different. Albert Shanker, previously mentioned, refuted the contention in a National Association of Manufacturers' report that school boards uniformly cave in to the political power of teacher organizations. Shanker claimed that the foot-dragging in settling disputes is by the school boards, not the teachers. Shanker wrote:

The NAM polemic tries to do just what the NAM is paid to do by its corporate members; promote anti-union legislation by conjuring up a phony monster—"union power." In doing this, it chooses to ignore the simple truth that in our pluralistic society neither labor nor management is all-powerful. . . . Both sides have power—but if at this point the power is unevenly distributed, it is that too much of it is wielded by school boards and too little by the teachers. [*New York Times,* October 7, 1973]

In much of the discussion of labor relations in government, the assumption seems to be that the picture is about the same in all parts of the country. Of course, the local situation is the one which people know best; since their experience is limited to that locality, it comprises their "world" and provides the basis for inevitable generalization. Unionism is an emotional issue, and emotions generate easy generalization. It is surprising that so many of the statements about public employee unions are based on New York City and a few other big cities. The quality of the political leadership if often overlooked. In small cities as well as big ones, examples of indiscriminate public employee vote-seeking politicians can be found. There is no reason why, with responsible public officials and union leaders, bargaining in government should damage the community. As the *UCLA Law Review* stated, it would be neither desirable nor constitutional to prohibit political activities of public employee unions. What the unions seek through political action is often desirable for the people as a whole, and other groups in society can oppose selfish union demands. The argument that collective bargaining is an inherently dangerous transplant in the political world of government is not sustained by the facts.

THE QUESTION OF PUBLIC PARTICIPATION

There is solid agreement by labor relations experts that if settlements are to be reached, negotiations must be conducted in closed sessions. Concessions simply will not be made if the press, members of the public, and any persons other than representatives of the parties are present. Because of the increasing criticism of secrecy in government, this argument does not convince everyone. The collective bargaining statute enacted in Florida in 1974 provides that "negotiations between a chief executive and a bargaining agent shall not be exempt" from that state's Sunshine Law. But Florida is an exception in this respect; in most of the country, negotiations are private.

It is easy to concentrate on preserving the conditions which permit agreement between the parties—but pretty much leave out the public. In fact, the participation of the legislative body itself may not amount to very much, and it may be presented with a *fait accompli:* the mayor may finally announce contract terms which the lawmakers are expected to finance to "end the emergency." Even when the chief executive is required by law to submit the contract to the legislature for approval, so many "tentative" commitments may have been made that it may not be in a position to make any changes.

Collective bargaining strengthens the role of the chief executive: leadership in the jurisdiction in labor questions is best provided by a single person. If necessary, union representatives will negotiate contract terms with legislative committees or the legislature as a whole, but this is an awkward arrangement. In the private sector, the unions bargain with a company representative who is authorized to make commitments for the firm. In government, both management and the unions much prefer trying to approximate the industrial collective bargaining model in this respect.

If the argument is made that the people can vote out of office elective officials who agree to exorbitant settlements, this does not repair the damage already done. Furthermore, it is often not realistic politically; by the time of the next election, much will have been forgotten and incumbents will be judged on the basis of many factors. Reviewing events of recent years, it is remarkable how chief executives, unpopular with the public and some of the unions for labor relations decisions, can recoup their standing with the public and even with the same labor leaders who swore vengeance on them.

Zagoria (1973) recommended trying to resolve impasses through *public referendum*. Specifically, he proposed that, if either party rejects a fact-finder's recommendations, the unresolved issues be placed on the ballot at the next regular or special election, whether it be for election of municipal or state officials or a bond issue or other referendum. If the impasse is over salaries and fringe benefits, what the employer was willing to grant could be put into effect immediately, and the disputed amounts could be considered in the referendum. The contract terms could be made retroactive so that "neither party is prejudiced by the referendum delay." Both parties would campaign to persuade the voters to accept their position on the referendum issues.

What better version of "finality" is there than referral of the disputed issues to the public for decision? Under some statutes, the final step in impasse resolution is to refer the matter to the legislative body for decision—a form of "finality" anathema to the unions. In other statutes, the final step is binding arbitration required by law—but in this case the final say is by arbitrators who often do not live in the community.

Based upon past experience, Zagoria judged that public employers are more likely to accept fact-finders' recommendations and that the union will usually be the party requesting the referendum. If the employer does reject the fact-finders' proposals and the union wins the referendum, the employer will have to "sign the agreement on the fact-finder's terms." If the union is the party rejecting the fact-finder's recommendations and then wins in the referendum, this is a "directive" to the employer to "up the ante in future negotiations."

Distinctly noticeable in the writings and public statements of many labor relations experts is lack of attention to schemes for public participation such as that suggested by Zagoria. This strongly suggests that they are too wedded to the private-sector collective bargaining model. Certainly, it is very difficult to develop workable plans for public participation, but it is unfortunate to leave any impression that the problem is not a high priority one.

Mention of fact-finding leads us to observe that its presumed advantage of involving the public may be more theory than reality. The theory is that public groups will be vitally concerned with fact-finders' reports and analyze them carefully. While it is unreasonable to expect the press to publish these reports in their totality, one would hope that it would give accurate summaries of them. Instead, the reporting in many newspapers tends to be very scanty; all the reader learns is that the fact-finders have supported one side or the other. Frequently, the newspapers support neither side on all impasse items, but do emphasize the recommendations on the economic issues and thereby give an incomplete and even distorted picture of the fact-finders' recommendations.

Nor do civic groups make intensive analyses of the reports and or take systematic efforts to educate the public accordingly. Of course, this is not a valid argument against fact-finding and is not intended as such. Fact-finders are neutrals, and their very appearance often induces the parties to adopt more flexible postures. We are simply calling attention to neglected possibilities of making fact-finding more effective as a stimulus to public participation. Taxpayer and business groups seem to apply themselves the most assiduously to informing the public on labor relations matters in government. Other groups should be more active in the same role.

That the public needs to be more vigilant in protecting its interests is seen in the obvious strategy of some employers to conceal or not fully reveal the *future* costs of contract settlements. Changes may have been agreed on in pension plans which do not increase costs in next year's budget or even those for some years to come but later do add greatly to the tax burden. A legal requirement that the employer make public its estimates of both *present* and *future* settlement costs is entirely reasonable. Even if it is decided that public disclosure should be required *after* rather than *before* the settlement is adopted, the effect should be to dissuade the employer from trying to conceal the true costs.

IMPACT ON LOCAL AUTONOMY

Starting in 1972, congressional committees gave active consideration to two bills requiring state and local governments to institute collective bargaining programs for public workers. One of these bills would have amended the Labor Management Relations Act to extend its coverage to state and local governments, bringing them under the National Labor Relations Board. The other would have established a National Public Employment Relations Commission to supervise collective bargaining in state and local government, just as the NLRB does for the private sector.

In its report, *Labor-Management Policies for State and Local Governments* (1969), the Advisory Commission on Intergovernmental Relations (ACIR) strongly opposed federal legislation of this kind, saying that, if it were enacted, "little would be left of the Federal principle of divided powers." The commission urged that all state governments adopt labor relations policies for public entities and viewed as distinctly undesirable the failure of so many of them to introduce order into what often was a chaotic situation in relationships with organized public employees.

The ACIR did not take a stand on whether the state legislation should require collective negotiations or meet-and-confer; under meet-and-confer the employer reserves the right to make unilateral decisions, and any agreements reached are not binding. It prepared two model bills for consideration by state legislatures, one providing for collective negotiations and the other for meet-and-confer, but both contained a local public agency option, as follows:

> This Act . . . shall be inapplicable to any public employer, other than the State and its authorities, which, acting through its governing body, has adopted by local law, ordinance, or resolution its own provisions and procedures which have been submitted to the Agency that such provisions and procedures and the continuing implementation thereof are substantially equivalent to the rights granted under this Act.

(In the collective negotiations bill, the words after "implementation thereof" are "do not derogate the rights granted under this Act.")

This provision is very similar to the one in the New York State Public Employees' Fair Employment Act passed in 1967. The experience in that state has been that relatively few local jurisdictions opt to come under this provision because of the expense and other difficulties of establishing and operating their own programs. Since they make this decision, the principle of local autonomy is preserved. Local governments naturally want state agencies to be flexible in their approach and take into account local conditions. In general, however, state labor relations legislation does not raise anywhere near the opposition that federal mandating does.

At its 1974 meeting the National Association of Counties adopted the position that labor relations policy should be decided "solely by each state." National legislation was "inappropriate because it usurps local prerogatives, it dictates use of revenues raised by state and local governments, and it violates intergovernmental partnership and smacks of federal paternalism" (LMRS Newsletter, October 1974). In August 1975, the United States Conference of Mayors approved a resolution that "these two bills and any similar bills could infringe on the constitutional rights of states and state authorized local government instrumentalities to govern their own affairs and would preempt the authority and responsibility of state and local governments to control, regulate, and manage labor relations programs for their employees" (LMRS Newsletter, August 1975).

Municipal officials are by no means all opposed to federal legislation; Mayor Young of Detroit and Mayor Kevin White of Boston supported it (U.S. House of Representatives, 1974). White said he thought the legislation would reduce labor strife because it would provide a framework for resolving disputes. He saw a "clear distinction" between the policy functions of state and local governments and their relationships with their employees. In the first case, Congress should not interfere; in the second, Congress had a "clear right" and "clear interest" to act to protect the "rights and privileges" of the employees. White rejected the constitutional argument as fallacious, citing congressional extension of the Fair

Labor Standards Act to some groups of state and local workers. The constitutionality of this extension was upheld by the U.S. Supreme Court in its decision in *Maryland* v. *Wirtz* (1968). (See Smith et al., 1974:46-59.)

In 1976, however, by a five-four vote, the Supreme Court reversed *Maryland* v. *Wirtz (National League of Cities* v. *Wirtz)*. The majority's basic argument was that it was not constitutional under the interstate commerce clause for the federal government to structure the way in which state and local governments make decisions about the wages and hours of their workers. Since collective bargaining would structure state and local decision making in very specific ways, most authorities concluded that the Court majority would not find either of the two above-mentioned bills constitutional. It appeared that the principle of state and local government right of free choice in deciding whether or not to require collective bargaining had won out.

EFFECT ON FINANCIAL MANAGEMENT

Statements abound as to the relationship between collective bargaining and the very substantial increase in public payrolls, but so little research has been completed on the subject that there is no basis for conclusions of wide application. Studies have been made which show that during certain recent time spans some groups of public employees like teachers, fire fighters, and police have received average annual increases exceeding those for the economy as a whole, but collective bargaining has been held only partly responsible (Hartman, 1971:192-197). Economists consider that the principal factor in annual salary increases are market forces and that the general nationwide rise in the level of salaries, plus inflation, would, in any case, have pushed up the compensation of government workers.

Undoubtedly, compensation of many public employees is substantially improved. So long as it is not excessive, the increases are justifiable in terms of equitable relationships with pay of private workers. This is the whole point behind "pay comparability" and collective bargaining; what has happened is that many taxpayers are not pleased with the financial costs of catch-up pay—they find the bill too high.

A U.S. Bureau of Labor Statistics survey during 1970 and 1971 of government and private enterprise pay in 11 large municipalities showed that municipal scales were generally better for equivalent work in office clerical, data processing, and maintenance and custodial work (Perloff, 1971:46-50). National surveys by the Labor Management Relations Service have shown that cities expend a higher percentage of working time pay for fringe benefits than private industry; for nonuniformed personnel, however, the difference is small. In New York City some union contracts provide for the city to make special annuity fund payments to several employee associations and unions in addition to those it makes to the particular retirement system. Some long-service employees in New York

City retire with benefits which, with social security, give them more income than before retirement (Permanent Commission on Public Employee Pension and Retirement Systems, 1973).

It is conceivable that salaries and benefits for most kinds of public jobs in most parts of the country could come to exceed those in the private sector, but there is no evidence that we are approaching such a situation. Certainly there is no justification for overpaying public employees, and there is also no justification for public employers to yield to excessive union demands.

Some municipalities are very much concerned about compulsory arbitration, particularly when required by state statute. They lose control over spending priorities and tax increase decisions; the funds to comply with the arbitral award must be found and new taxes levied if necessary. Of course, if compulsory arbitration eliminates or greatly reduces strikes, public services are not interrupted—which is the benefit side of the equation. Certainly the preferable arrangement is for the municipality itself to decide that it needs compulsory arbitration.

A reasoned argument can be made that collective bargaining forces public employers to make more effective use of total resources. A management that does not have to bargain collectively is not under the same pressures to scrutinize very carefully its expenditures for nonpersonal services. A school administration receives numerous budget requests to improve physical facilities, such as grounds, walks, athletic facilities, and classrooms. If it knows that, based on past negotiations, it must reserve a certain sum for the next contract settlement, it is more diligent in paring other parts of the budget. Of course, yielding to teacher demands could mean impairing the quality of public education. Yet a fact-finder will want to be sure that this is the case and that the school board is not neglecting what satisfied teachers contribute to educational quality.

The public employer sometimes goes to the bargaining table not knowing how much money it will have to finance settlements. Education and welfare are two good examples. In both cases the state and federal governments provide substantial grants-in-aid, but the precise amounts of these aids is difficult to predict accurately. Synchronization of financial decision making at all levels of government would solve this difficulty, but such synchronization is hardly likely under present governmental arrangements.

Of course, contracts can make clear that the financial terms are subject to receiving the necessary funds from the various sources, but if insufficient amounts are obtained from these sources the employer is apt to get most of the blame. Sometimes contracts are signed locally on the basis of tentative state allocations of funds which later are cut back when estimates of state revenues fall short. In such situations, the contract salary terms remain in effect, but some employees have to be laid off, perhaps quite a few. Of course, this could happen without collective bargaining, but the "union" will be held responsible for the extent of the layoffs.

With more experience, collective bargaining timetables, budget submission dates, and local financial decisions can be better arranged. Under Florida's collective bargaining statute, negotiations are supposed to be completed by the budget submission date, but, as in other states, they often have not been. Other state legislation requires local school boards to decide mileage rates by a certain date. While negotiations are still continuing, the school boards fix the mileage rate and thus the ceiling amounts available from local sources for contract settlements. Not surprisingly, some teacher representatives consider this a violation of good faith bargaining.

UNIONISM AND PRODUCTIVITY

Even without unions and collective bargaining, municipalities would have to develop productivity programs. Shrinking revenues and rising costs because of inflation make it necessary to increase productivity; public pressures for improving the delivery of public services and keeping costs down are intense.

Zagoria (Perloff, 1971:46-50) identified three kinds of productivity bargaining in the public sector. In one type, the objective is to induce the union to "negotiate out" traditional practices and/or contract provisions which make for less rather than more productivity. Examples are minimum crew requirements and work quota systems. The inducement to the union is more money; dollars are traded for "management flexibility." Zagoria pointed out that, since in these inflationary times workers would demand wage improvements anyway, the "contract cost for the changes may be less than indicated."

The second form is used when productivity gains can be measured in dollar benefits. The bargainers agree upon a formula for division of the gains, as in Detroit in the case of sanitation services. This obviously has great appeal to the workers because it means bonus money.

In the third plan, the contracts provide for joint labor-management committees to study work practices and procedures and recommend methods of improving productivity. As Zagoria stated, this approach "suggests that workers are interested in the work they do and bring a first-hand expertise to revamping the work process, that they have a major stake in the changes. It implies a management view of workers as people who would rather do a good and useful day's work than to log the hours and goof off as much as possible."

Unions shy away from management-initiated productivity plans when they suspect that the real purpose is speed-up, elimination of jobs, or other manipulation of the worker as a "cost" of operations. They want also to be sure that the measurement of productivity is valid and fair to the employee. For example, teachers accept the principle of "accountability," but not if their performance is to be measured exclusively by pupil reading attainment and other scores. The collective bargaining framework is an excellent one for dissipating these fears since the union does not have to agree to a particular productivity proposal. Of course, management has the right unilaterally to introduce changes in work methods; realistically, however, best results are obtained only when the workers accept the changes.

As financial stringencies continue, governments may follow the New York City example wherein pay raises are tied to demonstrable productivity savings. There is fear, however, that if the financial picture improves, unions will call for a repudiation of such arrangements, arguing that management neglects its own responsibility to make administrative and operating reforms which would greatly improve productivity.

THE TRANSFORMATION OF PUBLIC PERSONNEL MANAGEMENT

There has literally been a transformation of personnel management in government because of collective bargaining.

Traditional public personnel administration generally failed to deal comprehensively with personnel problems at the operating levels, i.e., where work activities are carried out. Lip service was given to the importance of line personnel management, but more time was spent on appointment, classification, and other routines required by law than on grievances and other problems causing difficulties in the relationships between management and the workers.

Binding grievance arbitration—now common in local governments—forces management to give more attention to the needs of employees and the conditions that cause worker dissatisfaction. When management knows that the arbitrator may decide in favor of the employee in a grievance case, it necessarily must reexamine the way in which its personnel program actually functions. Union stewards continuously monitor personnel decisions by supervisors; while they sometimes support unjustified grievances, they usually exercise a constructive function, as local officials themselves have often confirmed.

In theory, the public employer should *know* what is needed in a comprehensive personnel program and *provide* it. In practice, it is deficient in both respects; it is unaware of many problems at the operating level. Anyone who has analyzed any fair number of collective bargaining agreements in local governments quickly notes how numerous aspects of a complete personnel program, neglected in the civil service ordinance and regulations, are covered in the agreements. In some civil service jurisdictions, whole areas like discipline and grievances are in practice underdeveloped; matters of great concern to the employees are overlooked. The characteristic of union negotiators is that they pressure for inclusion in the agreement of all aspects of personnel policy and procedure which the workers consider important. What management believes that "good supervisors" can be expected to do the union wants made a contract requirement.

This is not to say that the union input is uniformly desirable. Some unions want seniority to be the key factor in promotions and in other personnel actions like reductions-in-force and reemployment. Unions may also insist on so many strictures on supervisors as to make it difficult for them to carry out effectively the work operations for which they are responsible. But management negotiators are not compelled to accept such demands.

By now it is clear that collective bargaining and the merit principle are not inherently contradictory and that they can be accommodated—as they now are being accommodated in many local governments. Bilateral decision making between management and the workers will not solve all personnel problems but it has greater potential for solving many of them than old-style paternalistic civil service systems.

CONCLUSIONS

Collective bargaining has contributed to the strain on local government budgets, but it also functions as a stimulant for closer scrutiny of expenditures for nonpersonal services and for efforts to improve productivity. Research on the relationship between increased payroll costs and collective bargaining has been meager, but the evidence does not support any conclusion that collective bargaining inherently causes the financial ruin of governments. While some unions have abused their new power, the benefits side of the bilateral process, in terms of improved employee morale and valuable worker inputs into policy making, tend to be overlooked. An example is personnel management which has been made much more meaningful and complete as the result of bilateralism.

For state and local governments which have adopted collective bargaining, the challenge is to adapt the bargaining process to the peculiar environment of government. Foremost is the need for development of proficiency by both management and the unions in the bargaining process and the administration of contracts. In the last analysis, both sides benefit when government services are rendered effectively. Contrary to the popular viewpoint, successful collective bargaining is more a cooperative than a conflictual relationship. For those jurisdictions which have not adopted collective bargaining, the challenge is to develop a viable substitute for the old-style and demonstrably ineffective unilateral paternalism. Finally, it need not be assumed that traditional bilateral collective bargaining is the *only* feasible alternative; to so limit the options would be a serious mistake.

REFERENCES

Advisory Commission on Intergovernmental Relations (1969). Labor-management policies for state and local governments. Washington, D.C.: U.S. Government Printing Office.
GERHART, P.F. (1974). Activity by public employee organizations at the local level: Threat or promise? Chicago: International Personnel Management Association.
GUYTON, T.L. (1975). Unionization: The viewpoint of librarians. Chicago: American Library Association.

HARTMAN, P.T. (1971). "Wage effects of local government bargaining." In Proceedings of the International Symposium on Public Employment Relations. Albany, N.Y.: New York State Public Employment Relations Board.

International Association of Chiefs of Police (1974). Guidelines and papers from the National Symposium on Police Labor Relations, 1974. Washington, D.C.: Author.

Labor Management Relations Service of the National League of Cities, United States Conference of Mayors, National Association of Counties (n.d.). Washington, D.C.

LIVINGSTON, F.R. (1972). "Collective bargaining and the school board." In S. Zagoria (ed.), Public workers and public unions. Englewood Cliffs, N.J.: Prentice-Hall.

NIGRO, F.A. (1970). "Labor relations in the public sector, I and II." Personnel Administration, (September, October, November 1970).

PERLOFF, S.H. (1971). "Comparing municipal salaries with industry and federal pay." Monthly Labor Review, 94(October).

Permanent Commission on Public Employee Pension and Retirement Systems (1973). Report. New York: Author.

SHAW, L.C. and CLARK, T.R., Jr. (1975). "The practical differences between public and private sector collective bargaining." in A. Anderson and H.D. Jascourt (eds.), Trends in public sector labor relations. Chicago: International Personnel Management Association.

SMITH, R.A., EDWARDS, H.T., and CLARK, R.T. (1974). Labor relations law in the public sector. Indianapolis, Ind.: Bobbs-Merrill.

U.S. Bureau of the Census (1974). 1972 census of governments. Washington, D.C.: U.S. Government Printing Office.

U.S. House of Representatives, Special Subcommittee on Labor (1972). Labor management relations in the public sector. Washington, D.C.: U.S. Government Printing Office.

U.S. Senate, Subcommittee on Labor and Public Welfare (1974). National Public Employment Relations Act, 1974. Washington, D.C.: U.S. Government Printing Office.

UCLA Law Review (1972). "Project: collective bargaining and politics in public employment." 19(August):887-1051.

WELLINGTON, H.H., and WINTER, R.K., Jr. (1971). The unions and the cities. Washington, D.C.: Brookings Institution.

ZAGORIA, S. (1973). "Referendum use in labor impasse proposed." LMRS Newsletter, 4 (9):2-3.

8

Collective Bargaining in
Municipal Governments:
An Interorganizational Perspective

CHARLES H. LEVINE
JAMES L. PERRY
JOHN J. DeMARCO

INTRODUCTION

□ THE UNEXPECTED GROWTH of public employee unionization and collective bargaining in American cities over the past 15 years has produced substantial changes in city governments and local politics. While racial politics monopolized much of the drama and media attention of the 1960s, the quieter changes brought about by the unionization of public employees during the same decade set the stage for equally significant shifts in local political power and governmental practices. For example, changes in public management brought about by collective bargaining have impacts on compensation patterns, service delivery systems, merit systems, supervisory authority, and intragovernmental resource allocation patterns (Lewin, 1973). Knowledge of these impacts and their primary and secondary effects on the fiscal and political well-being of cities and municipal governments is still rudimentary (Horton, et al., 1976); but if, as many observers have argued, these impacts threaten the long term viability of city governments, they clearly merit systematic investigation.

Even though the distinctive features of collective bargaining in public employment may be "exaggerated and even misleading" (Horton et al., 1976:503), collective bargaining is unlike any other bargaining process in American government. For example, negotiation episodes are more open to outside influences than conference committees in the Congress; they are usually of longer duration than plea bargaining in the courts; and the participant organizations usually have a greater divergence of goals than participants in the budgetary process. Therefore, as a legally constrained, semiclosed episode that involves a variety of third parties, public sector collective bargaining can yield insights into governmental bargaining processes in general and, in turn, bargaining in other political arenas can shed light on public employee collective bargaining.[1]

Like most emergent concerns in the social sciences, public sector collective bargaining suffers from preparadigmatic fragmentation, i.e., the failure of large groups of scholars to coalesce around common approaches, frameworks, and theories. Instead, the field has been dominated by journalistic accounts of crisis events, impassioned moralizing, and a pronounced melioristic orientation. To focus theory building and research about public employment collective bargaining, a framework must be constructed that captures the principal dimensions of collective bargaining episodes while retaining sensitivity to the distinctive features and commonalities between collective bargaining in the private sector and in government. To overcome preparadigmatic fragmentation and attract scholarly support, such a framework must have sufficient scope to encompass the various research concerns of a wide variety of scholars in the field, contain a familiar set of concepts, be operationalizable to allow for hypothesis testing. And, to provide a link between theory and practice, such a framework should account for the strategic behavior of participants in the bargaining process.

This essay presents a first approximation of a framework—a theoretical sketch rooted in interorganizational theory—which we believe captures much of the complexity of municipal collective bargaining. The framework is decomposed somewhat to allow brief discussions of the principal components of collective bargaining situations and to allow discussion of the major concerns of scholars who have studied public sector collective bargaining. We conclude by evaluating the state of knowledge about municipal collective bargaining and by suggesting some potentially high yield directions for future empirical research, theory building, and public policy.

AN INTERORGANIZATIONAL PERSPECTIVE ON MUNICIPAL COLLECTIVE BARGAINING

We consider municipal collective bargaining to be an interorganizational decision-making process. Proceeding from this perspective, we construct a framework for identifying and discussing the major components of the structure of municipal collective bargaining. The framework is presented in Figure 1. Although the framework has theoretical foundations, it is by no means complete and integrated. A great many of its empirical and practical implications need to be considered along with specific relationships between variables. The framework as it is developed here provides a "rough cut" instrument for organizing and generating hypotheses and guiding research.

The formal organization is treated here as the basic unit of analysis for studying both the employing organization and the union. Local governments are conceived of as being comprised of coalitions of semiautonomous organizations, groups, and individuals involved in competitive, conflictual, and cooperative relationships to influence the formulation and implementation of public policy. Within this coalitional arrangement, any single local government department or agency can also be understood as a coalition (or as a subcoalition within the more inclusive coalitional arrangement) whose "members include administrators, workers, appointive officials, legislators, clientele, etc." (Cyert and March, 1963:27).

The emergence of unionization and collective bargaining represents a basic transformation in the original configuration of a public organization's coalition (Perry and Levine, 1976). Two organizations, one representing employee interests and the other management interests, must jointly agree upon member contributions and organizational inducements (Barnard, 1938; March and Simon, 1958; Thompson, 1967). This change has implications for both the employing organization's and the local government's rule structure, resource allocation pattern, and service delivery behavior. The particular dyadic relationship between the two focal organizations "attempting to define or redefine the terms of their interdependence" (Walton and McKersie, 1965:3) through collective bargaining takes place within a context proliferated by third parties that form each organization's task environment. The presence and involvement of third parties suggests that bargaining processes are likely to be multilateral.

A primary consideration in approaching the analysis of the municipal collective bargaining relationship is the extent to which decisions about the inducements/contributions contract occur within a closed system bounded by the city government organization or an open system encompassing an interorganizational network. Typically, collective bargaining in the private sector is characterized by a closed system decisional structure with respect to items within the inducements/contributions contract. With the exception of constraints imposed by social legislation (minimum wage laws, equal employment opportunity legislation, health and safety laws), decisions about the inducements/contributions contract reside within the bilateral relationship. However, the bilateral relationship in the public sector frequently has superimposed upon it a number of lateral and hierarchical relationships which compete for control or literally control various inducements/contributions decisions.

Two areas of state and independent agency activity are particularly threatening to the closed system logic of bilateral negotiations: state control of the local government financial structure, and state and independent agency control of local government personnel practices. States control local government finances over a number of dimensions which together have an impact on local government flexibility to agree to particular employee organization demands. The establishment of legal debt limits, fixed tax rate levels, maximum allowable increases in local tax rates, and the procedures by which revenue increases are authorized have a considerable effect on a local government's flexibility to finance an agreement. For example, passage of a law requiring citizen approval of local property tax rate increases led a League of California Cities official to comment:

> the rate of the only major revenue source within the control of local officials which has the capacity to generate substantial revenues, the property tax, was frozen by statute. The balance of city and county revenue sources—sales tax, business license taxes, licenses and permits, fines and penalties, interest income, state and federal grants and subventions and service fees—are either comparatively minor in amount or fixed presently by statute. [Emanuels, 1973:8-10]

State mandating of particular practices for various functional groups, e.g., work week provisions for firemen and alternative health and pension plan systems for local government employees, presents additional problems. State mandating of inducements/contribution provisions provides an alternative channel for employee organization goal achievement and encourages the city employer to look toward state government as not only a "cost imposer" but a "resource provider." But state mandates reduce flexibility in constructing package settlements, and create differing expectations about what can be considered city controlled contractual changes. Although the "end-runs" of public employee organizations receive considerable attention, this type of behavior tends to be a strategy used by both parties in city government bargaining relationships. The propensity of the parties in city government bargaining to seek assistance and rewards at a level above them is traceable to their intensive dependence on state government.

Given these and other constraints on bilateral municipal collective bargaining, several recent studies suggest that there is doubt "whether a bilateral system can in fact be operationalized in this pluralistic environment" (Kochan, 1971:68). Kochan (1975a), for example, suggests that while the primary participants in bargaining may attempt to buffer the negotiation process from the more general political process on routine issues, it is unlikely that they will or can buffer the negotiation episode from multiple political interests on nonroutine issues. Nevertheless, this does not infer that the bargaining partners completely lose control of the negotiation episode to third parties. On the contrary, since both parties are semiautonomous, they exercise considerable influence in setting the agenda of negotiations and retain the latitude to affirm or reject a collective agreement. Furthermore, since most of the role structures imposed by state governments on local government collective bargaining exclude third parties from *direct* participation in deliberations, the dyad can still be meaningfully viewed as the "core" group.[2]

A FRAMEWORK OF THE MAJOR COMPONENTS IN
MUNICIPAL COLLECTIVE BARGAINING

Viewing municipal collective bargaining as a relationship between two organizations embedded in more inclusive social systems allows for the construction of a framework similar to the traditional bilateral structure of industrial bargaining but influenced from outside by a variety of direct and indirect forces. The components we have chosen to explicate are incorporated in Figure 1. While we will explore these components in greater depth below, it is appropriate to briefly introduce them now to define the framework's major components.

Like Warren (1967), we conceive of the bargaining environment as an *interorganizational field* in which legal, economic, technical, administrative, social, and political forces indirectly constrain and pattern the bargaining organizations, the negotiations episode, and the effects of settlements. Within this structure, the two focal organizations and the several peripheral organizations involved in the bargaining process are more directly linked through what Evan (1972) terms *organizational sets.* Each set is composed of organizations in proximal relationships to the focal organization. They constitute the focal organizations' "task environments" (Thompson, 1967) and are directly involved through exchanges in their goal-setting, goal-attainment, and decision-making processes. An organizational set includes a focal organization's coalition partners, its enemies, and neutral third parties.

Warren (1967:397-398) distinguishes between the two more inclusive levels of our framework—interorganizational fields and organization sets—in the following way:

The concept of interorganizational field is based on the observation that the interaction between two organizations is affected in part at least, by the organizational pattern or network within which they find themselves. . . . One need not reify this field structure into an entity independent of the behavior of its constituent parts in order to concede the importance of studying it at a more inclusive level than that of the exchange behavior of an individual organization with other actors in its environment.

The internal structure and decision processes of each of the two *focal organizations* influence goal formation and bargaining behavior. Like Thompson (1967), we conceive of organizations as "socio-technical systems" which mix social systems with technologies. We also accept Thompson's premise that "different types of jobs present different power situations and hence call for differences in political action" (1967:107). This suggests that technology and professionalization contribute significantly to the impetus to unionize, to the structure of the employing organization and the union, to the activities in which the organizations engage before and during negotiations and to the content of the inducement/contributions contract.

The *negotiation episode* may be described as a three-stage process.[3] The first stage involves the delimitation of the scope of bargaining from what Cobb and

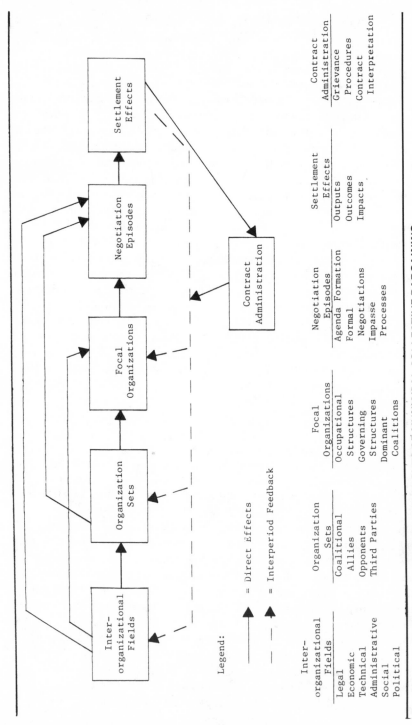

Figure 1: A FRAMEWORK OF MAJOR COMPONENTS OF MUNICIPAL COLLECTIVE BARGAINING

Elder (1972) call the "systematic agenda" to the "institutional agenda." The institutional agenda embodies a set of concrete, specific items scheduled for active and serious consideration by a given decision-making body while the systematic agenda refers to issues "more abstract, general, and broader in scope and domain than any given institutional agenda" (Cobb and Elder, 1972:14).

Walton and McKersie (1965) have differentiated institutional agenda items according to the underlying structure of their payoffs. Political action and bargaining form around these payoffs. Distributive items are associated with fixed-sum payoffs, and negotiation determines shares allotted to each organization; integrative items potentially increase the total payoff for both parties so that the distribution of shares is not a relevant consideration.

The second stage of the negotiation episode, formal negotiations, may be classified according to the nature of the organizational network in which they occur. The industrial relations literature focuses on a bilateral model of negotiations where two organizations bargain relatively unconstrained by elements in their interorganizational field and the activities of third party organizations (Chamberlain and Kuhn, 1965; Walton and McKersie, 1965; Dunlop, 1958). Kochan (1974a:526) proposes a more complicated framework, a multilateral model of negotiations, in which peripheral organizations as well as elements within the focal organizations "are involved in such a way that a clear dichotomy between the employee and management organizations does not exist." A third bargaining model, the intracoalitional model, describes a more structured and predictable situation in which the two organizations are subsumed in an overarching organizational network creating an arrangement that resembles interdepartmental bargaining within a single organizational hierarchy. In this case both bargaining partners would be subservient and allegiant to a third supraorganization like a political machine.

The third stage of the negotiation episode involves impasse processes which may develop if the bargaining partners fail to reach a negotiated settlement. These procedures involve a significant change in the structure of negotiations and encompass conflictual tactics like strikes and injunctions as well as third party conflict resolution and impasse procedures like binding arbitration.

The product of the negotiation episode consists of *settlement effects*: the substantive and symbolic outputs of the contract bargaining—the inducements/contributions contract and the concomitant attitudinal structure of the parties toward one another, the bargaining process in general, and the contract specifically; outcomes, the indirect implications for the economic, political, and administrative functioning of the community and local government; and impacts, the larger societal effects on the distribution of services, resources, and power within the community.

The final component in our framework, *contract administration*, encompasses the day-to-day interaction between the union and management organizations between negotiation episodes. It includes such activities as the handling of grievances and contract interpretation.

Each of these major components will be discussed in greater depth in the following sections.

INTERORGANIZATIONAL FIELDS

The previous section presented the broad outline of an analytic perspective for municipal collective bargaining. The objective of this and the following sections is to focus on the key structural properties of the basic framework. This section examines state legal policy, local labor markets, fiscal capacity, and community attitudes as elements of the interorganizational field that influence municipal collective bargaining.

THE SUPERORDINATE RULE STRUCTURE: STATE LEGAL POLICY

Since the passage of the Wisconsin Public Employment Relations Act in 1959, more than 25 states have adopted legislation governing collective bargaining for public employees. As a major factor structuring municipal labor relations, state legal policies may be conceived as: (1) embodying a set of rules about the legitimacy of joint decision making and the instrumentalities of the process; (2) creating rules which systematically influence the distribution of power to various parties; and (3) establishing the boundaries of acceptable behaviors and outcomes. The general provisions of public employee collective bargaining laws and the manner in which they serve to structure each of these dimensions are discussed below.

State Statutory Provisions for Municipal Bargaining

From an ongoing review of legislation, attorneys general opinions, and court decisions, the Labor-Management Services Administration of the Department of Labor has developed a classification system for the provisions of state public employment collective bargaining law. The classification scheme provides a useful tool for description and comparison of the state regulation of municipal collective bargaining. Table 1 provides a list of the 11 provisions and a brief explanation of each.

Legitimacy of Joint Decision Making

Although use of the term collective bargaining generates widely accepted denotative and connotative definitions based on over four decades of experience in the private sector, frequently some of the key tenets of these private sector definitions are absent from the state statutory authority for collective bargaining in municipal governments. The two most significant departures from private sector practice revolve around the right to bargain collectively and the

TABLE 1

AREAS OF STATE LEGAL REGULATION OF
MUNICIPAL COLLECTIVE BARGAINING

Administrative body	Designates an existing agency or creates a new agency to administer statutory provisions and develop policy. Examples: Public Employment Relations Board, State Department of Labor.
Bargaining rights	Defines the extent to which the labor-management relationship is a joint or mutual relationship. This provision may require a "mutual duty to bargain" or merely require an employer to "meet and confer" with an employee organization.
Recognition	Establishes the type of recognition accorded an employee organization (e.g., exclusive, proportional) and the procedures for designating an employee representative.
Unit determination	Specifies the criteria and procedure for selecting the appropriate unit of employees for bargaining.
Scope of bargaining	Specifies the items, i.e., wages, hours, terms and conditions of employment, which may be bargained.
Impasse procedures	Designates the procedures to be followed to achieve a settlement in the event of an impasse between the parties.
Strike policy	Establishes the legality of and penalties for strikes and work stoppages.
Management rights	Establishes the prerogatives of the public employer.
Unfair labor practices	Defines actions which violate basic requirements of the labor-management relationship, including failure to bargain or participate in impasse procedures in good faith, refusal to comply with an arbitration award, and violation of election provisions.
Grievance procedure	Defines the requirements for submission and arbitration of employee grievances.
Union security	Spells out policy on the legality of dues checks off, maintenance of membership, agency shop, and union shop clauses in the collective bargaining agreement.

right to strike. State statutes provide a variety of alternatives to the private sector model which requires a "mutual" duty to bargain. Statutes may permit, but not require, collective bargaining, may merely require the city employer to meet and confer with employee representatives, or may prohibit entirely any form of joint decision making. While state bargaining rights policy does not necessarily control overt interorganizational behavior at the local level, it does establish a potentially enforceable definition of the legitimate bounds of joint decision-making behavior for the participants and third parties.

The almost blanket prohibition of municipal employee strikes serves to create additional potential threats to the legitimacy of joint decision making. First, in the absence of a consensus on an alternative means for achieving finality, considerable doubt can be cast upon the legitimacy of the instruments and outcomes of present practices for terminating negotiations. Imposition of a settlement by the employer or a third party neutral is potentially unacceptable to a number of participating and nonparticipating actors. Second, if a strike should occur, whatever the merits of the case, it threatens the legitimacy of the joint decision-making process.

Systematic Influences Upon the Distribution of Power

Collective bargaining serves as a mechanism for both redistributing power within existing city political relationships and distributing power in the restructured employment relationship between employee organizations and municipal governments. Comparing the structural changes that take place in moving from a management administered system to a bargained system highlights the implications of collective bargaining for existing city political relationships (Taylor, 1957). In addition to the basic shift of authority between city government and its employees, the structure of two external-internal relationships undergo significant alteration through the collective bargaining relationship: municipal employees vis-à-vis the public and the city government vis-à-vis the public.

The change in the structure of the relationship between public employees and the general public occurs in two areas. First, the organization of public employees is facilitated and the subsequent organizational arrangements are accorded status as legitimate interest groups. This change, although pointed to by some as the primary structural shift stimulated by collective bargaining, does not differ substantially from early ad hoc efforts to accord public employees "representative" status. The second, and most important shift in the distribution of power between public employees and the public involves the exclusivity in decision-making authority. Decisions about wages, hours, and conditions of employment become more centralized and less susceptible to public access and manipulation. An analogous shift occurs in the structure of the relationship between the city government and the public. The taboos placed on "fish bowl" bargaining and similar arrangements by the imperative of a bilateral decision-making structure are indicative of the removal of items from public scrutiny and potential control.

In terms of distributing power between employee organizations and municipal governments, state legal policy exercises considerable control over the structure and goals of the contending organizations. Two of the primary statutory instruments for influencing the relative power between employee organizations and municipal governments are the criteria for unit determination and the provisions for union security. Thus far the debate over the bargaining unit issue in the public sector has been cast into economic and prescriptive arguments, but the underlying political implications are difficult to disguise. Jones (1975:22) formulates the traditional view:

> The central issue in unit determination in the public sector is how best to balance the objectives of the employees' right to self-determination (which

may justify a large number of small units) and the need for employer efficiency in bargaining.

Beyond this prescriptive argument for balance between employee rights and management efficiency lies another set of variables inextricably linking unit determination to the relative distribution of power. For example, prior to the formalization of collective bargaining in New York City, Mayor Wagner encouraged the proliferation of bargaining units "to satisfy as many different unions as possible and to prevent one or a few unions from growing too large and too strong" (Horton, 1973:37). Unit determination has likewise been linked to the capability of achieving industrial democracy, the ability of the employee group to achieve intragroup consensus, the ability for the employee group to secure recognition, intergroup equity, and the ability for bargaining to take place near the locus of power in the governmental organization.

While size and related aspects of unit determination may be conceived of as factors which cut horizontally across the city government organization, an analytically separable dimension of unit determination concerns the vertical structure of bargaining. What roles and positions from the hierarchy of the government organization are potentially legitimate participants in collective bargaining? Unlike the policy established by the National Labor Relations Act, supervisory employees are ordinarily permitted to bargain collectively with city government organizations. Given a range of possible decisions which may flow from this policy, there is a high probability that supervisory participation in collective bargaining enhances the vertical integration of employee interests.

An issue on which state policy at this point plays a less decisive or deterministic role, but for which the political implications are more conclusive, is that of union security. State policies on union security range from mandating particular forms, to permitting the negotiation of security provisions, to prohibiting any form of security provision. The rationale for arguing that some form of either mandated or negotiated security provision is particularly important for enhancing the absolute power of the employee organization is cogently set out by Olson (1968:80):

It follows that most of the achievements of a union, even if they were more impressive than the staunchest unionist claims, could offer the rational worker no incentive to join; his individual efforts would not have a noticeable effect on the outcome, and whether he supported the union or not he would still get the benefits of its achievements.

The attainment of some form of union security therefore assures organizational stability through guaranteeing a limited—but important—level of participation.

THE EXTERNAL ECONOMY

While collective bargaining in municipal government is often perceived as having a substantial political content, it is also subject to the forces of the external economy, such as macroeconomic fiscal and monetary fluctuations, which shape interactions during the collective bargaining process. For purposes of exposition, these forces are organized into two categories: input characteristics and output characteristics.

Input Characteristics: Local Labor Markets and the Ability to Pay

One of the most widely discussed economic aspects of municipal government bargaining surrounds the notion of fiscal capacity or "ability to pay." Although ability to pay has a perceptual component, its roots are grounded in a number of community demographic characteristics. Crouch (1971:2) notes:

> the city's economic ability, in simple terms, relates to its (1) property values as reflected in land use, (2) volume of trade, (3) type of business and industry within its borders, (4) the average income of its people, and (5) their need and desire for services.

These demographic characteristics enter the collective bargaining process not only indirectly in shaping union and management assessments of fiscal capacity but often directly as criteria for third party assessments (Ross, 1969; Baker and Colby, 1976).

The significance of the ability to pay concept in municipal collective bargaining lies in the uncertainty surrounding its definition. Unlike a corporation's profit which can be defined from a corporate income statement, a city government's ability to pay is based upon two residuals of a city's political system. First, any settlement which may require additional financing is dependent on public willingness to accept a tax increase. Second, any attempt to increase taxes for the purpose of financing a settlement may also create disputes over allocational priorities and stimulate demands to serve other city needs.

Supply and demand characteristics of local and national labor markets also serve to structure the interorganizational decision-making process. For example, in a study of market influences on municipal employee wages, Freund (1974) concluded that four labor market forces did have a significant impact on wage decision making: the average SMSA unemployment rate, the percent change in nongovernment wages, relative municipal wages, and the percent change in the cost of living in the SMSA.

COMMUNITY ATTITUDES

Attitudes of the citizenry of a community toward organized labor in general and specific unions and occupations play a key role in municipal collective bargaining. Public attitudes can affect union power, bargaining strategies, and management's assessment of settlement options. An illustration of the importance of the structure of community attitudes is provided by Craft (1971:67-68) in his analysis of fire fighter strategies:

> The union emphasizes building a reservoir of goodwill and developing an "image" of the fire fighter as a community servant. . . . The image and goodwill are subsequently used in garnering community and city officials' support for wage proposals. At best, it is hoped that these parties will respond positively to wage proposals at budget decision time. At worst, it means that they will be neutral, and not hostile to fire fighter wage proposals.

There is also evidence that public attitudes toward unions change dramatically over time due to the events and outcomes of previous negotiation episodes. For example, the response of San Francisco residents to the 1975 police and fire strike was to change the "rules of the game" by rescinding City Charter provisions they had once found acceptable (Los Angeles Times, 1975:1). Community attitudes had a similarly "toxic" affect on the 1968 strike by New York City sanitation workers (Raskin, 1968:VI-34).

ORGANIZATION SETS

Some of the previously cited literature on labor market conditions and community demographic characteristics is relevant for understanding the role and significance of the task environments of the union and public organizations in structuring the negotiations episode and influencing outcomes. Community needs and desires for services and resource limitations are likely to be aggregated and articulated by interest groups and elected officials all of whom play potentially important roles in the bargaining process. Similarly, not only is the extent of employee organization a significant element of the interorganizational field, but the control of competition by a particular union is also significant in the negotiation episode.

Viewing municipal collective bargaining as a joint decision-making process which involves two focal organizations with intersecting task environments suggests that the structure of the organizational sets plays an important part in the decision to engage in negotiations. Warren (1972:23) states that:

> Organizations enter voluntarily into concerted decision making processes only under those circumstances which are conducive to a preservation or expansion of their respective domains . . . concerted decision making is possible only if trade-off inducements can be offered to ensure voluntary participation or if coercive power may be exercised by one or more of the parties over the other(s), or by a third party, or by a legitimate authority

within a hierarchical structure of which the organizations are an integral part.

To the degree that elements of the intersecting task environments require interorganizational decision making for the realization of goals and to the degree that interdependence and the ability to coerce are woven into the organizational set relationships, a form of bargaining will characterize interorganizational relations.

Elements of the intersecting task environment not only have great influence on the degree to which interorganizational decision making takes the form of a bargaining relationship, but they also enter indirectly and directly into the formal negotiations process (Kochan, 1974a; 1975b). Thus, third parties included in the organization sets of the focal organizations influence both the formation of the bargaining agenda and the structure of negotiations.

Two organization sets are of particular importance in city government collective bargaining: (1) civil service commissions and independent personnel boards and (2) political parties and interest groups. Both sets play important roles as proximal actors in the bargaining process.

CIVIL SERVICE COMMISSIONS AND INDEPENDENT PERSONNEL BOARDS

What some view as a transitional issue in city government bargaining, but one that remains salient in many jurisdictions, is the division of prerogatives between the civil service commission and the collective bargaining process. Where de facto collective bargaining and the civil service system coexist without recognition of their overlap, civil service regulations supersede collective bargaining. Jones (1975:150) summarizes the consequences of this arrangement:

> This leads to cases where, for example, employees have the opportunity to use grievance procedures established by both the civil service and collective bargaining agreements. The result is an unstable coexistence, creating frustration particularly for unions because management can use the existing civil service laws to restrict the scope of bargaining or as an excuse for its failure to implement negotiated agreements.

Even where some accommodation has been made between the civil service system and collective bargaining, the scope of bargaining has been restricted to maintain some limited powers for the civil service commission. This accommodation still leaves open opportunities for manipulation and conflicts over jurisdiction.

POLITICAL PARTIES AND INTEREST GROUPS

Although explicit examination of the role of political parties and interest groups in city government collective bargaining is limited, some indirect assessments of the affects of party structure and interest group activity are possible. Banfield and Wilson (1966:214) suggest one general relationship between political party structure and the influence of union organizations in city government:

> The political weight of organized city employees in a particular city depends largely upon the nature of that city's political structure, and especially upon the degree to which influence is centralized. Where party organization is strong, the city administration is in a relatively good position to resist the demands of the organized employees. . . . On the other hand, where party organization is weak or altogether absent, the political weight of organized employees is relatively large and might be decisive.

Several "qualifications" should be appended to Banfield and Wilson's general proposition. First, some consideration must be given to the independence of the party organization from the municipal employee organization. Integration or overlap in membership may make it more difficult to control employee influence. Second, although the ability to resist the demands of organized employees may vary with the centralization of political structure, the willingness of those at the apex of the party organization to resist employee demands may not coincide with the structural prediction. In this case, organized public employees are still likely to wield significant influence.

While some plausible linkages can be drawn between political party structure and the generalized influence of public employee organizations, the impact of the structure of interest group activity is less easily discernible. Central Labor Councils in larger municipal jurisdictions and taxpayer lobbies have been occasional participants. Constituent interest groups of various functional departments, such as welfare recipient and parent-teacher organizations, are also occasionally drawn into the process. Nonetheless, unlike many policy arenas, collective bargaining in city government has not generated a unique or substantial set of differentiated and vocal interest groups.

On the other hand, the general level of interest group activity within a municipality does play a significant role in shaping the process. Since organized groups are the primary means of demand articulation in city governments, the level of interest group activity is likely to influence the degree of competition for resources, available counter-balances to employee organizations, and the level of third party interest in negotiations.

FOCAL ORGANIZATIONS

THE EMPLOYEE ORGANIZATION

Three variables tend to account for the majority of variation among internal structures of employee organizations: professionalization, differentiation, and stratification.

The degree of professionalism of the employee groups has clear implications for the composition and priorities of goals sought in bargaining. Kleingartner (1973:165) notes that "there is embodied in the idea of professionalism . . . certain imperatives in bargaining and in the employment relationship." He indicates that all employee groups pursue goals related to "fairly immediate, closely job-centered gain." What distinguishes the demands of organizations of increasingly professionalized employees is the pursuit of goals "encompassed by the concepts of (1) autonomy, (2) occupational integrity and identification, (3) individual satisfaction and career development, and (4) economic security and enhancement."

The degree of differentiation has a decided impact on the structure of internal organizational decision-making processes. One source of internal differentiation—occupational—and its general implication is illustrated by District Council 37, AFSCME:

> Over *1200* titles are represented. What the union leaders continually emphasized in interviews, and what was clearly observable in day-to-day contact with Council operations, was the need to appreciate the difference in bargaining for and servicing this variety of titles versus other unions like the UFT and PBA which represent basically one title or any of the other public service unions in the city representing limited occupational groups or relatively few titles. [Mann and Mann, 1975:28]

The need to service a substantial number of occupational groups requires a highly differentiated structure for the employee organization, a substantial number of boundary positions, and decentralization of discretion which, in turn, shape internal political processes.

A potentially more disruptive form of differentiation is stratification which can occur along racial and ethnic lines. Lincoln (1967:273-274) writes of the internal structure of New York City's Transport Workers Union:

> The TWU still has an all-Irish officialdom and retains a distinctly Irish air. . . . There have, however, been significant changes in the composition of the workforce. Substantial numbers of Negroes and Puerto Ricans have moved into the lower-paid jobs, and there are signs of considerable friction between them and the Irish leadership of the union. In 1963, several hundred Negro TWU members formed the Transit Fraternal Association to fight job discrimination by the union and the employers.

This particular situation required the union leadership to establish a high level of perceived external threat (through continuously high levels of conflict) among its membership in order to maintain cohesion and overcome the stratification problem.

Both of these forms of differentiation are causal factors in shaping the internal political structure of the employee organization. Not only do occupational and social differentiation and stratification affect the optimum degree of goal consensus which can be achieved, but they also stimulate demand aggregation and membership cooptation mechanisms so that rank-and-file discontent does not seriously impede instrumental activities undertaken by the dominant coalition. These demand aggregation and cooptation mechanisms require the dominant coalition to balance their needs to maintain control against the threats posed by the differentiation problems. In the case of AFSCME District Council 37 cited earlier, organizational size and occupational differentiation led to the following pattern of demand aggregation:

Demands are solicited from the membership rather than their being made up solely by the committee. . . . Before the final package of demands is made up (usually before the council staff and the committees begin discussions), the research and negotiation department and the executive director's office meet and decide what should be the major demands or thrusts of the negotiations. The key policy decisions come at these "premeetings." [Mann and Mann, 1975:28]

This procedure allows the dominant coalition both the opportunity to control key decisions as well as the means of managing a source of internal discontent. New York City's TWU developed a federative pattern of demand aggregation and membership cooptation in the face of more serious problems presented by occupational differentiation and racial stratification. Following considerable friction among groups and bitterness against the union leadership, TWU adopted the following procedures:

TWU Local 100 agreed to pay all administrative costs, but each such division was allowed to have its own meetings, bylaws, officers, and organizers. Each represents its own members in grievance processing and contract administration, and has proportional representation on negotiating committees, the Local 100 Executive Committee, and delegations to International and Federation conventions. [Lincoln, 1967:278]

MUNICIPAL GOVERNMENT ORGANIZATIONS

In their behavioral theory of labor negotiations, Walton and McKersie (1965:4) identify four systems of labor negotiation activity. Of these four systems, *intra*organizational bargaining is set apart from the other systems of activity because of its focus on the "function of achieving consensus within each of the groups." The divergence of the structure of municipal governments from that of private organizations suggests that the nature and extent of intraorganizational bargaining may be significantly different from private sector models. Two of the key structural features of municipal governments which are likely to influence the nature and extent of intraorganizational bargaining, and its activities within the interorganizational field, are the distribution of managerial authority and succession systems for executive personnel.

Distribution of Managerial Authority

For many years the authoritative rule prohibiting collective bargaining by city government employees was the sovereignty doctrine. The concept is defined as the "supreme, absolute, and uncontrollable power by which any independent state is governed." Its use to prohibit collective bargaining was summarized by Spero (1948:1): "Government insists that, in order to preserve the integrity of public authority, it must posssess the right of final determination in all its employment relations." The passing of the sovereignty doctrine has served to place residual rights of determination into various roles and positions of the city government organization. The relevant roles and positions can be identified as: the local governing body, the elected or appointed chief executive, labor relations specialists, including the chief negotiator, and department and agency heads.

Although collective bargaining no longer permits local legislatures complete latitude over local personnel matters, legal authority to bind the city government to a labor agreement still resides with the local governing body. Burton (1972:132) describes both the initial and long-term implications of the legislature's status:

> A common initial response of local governments is to impose a system of collective bargaining on the existing structure of authority with little or no modification. Primary responsibility for negotiations can be assigned to either the executive or legislative branch, but no pre-existing center of authority is subordinated or greatly diminished in influence. . . . Usually, however, the participation of elected legislative officials in labor negotiations is shortlived and bargaining responsibility is soon transferred to the executive branch.

From the legislative perspective, two aspects of this transfer of responsibility have important processual implications. First, it is a transfer of responsibility for negotiations and not the authority to commit the city government to a binding agreement. Second, Burton (1972:134) notes that "institutional or informal arrangements must be developed to insure the effective delegation of authority from the legislative branch. . . . The delegation problem has not yet been solved in many cities."

The transfer of responsibility to the executive branch poses an opportunity for greater centralization of authority within the office of the elected or appointed chief executive. With this shift in locus of control comes a related shift in the level of specialization and delegation of functions to staff agencies. The establishment of staff specialization in labor relations poses a threat to budget and personnel officers. The role of chief negotiator, in turn, represents a threat to executive control and the legitimacy of the interests of other city government officials in the bargaining process.

Succession System for Executive Personnel

The characteristic succession system for city government chief executives often involves balancing behavior and this has significant implications for the collective bargaining process. Nowhere was this more evident than in the events which preceded the New York City transit strike of 1966. Lincoln (1967:288) recounts:

> John Lindsay had built his successful electoral campaign on the issues of financial irresponsibility and "back door deals." The year-end transit negotiations thus became an enormous political liability for Wagner: he would be blamed if a strike occurred or if he prevented one the only way he could, by a "back door deal." Relatively early in the negotiations Wagner made the statement that "no agreement can be reached without the availability of the incoming Mayor to consider, accept, modify or reject the eventual proposal for an additional subsidy to the TA—or whatever formula he might want to suggest as a counter-proposal;" Lindsay took the position that "I'm not Mayor until January 1, and I do not have the power or responsibility as Mayor."

Although probably less volatile, the succession of top-level appointed personnel, such as city managers, can also be a destabilizing influence on collective bargaining relationships. Saltzstein (1974:217) reports from a survey of city managers' attitudes toward organized public employees that "many managers ... expressed intense hostility, as evidenced by numerous unprintable marginal comments and the fact that over 22 percent indicated that they do work actively to prevent or control employee organizations." While we might expect professionalism to moderate organizational influences upon city managers' attitudes, the evidence is quite to the contrary. Saltzstein (1974:227) concludes from his study:

> When faced with community or organization pressures that are contrary to professional direction, a manager is likely to forego professional guidance. Given the structure of the job, it would be foolish for him to do otherwise.

The traditional recruitment and socialization process for lower level managers and supervisors also has impacts upon the orientation and cohesion of the governmental coalition. Three factors tend to promote some identification between lower level managers and the interests of employee organizations. First, it is possible, as we noted earlier, that supervisors may have been or may be members of a union or association. These attachments enhance the identification between managerial and supervisory employees and the interests of members of the employee organization. Second, the tenure and selection systems of city government organizations tend to favor internal recruitment so that a manager or supervisor's vertical movement within the organization may provide him with experiences similar to the employees whom he supervises. This experiential bond may also contribute to lower management cohesion. Finally, and perhaps most important, the rewards of lower managerial and supervisory employees tend to be pegged to those of the employee organization. Thus, success for the employee organization often serves the self-interests of managerial and supervisory employees.

NEGOTIATION EPISODES

The idea of isolating and placing analytical boundaries around the negotiation episode serves as a useful temporal distinction between the structure of a collective bargaining situation and the joint decision-making process. The negotiation episode encompasses those activities which are oriented toward the attainment of a settlement; the episode begins with the initiation of negotiations (e.g., the filing of union demands) and ends with the termination of negotiations either through the acceptance by the parties of some form of collective agreement or the dissolution of the union.

This conceptualization serves several purposes in addition to distinguishing the structure of the situation from the process. First, while the structural properties of the urban interorganizational field and organization sets are persistent factors in decision-making, the concept of negotiation episode serves to focus attention—both temporally and functionally—on the participants in a particular episode. Second, the negotiation episode provides a locus for the analysis of bargaining strategies separate from longer term coalition building activities. Thus, it allows for differentiation between properties of the collective bargaining context, behavioral strategies of the negotiating parties, and outcomes. Finally, it tends to shift the focus of analysis away from the group of labor and management representatives created by the negotiation episode. This group is itself analyzable at less inclusive systematic levels for the affects of variations in small group processes, individual negotiating skills, and personality factors (McGrath, 1966). Although these individual and group dimensions may contribute to outcomes, we contend: (1) that prior structural features are of much greater significance in determining settlement outcomes; (2) that during the negotiation episode more inclusive systemic levels continue to provide the major inputs for collective decision making (Turk and Lefcowitz, 1962); and (3) that collective bargaining in the public sector is less distinguishable from its private sector counterpart at these less inclusive systemic levels.

AGENDAS

The importance of agenda formation to decision making in general and political decisions in particular is discussed by Cobb and Elder (1972:12); they write:

Pre-decisional processes often play the most critical role in determining what issues and alternatives are to be considered . . . and the probable choices that will be made. What happens in the decision-making councils of the formal institutions . . . may do little more than recognize, document, and legalize, if not legitimate, the momentary results of a continuing struggle of forces in the later social matrix.

While not all battles are won prior to negotiations, the agenda formation stage can be crucial in that some items are included, while others never reach the bargaining table. Bachrach and Baratz (1970:44) observe that agenda manipulation is a formidable tactical weapon and identify a "nondecisional" process as one where "demands for change in the existing allocation of benefits and privileges in a community can be suffocated before they are even voiced; or kept covert; or killed before they gain access to the relevant decision making arena."

The properties of the institutional agenda can be examined in terms of Walton and McKersie's (1965) classification of agenda items outlined earlier—distributive, integrative, and mixed. Distributive items, those concerning the allocation of a fixed-benefit, and integrative, pure problem-solving items represent polar types. We would expect items of each type to be present in the majority of negotiation episodes.

AGENDAS AND "BOUNDARY POLITICS"

In municipal collective bargaining the employing organization and the union organization normally share a joint problem-solving task in which *both* can be made better off by increasing the *total* value of resources available to the employing organization. Critics have frequently leveled the charge that because public organizations operate in a nonmarket context there is no incentive for management to deny the union increased benefits (Wellington and Winter, 1971). Indeed, because profits are unavailable as indicators of performance, it is in the interests of management and the union to collude to increase the organization's total budget irrespective of the effects on performance, the total municipal budget, or the community's financial condition.

In fact, however, if one begins to explore the literature, the qualifications to a monolithic model of unrestrained municipal union power become the rule rather than the exception. For example, Wellington and Winter (1971:17), following exposition of a public sector model positing almost unrestrained union power, note: "at any given time, taxpayer resistance or the determination of municipal government, or both, will substantially offset union power even under existing political structures." Thus there are limits to the extent to which positive sum solutions can be generated through collusion between the union and employing organization. These limits can be differentiated into three conceptually distinct negotiation "spaces" bounded by political, economic, technical, and legal constraints: bargaining space, feasibility space, and legitimacy space.

The three "spaces" overlap one another so that the area common to all three contains the most probable solutions or outcomes of the bargaining process. *Bargaining space* is conceived as the set of possible solutions to a negotiation episode as indicated by the scope of the cumulative goals of the employee and the municipal organizations. Empirically, the bargaining space consists of the agenda items introduced by the parties at the initial negotiation conferences (Stevens, 1963). *Feasibility space* refers to the set of potential solutions given such constraints as limitations on fiscal resources, taxing ceilings, technology, administrative capacity, and the dictates of statutory law. *Legitimacy space* is defined as the set of legitimate tactics and solutions or outcomes to the negotiation episode given the dominant symbols, norms, and values of the community.

These three dimensions of collective bargaining situations provide useful concepts with which to describe the unusual "boundary politics" of municipal collective bargaining. Two aspects of "boundary politics" are especially significant. First, the manipulation of constraints has strategic implications. For example, when goals are feasible but illegitimate, actions of environmental actors may serve to limit the scope of actual decision making by manipulating the dominant community values, myths, political institutions, procedures, and decision rules within which collective bargaining takes place. Typically, this process would be initiated through threatened or attempted issue expansion by the member of the bargaining dyad which views its position as legitimate. The threat of issues expansion may be sufficient to achieve withdrawal of a demand. On the other hand, when goals are incompatible, infeasible, but legitimate, issue expansion is likely to be unconvincing as a threat and unsuccessful as a strategy. In this case, strategies are likely to be confined to the mobilization of political support and the utilization of force.

Second, the analyses of bargaining, feasibility and legitimacy spaces leads to the search for the locus of control over these constraints, and to the conclusion that a radical expansion of the scope of municipal bargaining is unlikely to occur. In most American municipalities constraint analysis leads to the conclusion that a variety of different actors located in different organizations at different levels of government (and private industry in the case of municipal bond ratings) control different constraints on negotiated settlements. For example, the relaxation of fiscal constraints on bargaining in many cities that have reached state tax ceiling requires either an act of the state legislature, a state constitutional amendment, or a reassessment of local property following a local referendum. All three procedures are difficult and require at least the tacit support of the media, state political leaders, and the passivity of local taxpayers.

These obstacles to increasing the scope of collective bargaining outcomes can only be appreciated through the analysis of constraints and interorganizational politics.

The impact of constraint manipulation has been reported in Horton's (1976) examination of the collective bargaining environment and processes in New York City during the 1960s and 1970s. He notes that in the period between 1965 and 1970 collective bargaining occurred in a relatively prosperous economic environment and the negotiations process was one "wherein resources were considered susceptible to almost infinite distribution" (1976:201). The city's financial resources were increasing for several reasons: tax rates and the economic base were growing; intergovernmental aid, mostly federal, was rising; and the city's substantial borrowing power and ability to obscure financial facts through a complex budgetary and accounting system gave at least the impression that money was not in short supply (Horton, 1976:188). The political position of the city unions rose during the Wagner-Lindsay years due to "two important political resources largely controlled by municipal union leaders—labor peace the year before the campaign and labor support during the campaign" (Horton, 1976:187). Thus, by manipulating community attitudes through the use of strikes or strike threats, union leaders were able to include their demands in the expanding legitimacy space for negotiations. A similar manipulation of the feasibility space by the city administration through borrowing and increases in pension rights also occurred.

The feasibility space for negotiations shrunk considerably during the early 1970s because New York City never fully recovered from the recession of 1969-1970; intergovernmental aid (measured in constant dollars) declined; and the long-run effects of the city's earlier budgetary practices began to be widely publicized. The political leadership of the city, including the union leadership, were "either slow to appreciate or unwilling to address the significance for municipal labor relations of the city's changing economic condition" (Horton, 1976:193). The inability of New York City to sell its securities and the spector of imminent default became public in 1975 with considerable impact on the legitimacy space for bargaining. Various state and federal officials became directly involved in the city's labor relations and the establishment of a special state agency, the Municipal Assistance Corporation (MAC), severely reduced the city administration's discretion. A change, though not as drastic, in community attitudes toward the city's bargaining practices coincided with these legal administrative developments.

The contraction of both the feasibility and legitimacy spaces directly affected the scope of solutions. The initial results were large-scale layoffs of city workers and an eventual decline in (real) wages. In late 1975 the city negotiated wage deferral agreements postponing 1976 increases, and the state legislature imposed a three-year ban on pension bargaining (Horton, 1976:196). Thus, through the manipulation of constraints, the outcomes and impacts of collective bargaining in New York City were tightly controlled.

NEGOTIATIONS

The formal negotiations process may be examined in terms of the structure of the interorganizational relationship.[4] Dunlop (1958) and others have developed bilateral models of negotiations to describe private sector collective bargaining. In actual practice a purely bilateral arrangement is rare. Clients and suppliers of the management organization can directly influence negotiations; national labor organizations as well as rival unions are also potential third parties to the bargaining process. And, of course, government entities can and do enter the formal negotiations process as neutral or not-so-neutral third parties (Thompson, 1967:110). The bilateral model, then, is a polar type representing the nearly complete closure of the interorganizational decision-making process.

Most formal negotiation structures in both the public and private sectors may be more accurately characterized as multilateral involving a variety of third parties and intraorganizational coalitions. One model which describes the multilateral nature of municipal collective bargaining has been developed by Kochan (1974a). Kochan (1974a:526) describes the process as highly multilateral because of the "political and organizational characteristics of city governments." Elements from within the community are motivated and are able to intervene in negotiations because of the "public" nature of the agenda items and the political accessibility of the government employer. Therefore, the degree to which elements from the intersecting task environments enter the negotiations varies widely across communities and across public services.

Kochan's model also includes an intraorganizational dimension that recognizes the involvement of subgroups from within each of the focal organizations:

> the occurrence of internal conflicts within a city management provides the motivation for various city officials to directly pursue their interests in the interorganizational bargaining process ... [and] because decision making power is so often shared among public officials they can usually intervene in the bargaining process (if they have incentive to do so) far more easily than can management officials in a private company who are not designated negotiators. [Kochan, 1974a:528]

Though Kochan's (1974a:529) analysis focused on multilateral bargaining resulting from features of the municipal employing organization, he states that the model is symmetrical in nature and that "there is no theoretical reason for rejecting intraunion conflict as a determinant of multilateral bargaining."

More support for a multilateral model has been provided by Feuille (1974). He examined police labor relations in 22 cities and found evidence that both the police unions and city officials work through and with other groups to effect their goals. He documents cases where union leaders met with city councilmen and mayors away from the bargaining table in an attempt to push their demands. There were instances of political candidates receiving election endorsements from the police associations as well as lobbying activities by the unions to promote favored legislation. City councils had, in some cities, passed legislation which bypassed or overruled negotiated and arbitrated agreements, usually in favor of union interests. And there were numerous examples of work slowdowns which directed citizen pressures on the city administration to accede to union demands.

These devisive strategies are not directed at city governments exclusively; often rival unions become the target of coalitional realignments, further adding to a multilateral structure of negotiations. For example, Blackwell and Haug's (1971) probe of the circumstances and events leading to the Cleveland Water Works Strike of 1969 demonstrated how the rivalry between the city's industrial and craft unions over jurisdictional domain and political leadership were a complicating factor in the negotiations process.

Because actors within and outside of the two focal organizations have some autonomy to align themselves in coalitions outside of the formally designated channels, a coalition structure that overlaps or encompasses the formal negotiating dyad can emerge. Warren (1972:24) uses the term "federative" to describe decision making by two organizations subsumed in a larger multiorganizational coalition; that is, to describe intracoalitional negotiations.

> Here there is not only a partial concerting of decisions through ad hoc interaction but also a special organization set up for such concerted decision making, involving strictly delimited parts of the total decision making scope of the organizational units involved. And the occasion for the concerted decision making is less likely to be an ad hoc, episodic issue than to be a continuous interaction around certain issues over an extended period of time.

Collective bargaining in such a context resembles negotiations between public employee unions and a city government when the urban environment is dominated by a powerful political machine (Douglas, 1976).

BARGAINING BEHAVIORS

A multilateral theory of collective bargaining has not, as yet, been fully developed. The bulk of effort, at this point in the development of the theory, addresses the processes involved in multilateral bargaining situations. The concept needs to be more specifically defined and particular real-world strategies should be expressed as general concepts. While not pretending to have developed a formulation that is mutually exclusive or exhaustive of multilateral processes, the interorganizational perspective suggests a typological scheme within which to classify the wide variety of negotiation situations and strategies found in the public sector.

In essence, we attempt to break down the identifiable components of the process. Our framework for doing this includes two formal organizations operating within intersecting domains and task environments. The focal points within each organization are their dominant coalitions (or bargaining-unit control factions) and not the boundary unit; boundary-role conflict is assumed to be zero. Bargaining will be operationally defined as any communication, direct or indirect, between the two organizations for the purpose of influencing the nature of the resulting inducements/contributions contract. Within such a framework, we identify eight distinct bargaining behavior sets.

Bilateral bargaining: All contact between the two organizations takes place through the designated boundary units which are extensions of the dominant coalitions and subject to environmental constraints only through the dominant coalition.

Formal multilateral bargaining: A third party enters the bilateral process. All bargaining takes place through the designated boundary units under complete control of the dominant coalitions. We suspect that a tendency to form an interorganizational coalition will develop though the model does not require it.

Intraorganizational bargaining: Minor factions within each organization affect the bargaining process at the boundary points through their pressures on their respective dominant coalitions. This is "factional bargaining" in the terminology of Walton and McKersie (1965).

Boundary unit bypass bargaining: The formal bargaining channels are by-passed by one of the dominant coalitions. The result is a form of collusion, conspiracy, or subversion. The public nature of the formal negotiations or a desire to fragment the opposing dominant coalition provide the motivation.

Organizational end-run bargaining: One of the dominant coalitions attempts to apply pressure to its counterpart from within the counterpart's own organization by directing pressure to one or more of its minor factions.

Domain end-run bargaining: One of the dominant coalitions mobilizes forces in the intersecting domain to apply pressure to the opposing dominant coalition either directly or through a minor faction.

Factional interorganizational bargaining: Minor factions try to subvert their own dominant coalition either directly by applying pressure to the opposing dominant coalition or indirectly through its minor factions.

Factional domain bargaining: Minor factions mobilize units within their domain to pressure the opposing dominant coalition either directly or through its minor factions.

In all of these cases, except the first, the current definition of multilateral bargaining is satisfied: i.e., a "process in which more than two distinct parties are involved . . . [and violation of the criterion that] all interactions between management officials and the union organization are channeled through the formally designated negotiators" (Kochan, 1974a).

To outline the multilateral process, it is proposed that this breakdown may provide a starting point. The first three multilateral situations have been discussed specifically in the literature (Caplow, 1968; Walton and McKersie, 1965, and the large volume of literature on bilateral negotiation). These discussions need to be incorporated into the multilateral framework more precisely and, indeed, these processes might be classified as distinct bargaining behavior sets. The remaining models may then form a possible starting point for further research on the multilateral process and the effects of different bargaining configurations on outcomes.

IMPASSE PROCESSES

The third stage of the negotiation episode is contingent upon the failure of formal negotiations to produce a settlement. The literature on public sector impasse processes deals, for the most part, with the effects of strikes and strike-related activities by the unions and the use of court injunctions and other legal tactics by government employers. In addition, third party solution mechanisms, such as mediation, fact-finding, and arbitration, are invariably included in any discussion of the impasse question.

The use of strikes, strike threats, or "slowdowns" which do not result in a total work stoppage, are important, though usually illegal, in municipal labor negotiation impasses. Wellington and Winter (1971) observe that strikes against city governments differ from industrial strikes because they do not function as economic weapons against the employing organization since city revenues are not interrupted during the work stoppage. Instead, strikes deny services to the community and hence bring citizen groups into the dispute. These groups apply political pressure on the city administration to settle. The most common reaction of city officials to work stoppages or strike threats is to bring the judiciary directly into the dispute by obtaining a court order or injunction ordering union members to resume normal work activity. Wellington and Winter point out that the legal strategy of the injunction is, however, seldom effective in meeting the strike weapon. Sanctions imposed by the courts on the unions, such as fines or incarceration for union leaders, frequently result in "strikes-for-amnesty" which usually continue beyond the period when the parties are unable to settle on contract terms alone. These union efforts are generally successful as evidenced by numerous cases where jail sentences were never served and fine obligations canceled.

The second major area of study on impasse processes concerns the formal inclusion of a third party into the negotiation process. Mediation, fact-finding, arbitration, and their variants have been examined extensively as to their effects on outputs, their ability to reduce the incidence of strikes and their effects on the interorganizational decision-making process itself.

Mediation does not result in binding settlements; and for this reason, it is difficult to isolate its effects on outputs. Consequently, most discussions of the mediation process concern the degree to which settlements are achieved as a direct result of the third party intervention without the occurrence of strike activity (cf. Rehmus, 1965; Simkin, 1971; Stevens, 1967).

Several studies indicate that of all the formal impasse procedures commonly employed, most negotiation impasses are settled at the mediation stage (McKelvey, 1969; Pregnetter, 1971). However, this fact says nothing about how mediated settlements are effected by confounding influences. Kochan and Dick (1976) have conducted a systematic empirical examination of the public sector mediation process that recognizes that the effectiveness of the process must be separated from the effectiveness of the particular mediator in question. Consequently, they test a model of mediation effectiveness which posits the following determinants: (1) the nature of the dispute causing the impasse; (2) the strategies employed by the mediator; (3) environmental influences beyond the mediator's control; and (4) personal characteristics of the mediating individual. The results of their analysis suggest that while the skill and strategies employed by the mediator do have some influence on the effectiveness of the mediation process, the nature of the impasse, forces from within the interorganizational field, and organizational characteristics appear to be the most significant determinants of mediation effectiveness (Kochan and Dick, 1976:24).

Like mediation, the results of fact-finding are not binding on the negotiating parties. Thus, research has tended again to focus on fact-finding affects on the process of negotiations. For example, various studies indicate that fact-finding has "played a significant role in reducing the incidence of overt conflict in public sector labor-management relations" (Yaffe and Goldblatt, 1971:62). However, Kochan et al. (1977) suggest that the effectiveness of the fact-finding procedure is limited in the presence of arbitration provisions. They argue that fact-finding has not significantly reduced the number of impasses which go on to arbitration nor resolved more than a minority of the disputes it considers. These findings would seem to indicate that, in many cases, the fact-finding procedure is being by-passed in favor of more definitive arbitration procedures.

Several studies have voiced the concern that both fact-finding and arbitration produce substantial changes in the negotiation process itself (cf., McKelvey, 1969; Zack, 1970; Gilroy and Sinicropi, 1972). McKelvey (1969:543) argues that resorting to formal impasse procedures may "become . . . an addictive habit, the first and not the final step in collective negotiations." However, several empirical analyses provide conflicting evidence for this claim (Yaffe and Goldblatt, 1971; Anderson, 1970a; Anderson, 1973; Word, 1972). There is also concern that fact-finding and arbitration produce a "chilling effect" on the bargaining process. Each party may harden its demands in anticipation of the third party's "splitting the difference." Again, the empirical evidence is inconclusive (Thompson and Cairnie, 1973; Loewenberg, 1970; Anderson, 1970b; Kochan et al., 1977).

Since the arbitration procedure produces a set of settlement provisions which are legally binding on both parties, there has been considerable research on its effects on the outputs of the negotiation episode. Most studies have failed to show a significant difference in contract wage levels between arbitrated settlements and those resulting from negotiated ones (cf., Howlett, 1973; Thompson and Cairnie, 1973; Loewenberg, 1970). On the other hand, while arbitration seems to be effective in reducing conflictual behavior (Howlett, 1973), there are many obvious examples of its failure (Bowers, 1974). Like its effects on contracts, researchers have not yet untangled how the ability of the arbitration procedure to reduce strike activity is related to the larger structural and dynamic features of the collective bargaining process.

SETTLEMENT EFFECTS

OUTPUTS

The negotiation episode is viewed as a joint decision-making process which produces two immediate results—the inducements/contributions contract and the modification or reinforcement of the attitudinal structure of interorganizational relations.

Thompson (1967), expanding the previous work of Barnard (1938), Simon (1957), and March and Simon (1958), describes the inducements/contributions contract as defining (at least the parameters of) the individual's relationship to the organization. Expanding upon this concept, the formal bargaining agreement can also be seen as defining, to a large extent, the relationship between the two focal organizations in the context of jointly established domains (Perry and Levine, 1976). Kochan and Wheeler (1973) have developed an empirical measure of bargaining outputs by analyzing items from a sample of municipal labor contracts and by conducting a survey of various public union officials to identify those items not usually found in formal agreements but thought to be important by the union negotiators. They constructed an output scale containing such items as wage levels, fringe benefits, union security provisions, control of the physical and health related environment of the workplace, and such management discretion items as length of work week, grievance, procedures, job descriptions, disciplinary procedures, and residual rights of management. In a subsequent analysis of the determinants of bargaining outputs, Kochan and Wheeler (1975) related environmental characteristics, union and management organizational characteristics, and features of the bargaining process to the resulting contract. The results of their regression analysis suggest that the legal structure of municipal collective bargaining is positively related to union "success;" the comprehensiveness of bargaining laws and the presence of impasse provisions have the strongest effect of all environmental variables on outputs. Also, aspects of the political environment indirectly affect outputs by influencing the internal management structure, the goals of the union and management, and the dispersion of organizational power. These factors contribute to the multilateral structure of negotiations and influence the decision of the union whether or not to engage in militant and overtly political strategies.

Perry and Levine (1976) employed a modified version of the Kochan-Wheeler scale to measure the degree of contractual change. The results of that analysis suggest: that the degree of goal divergence between the two focal organizations is positively related to the degree of conflictual behavior which, in turn, shows a positive relationship to the amount of contractual change for certain nonsalary items; and, the asymmetry of relative dependence on task environment elements favoring the employing organization (asymmetry of organizational "power") appears to have a negative influence on the degree of change in the inducements/contributions contract.

Gerhart (1976) developed a similar scaling device to measure contract provisions. He related a host of environmental characteristics and organizational features to the content of resulting settlements. His theoretical model and empirical analysis suggest that economic, demographic, and political variables directly affect the relative bargaining power of the two focal organizations which shape the outputs of the negotiation episode. In addition, interorganization field forces indirectly shape outputs through their influence on goal intensity, which is related to both agenda formation and relative bargaining power. Gerhart's model is similar to our conceptualization of negotiation spaces in the formation of the institutional agenda and to the concepts of "degree of goal divergence" and "asymmetry of relative dependence" in the Perry and Levine analysis.

In addition to the inducements/contributions contract, there is an attitudinal component of the outputs to the negotiation episode. Walton and McKersie (1965:184) identify attitudinal structuring as being an important subprocess of the collective bargaining relationship:

[another] result of the negotiations process is a maintenance or structuring of the attitudes of the participants toward each other. . . . The issues in labor negotiations involve important human values; fulfillment of the contract terms is strongly contingent upon attitudes; instruments in the contest involve social ideologies and psychological tactics as well as economic sanctions; and the relationship between the parties is an exclusive and continuing one.

The attitudinal component, though difficult to identify and describe, has obvious importance to the enduring relationship (Selekman et al., 1958). Walton and McKersie (1967:188) describe attitudinal outputs in terms of "action tendencies" which range from conflictual to collusive. These tendencies influence both the administration of the present contract and the purposeful behavior of the focal organizations in subsequent negotiation episodes.

OUTCOMES

The outcome of the negotiation episode includes the indirect implications of the contractual and attitudinal outputs for the economic, political, and administrative subsystems. Our distinction between outputs and outcomes is similar to that of Levy, Meltzner, and Wildavsky (1974); outcomes are the consequences for individuals and organizations of collective bargaining relationships.

Since the dramatic near-default of New York City in 1975, there has been much discussion concerning the effects of collective bargaining on the economic health of American municipalities. Ross (1976) examined the local government budget crisis and argued that while high wage-benefit packages resulting from collective bargaining may add to financial distress, more macroeconomic factors are the determining elements.[5] These features of the interorganizational field include: white flight from the central city and the resulting decline in the tax base; the inability of the inflexible property tax to respond to economic needs; the failure of intergovernmental aid to keep pace with inflation; the increasing cost of borrowing; and the effects of regional and national recessions.

The principal effect of increased labor costs in economically constrained bargaining environments is not a reduction in municipal expenditures, which are highly inelastic, but a reduction in employment through attrition and in some cases layoffs (Horton, 1976; Ross, 1976). The result tends to be a decline in the amount of government services provided without a corresponding decrease in tax rates.

Collective bargaining and public employee unionism appear to have a great effect on the urban political environment. Foley (1975:i) examined union influences on local power distributions in Philadelphia. The results of his case study suggest that:

> Union activity has fundamentally altered the structure of city government by undermining the political independence of the Civil Service Commission and the City Council, both of which have delegated extensive powers over pay scales, fringe benefits, and the raising and distributing of city revenue . . . and by consolidating responsibility for negotiating financial matters in the Office of the Mayor.

Horton, Lewin, and Kuhn (1976:511) noted a similar concentration of authority in the new administrative agencies responsible to the major for negotiations. Their examination of Chicago, New York, and Los Angeles further suggests that mayors are aware of "both the potential managerial problems and political opportunities posed by public unionism."

Lewin and Horton (1975) examined outcomes of collective bargaining on the administrative environment and, more specifically, on the merit system in local government. They concluded that "collective bargaining will produce mixed consequences for merit, consequences which are highly dependent upon particular management-union goals and power relationships in collective bargaining" (1975:205). Thus collective bargaining has the potential to supplant many of the traditional aspects of local personnel administration, but its outcomes are not preordained. For example, Lewin and Horton (1975) noted that the use of seniority in promotions and pay differentials is gradually being implanted in local personnel systems where the union possesses substantial bargaining power. At the same time many of the unions which support a seniority system also bargain for and often obtain a rule-of-one appointment and promotions procedure. This is clearly more consistent with the merit principle than the traditional rule-of-three (and also serves to limit management discretion concerning hiring and promotions).

Horton (1976) examined the effects of collective bargaining on production and productivity. Productivity bargaining, "the negotiation and implementation of formal collective bargaining agreements which stipulate changes in work rules and practices with the objective of achieving increased productivity and reciprocal worker gains," would seem to offer substantial benefits to both the city administration and the workers (Newland, 1972:808). The results of Horton's study, however, suggest that in most cases even if increased productivity can be "bought" by salary changes, the resulting increase in unit labor costs will neutralize most or all of the productivity gain.

In what is perhaps the most ambitious and comprehensive treatment in the literature, Horton, Lewin and Kuhn (1976) are examining the consequences of collective bargaining for five selected areas: compensation levels; service delivery; personnel administration; formal government structure; and informal politics. Their general orientation is "to hypothesize that only one pattern of union impact on [the] public sector . . . will occur, overlooks the diversity of organizations, relationships, and impacts that may occur even within a single jurisdiction" (1976:504). Their approach is in agreement with the general statements contained in our model. Outcomes are the result of the interaction between negotiations, outputs, and elements of the interorganizational field and organization sets. To the extent that there is variability in any of these factors across communities, outcomes are likely to vary.

IMPACTS

The aftermath of the negotiations episode has important distributive consequences for society in general. The outputs and outcomes of the collective bargaining process impact on individual citizens, institutions, and societal groups. Levy, Meltzner, and Wildavsky (1974) discuss ultimate consequences or impacts in terms of *who benefits*; similarly we are concerned with "who gets what" as a result of collective bargaining. Research in this area is understandably sparse. "The immense problems in figuring out what these consequences might ultimately be" (Levy et al., 1974:8) are a product of the complexity of social interactions, the unmanageable number of relevant variables, and the limitations of social science methodology.

Wellington and Winter (1971:65) argued that public employee unionism and the outcomes of negotiation episodes are forces "restraining expansion of the role of government. . . . Other considerations being equal, [they] require government to decline new functions or to shed those it already has." Those groups in society, the poor, the elderly, and so on, that depend disproportionately on government resources to meet the needs of daily life are, if Wellington and Winter are correct, losing actual or potential benefits to public employee union members.

Jones (1975) examined the impact of "task performance rules" on local service delivery agencies and related them to the actual services provided. These rules structure the repetitive decisions made within the organization and have distributional consequences for the citizenry. To the extent that these rules result from negotiated settlements, collective bargaining clearly affects the level of government services provided to different groups within the city.

There has been no significant research to date on the impact of collective bargaining settlements on racial groups in American cities. Much attention has been given to the effects of judicially imposed affirmative action programs on the provisions of negotiated agreements in that they often conflict with negotiated seniority provisions for promotions. Yet the degree to which the outcomes of negotiation episodes impact on the racial and ethnic composition of public work forces has received little systematic attention.

CONTRACT ADMINISTRATION

The implementation of the formal provisions contained in the inducements/ contributions contract is an obvious link between the outputs and outcomes of the negotiation episode. Yet it is useful to distinguish, conceptually, between these indirect influences on the community, and the day-to-day administration of the formal agreement in the period between negotiation episodes.

The literature concerning contract administration is sparse in comparison to the treatment of other aspects of municipal collective bargaining. But interperiod developments can greatly affect future negotiations. Heisel (1971:1) argues:

> Labor relations is a year-round task. . . . Negotiations are but one phase of the total relationship. . . . Union and union members contact every representative of management on a continuing basis. . . . If these contacts are generally harmonious, then negotiations can be conducted .with a minimum of rancor. If, on the other hand, they are not harmonious, then it can be expected that negotiations will be bitter.

The results of the Perry and Levine (1976) study are supportive. The number of grievances filed between negotiation episodes were used as an indicator of the stability of the relationship between the two parties. Their findings suggest that the greater the number of grievances the less stable the relationship and the greater the potential for conflictual behavior in the next negotiation episode.

Disputes and grievances can arise from a variety of sources. Ferris (1975) notes that subjective interpretation of the contract language, which occurs in connection with any written instrument, contributes to ambiguity. Apart from inherent features, the quality of draftsmenship of the legal document is an additional factor. Furthermore, disputes may result from purposeful agreement during the negotiations to unclearly worded items in the contract in order to avoid immediate impasse. The result is a postponement of the confrontation until the administration phase. The postponement of the decision may also be a tactical maneuver. Either party may believe it to be advantageous to their position, as circumstances change, to delay the decision-making process.

Disputes arising during the administration of the contract are seen by both the management and the union as being inevitable. Consequently, most negotiated labor agreements contain formal provisions for handling grievances (Heisel, 1971). The procedures not only serve to mitigate conflict and contribute to the stability of the interorganizational relationship, but they also impact on the internal dynamics of city administration. Stanley (1972) notes that grievance procedures are increasingly being called into use by public unions. To a certain extent, they have supplanted traditional management disciplinary mechanisms as well as many quasi-judicial responsibilities of Civil Service organizations. This suggests that the interorganizational nature of the decision-making process in collective bargaining extends into the work situation and the management of the technical core of public agencies.

Similarly, the implementation of contract provisions specifying work assignment rules, workload and manning procedures, and so on are constantly being monitored by union personnel (Stanley, 1972). The result is a limitation of management discretion and an increase in the bilateralism of decisions which affect day-to-day work activities.

CONCLUSIONS: SOME RESEARCH QUESTIONS AND POLICY IMPLICATIONS

Each of the six major components in our framework contains numerous, potentially high-yield research questions with implications for public policy. Rather than attempting to explicate all these questions, we will instead illustrate the utility of the framework in this conclusion by identifying one research question in each category that is especially important for the conduct of municipal collective bargaining.

Interorganizational Fields

As the strength of unions increases and the fiscal solvency of more cities weakens, will the federal government play an increasingly more important and direct role in municipal collective bargaining? So far the federal government seems content to allow the states to tackle the major problems that stem from municipal collective bargaining and has chosen to play an indirect role through means such as revenue sharing that strengthens sagging city finances. If these funds are increased as a partial result of municipal union lobbying in Washington, then we might expect pressure to build for an increased role for the federal government in municipal labor relations and for a national public employment labor relations policy. While we have scraps of evidence to support this line of reasoning, a well-documented historical and developmental study that also taps the experiences of other nations needs to be mounted to build support for this contention/projection.

Organization Sets

Has the configuration of power in local politics changed since unions have come to occupy a more central role as financers of city debt through the purchase of municipal bonds? Also, with this new stake in the functioning of local government, have unions trimmed demands and become more "responsible" in contract negotiations? The implications of this question for public policy are reasonably straightforward; if unions in financially troubled cities own a large portion of the city's debt through their pension fund, then, it is frequently argued, they will take a partnership-like approach to the management of the city and its labor relations policy. The validity of this argument remains to be fully tested.

Focal Organizations

Is the governance of municipal unions democratic or oligraphic? How does this vary across communities and across unions? What impact does this have on union demands and service delivery? While we have some evidence that local government management tends to be divided during collective bargaining negotiations and that this fragmentation weakens the government's position, we have little similar insight into the internal politics of municipal unions. If municipal unions are similarly fragmented by internal competition, we would expect them to be weaker in negotiations but more responsible to the needs of their membership. The links between internal conflict and external conflict, therefore, need to be much more thoroughly investigated. If municipal unions are oligarchic and unresponsive to their membership, then states and the federal government need to forge policy that encourages the development of greater levels of union democracy.

Negotiation Episodes

To what extent are the outcomes of negotiation episodes determined by the political and economic structure of the community or by the strategies employed by the two organizations? How does this mix of determinants vary across communities? We have some evidence that the malleability of contracts is a function of the feasibility and legitimacy space provided by the community. Unfortunately, most of this information has been developed from case studies; and there has not been developed a systematic categorization of communities that include economic *and* political variables that has allowed this proposition to be validly tested. If valid, however, the policy implications are obvious: In cities with little negotiation space, city leadership is confronted with the cruel tradeoff between reducing service levels (and incurring intense political pressure) or threatening the city's financial solvency as municipal union demands grow (even if only slightly). Across communities, this tradeoff has substantial implications for economic development policy because, left without aid from state or federal sources, rich communities will get more industrial development and poor communities will be less able to compete; thereby enhancing the relative advantage of rich communities.

Settlement Effects

Does increased pressure for affirmative action for women and minorities collide with negotiated contract provisions in such a way as to make unions the "natural enemy" of affirmative action? To our knowledge, there has been no major research that examines the conflictual or complementary status of municipal collective bargaining and affirmative action programs. Recent court decisions in several cities have nullified negotiated provisions with respect to seniority in order to change the composition of various hierarchical levels to include more women and minorities. The impetus for these changes, the degree to which the objectives are achieved, and the impact on bargaining relationships and community politics needs to be systematically explored.

Contract Administration

How do patterns of grievance activity and its management affect subsequent negotiation episodes and contract changes? Evidence that provides links between contract administration, bargaining behavior, and contract change is scant. One can only hypothesize that the type of grievance patterns that emerge between bargaining episodes contributes substantially to defining the bargaining agenda. The issues raised in piecemeal fashion between bargaining episodes may be important predictors of future settlements. The implications for understanding the contributions of the management of settlements for both union solidarity and municipal government personnel relations are an important and neglected area within the municipal collective bargaining literature.

The analytic differentiation in our framework between interorganizational fields, organization sets, focal organizations, negotiation episodes, settlement effects, and contract administration is instrumental to understanding our central contention: that is, *that despite the dramatic and attention-attracting dynamics of the negotiation episode, structural features of more inclusive systemic levels provide major determinants of both the strategies and outcomes of the collective bargaining process.* This position recognizes that local governments, their agencies, and public employee organizations are linked to external actors by a multiplicity of direct and indirect influences. These actors and influences possess the capacity to initiate changes in the structure of opportunities and incentives facing the employing organization and its union. There is substantial support for this interorganizational perspective woven into the literature on public sector collective bargaining. A review of some of this literature has served to identify key variables and their relationships within our framework.

NOTES

1. For a discussion of some of the contextual differences between collective bargaining in the public and private sectors, see Hildebrand (1967), Stanley (1972), and Wellington and Winter (1971).

2. For a similar use of the conceptualization of "core" groups in the political science literature, see Sayre and Kaufman (1965:710).

3. The negotiation episode can be understood as a generalization of less abstract and more traditional approaches for meaningfully distinguishing between significant temporal points in the collective bargaining process. For two approaches that employ the budgetary cycle as a theoretically and practically significant point in the collective bargaining process see Derber et al. (1973) and Craft (1971).

4. In this section we focus on behavioral approaches to collective bargaining and ignore approaches rooted in analytic theory like game theory and economic theory. While some interesting formulations of collective bargaining have been developed in this literature (Cross, 1969), they have as yet been unable to solve the problem of "strategic interaction" (Young, 1975: chapter 1) and therefore, predicting the behavior and outcome of two or more purposeful actors involved in multistage bargaining using these models is imprecise.

Briefly, the strategic interaction problem is the most formidable obstacle to the development of a determinant theory of bargaining applicable to real-world situations. It is rooted in the propensity of all parties in a bargaining process to think ahead and outguess their opponents. This leads to an infinite regress; what has been called the "outguess-regress problem": i.e., "if he thinks that I think that he thinks that I think...," etc. It is such a difficult methodological problem that it limits the applicability of game theoritic and economic models of bilateral monopoly because they do not overcome the outguess-regress problem. Instead, they either remove it from their analysis by means of prior assumptions or try to incorporate a learning component and a time element by adding more and more variables. But, as Young (1975) argues, assuming away strategic interaction tends to emaciate the bargaining process since strategic interaction is the heart of the process. Similarly, reducing learning and other cognitive processes to a cluster of variables with no theoretical coherence is a methodologically indefensible practice from a behavioral standpoint. As a result, unless the outguess-regress problem can be solved, economic and game theoretic models may have reached the limit of their explanatory usefulness and their applicability may be limited to their heuristic value.

5. Numerous studies indicate that the presence of collective bargaining leads to higher wage rates than in communities where it is absent but that the differential is, in most cases, small. See, among others, Lipsky and Drotning (1973), Ashenfelter (1971), Ehrenberg (1973), and Freund (1974).

REFERENCES

ANDERSON, A. (1970a). "The use of fact-finding in public employee dispute settlement." Proceedings of the Twenty-Second Annual Meeting, National Academy of Arbitrators. Washington, D.C.: Bureau of National Affairs.

——— (1970b). "Compulsory arbitration under state statutes." Proceedings of the New York University Twenty-Second Annual Conference on Labor. New York: Mathew Bender.

——— (1973). "The impact of public sector bargaining: An essay dedicated to Nathan P. Feinsinger." Wisconsin Law Review, 1973:986-1025.

ASHENFELTER, O. (1971). "The effects of unionization on wages in the public sector: The case of fire fighters." Industrial and Labor Relations Review, 24(January):191-202.

BACHRACH, P., and BARATZ, M. (1970). Power and poverty. New York: Oxford University Press.

BAKER, D., and COLBY, D. (1976). "The politics of municipal employment policy: A comparative study of U.S. cities." Prepared for delivery at the annual meeting of the Midwest Political Science Association, Chicago.

BALES, R.F. (1950). Interaction process analysis: A method for the study of small groups. Reading, Mass.: Addison-Wesley.

BANFIELD, E.C., and WILSON, J.Q. (1966). City politics. New York: Vintage Books.

BARNARD, C.I. (1938). The functions of the executive. Cambridge, Mass.: Harvard University Press.

BLACKWELL, J.E., and HAUG, M.R. (1971). "The strike by Cleveland Water Works employees." In W.E. Chalmers and G.W. Cormick (eds.), Racial conflict and negotiations. Ann Arbor: Institute of Labor and Industrial Relations, The University of Michigan-Wayne State University.

BOWERS, M.H. (1974). "Legislated arbitration: Legality, enforceability, and face-saving." Public Personnel Management, (July/August):220-278.

BURTON, J.F. (1972). "Local government bargaining and management structure." Industrial Relations, 11(May):123-139.

CAPLOW, T. (1968). Two against one: Coalitions in triads. Englewood Cliffs, N.J.: Prentice-Hall.

CHAMBERLAIN, N.W., and KUHN, J.K. (1965). Collective bargaining. New York: McGraw-Hill.

COBB, R.S., and ELDER, C. (1972). Participation in American politics: The dynamics of agenda-building. Boston: Allyn & Bacon.

CRAFT, J.A. (1971). "Fire fighter strategy in wage negotiations." Quarterly Review of Economics and Business, II(autumn):65-75.

CROSS, J.C. (1969). The economics of bargaining. New York: Basic Books.

CROUCH, W.W. (1971). Municipal revenues. Washington, D.C.: Labor Management Relations Service, United States Conference of Mayors, National League of Cities, National Association of Counties.

CYERT, R.M., and MARCH, J.G. (1963). A Behavioral Theory of the Firm. Englewood Cliffs, N.J.: Prentice-Hall.

DERBER, M., JENNINGS, K., McANDREW, I., and WAGNER, M. (1973). "Bargaining and budget making in Illinois public institutions." Industrial and Labor Relations Review, 27(October):49-62.

DOHERTY, R.E. (1971). "Public employee bargaining and the conferral of public benefits." Labor Law Journals, 22(August):485-492.

DOUGLAS, J.D. (1976). "Urban politics and public employee unions." In A.L. Chickering (ed.), Public employee unions. Lexington, Mass.: Lexington Books.

DUNLOP, J.T. (1958). Industrial relations systems. New York: Holt, Rinehart & Winston.

EHRENBERG, R.G. (1973). "Municipal government structure, unionization and the wages of fire fighters." Industrial and Labor Relations, 27 (October):36-48.

EMANUELS, K. (1973). "Impact of property tax rate limitations on the meet-and-confer process." California Public Employee Relations. 19(December):8-10.

EVAN, W.M. (1972). "An organization-set model of interorganizational relations." In M. Tuite, R. Chisolm, and M. Radnor (eds.), Interorganizational decision making. Chicago: Aldine.

FERRIS, F.D. (1975). "Contract interpretation-A bread and butter talent." Public Personnel Management, 4(July/August):10-17.

FEUILLE, P. (1974). "Police labor relations and multilateralism." Journal of Collective Negotiations, 3(summer):209-220.

FOGEL, W., and LEWIN, D. (1974). "Wage determination in the public sector." Industrial and Labor Relations Review, 27(April):410-432.

FOLEY, F.J. (1975). "Public employee unionism in Pennsylvania: Impact on local power distributions." Prepared for delivery at the 1975 Annual Meeting of the American Political Science Association, San Francisco Hilton, San Francisco, September 2-5.

FREUND, J.L. (1974). "Market and union influences on municipal employee wages." Industrial and Labor Relations Review, 27(April):391-404.

GERHART, P.F. (1976). "Determinants of bargaining outcomes in local government labor negotiations." Industrial and Labor Relations Review, 29(April):331-351.

GILROY, T.P. and SINICROPI, A. (1972). Dispute settlement in the public sector: The state of the art. Washington, D.C.: U.S. Government Printing Office.

HEISEL, W.D. (1971). Day-to-day union management relationships. Public Employee Relations Library Report No. 31. Chicago: Public Personnel Association.

HILDEBRAND, G.H. (1967). "The public sector." In J. Dunlop and N. Chamberlain (eds.), Frontiers of collective bargaining. New York: Harper & Row.

HORTON, R.D. (1976). "Economics, politics, and collective bargaining." In A.L. Chickering (ed.), Public employee unions. Lexington, Mass.: Lexington Books.

――― (1976a). "Productivity and productivity bargaining in government: A critical analysis." Public Administration Review, 4(July/August):407-414.

――― (1973). Municipal labor relations in New York City: Lessons of the Lindsay-Wagner years. New York: Praeger.

HORTON, R.D., LEWIN, D., and KUHN, J.W. (1976). "The impact of collective bargaining on the management of government: A diversity thesis." Administration and Society, 7(February):497-515.

HOWLETT, R.G. (1973). "Contract negotiation arbitration in the public sector." Cincinnati Law Review, 42:47-75.

JONES, B.D. (1976). "Distributional considerations in models of government service provision." Prepared for delivery at the annual meeting of the Southwest Political Science Association, Dallas, April 7-10.

JONES, R.T. (1975). Public sector labor relations: An evaluation of policy-related research. Washington, D.C.: National Science Foundation.

KLEINGARTNER, A. (1973). "Collective bargaining between salaried professionals and public sector management." Public Administration Review, 33(March/April):165.

KOCHAN, T.A. (1971). City employees bargaining with a divided management. Madison, Wis.: Industrial Relations Research Institute.

――― (1974a). "A theory of multilateral bargaining in city governments." Industrial and Labor Relations Review, 27(July):525-542.

――― (1974b). "Internal conflict and multilateral collective bargaining in city governments." Prepared for delivery at the annual meeting of the American Psychological Association, September 2.

――― (1975a). "City government bargaining: A path analysis." Industrial relations, 14(February):90-101.

――― (1975b). "Determinants of the power of boundary units in an interorganizational bargaining relation." Administrative Science Quarterly, 20(September):434-452.

KOCHAN, T.A., and DICK, T. (1976). "A theory and empirical examination of the public sector mediation process." Prepared for delivery at the annual meeting of the Society of Professionals in Dispute Resolution, October 24, Toronto.

KOCHAN, T.A., EHRENBERG, R.G., BADERSCHNEIDER, J., DICK, T., and MIRONI, M. (1977). "An evaluation of impasee procedures for police and firefighters in New York State: Final report submitted to the National Science Foundation." New York: Cornell University, New York State School of Industrial and Labor Relations.

KOCHAN, T.A., and WHEELER, H. (1973). Contract analysis of fire fighter labor agreements. Washington, D.C.: U.S. Department of Labor, Labor Management Services Administration, Division of Public Employee Relations.

——— (1975). "Municipal collective bargaining: A model and analysis of bargaining outcomes." Industrial and Labor Relations Review, 29(October):46-66.

LEVY, F., MELTSNER, A.J., and WILDAVSKY, A. (1974). Urban outcomes. Berkeley: University of California Press.

LEWIN, D. (1973). "Public Employment relations: Confronting the issues." Industrial Relations, 12(October):309-321.

LEWIN, D., and HORTON, R.D. (1975). "The impact of collective bargaining on the merit system in government." The Arbitration Journal, 30(September):199-211.

LINCOLN, A.A. (1967). "The New York City transit strike: An explanatory approach." Public Policy, 16:273.

LIPSKY, D., and DROTNING, J. (1973). "The influence of collective bargaining on teacher's salaries in New York State." Industrial and Labor Relations Review, 27(October):18-35.

LOEWENBERG, J.J. (1970). "Compulsory arbitration for police and firefighters in Pennsylvania in 1968." Industrial and Labor Relations Review, 23(April):367-379.

Los Angeles Times (1975). "S.F. supervisors hit back, attempt to trim benefits." August 23, p. 1.

MANN, S.Z., and MANN, S. (1975). "Decision dynamics and organizational characteristics: The AFSCME and NYC District Council 37." Prepared for delivery at the 1975 Annual Meeting of the American Political Science Association, San Francisco, September 2-5.

MARCH, J.G., and SIMON, H.A. (1958). Organizations. New York: John Wiley.

McKELVEY, J.T. (1969). "Fact-finding in public employment: Promise or illusion?" Industrial and Labor Relations Review, 22(July):528-543.

McGRATH, J.E. (1966). "A social-psychological approach to the study of negotiations." In R.V. Bowers (ed.), Studies on behavior in organizations. Athens: University of Georgia Press.

NEWLAND, C.A. (1972). "Personnel concerns in government productivity improvement." Public Administration Review, 32(November/December):807-814.

OLSON, M. (1968). The logic of collective action. New York: Schocken Books.

PREGNETTER, R. (1971). "Fact-finding and teacher salary disputes: The 1969 experience in New York State." Industrial and Labor Relations Review, 24(January):226-242.

PERRY, J.L., and LEVINE, C.H. (1976). "An interorganizational analysis of power, conflict, and settlements in public sector collective bargaining." American Political Science Review, 70(December):1185-1201.

RASKIN, A.H. (1968). "How to avoid strikes by garbagemen, nurses, teachers, subwaymen, welfare workers, etc." New York Times Magazine, (February 25).

REHMUS, C. (1965). "The mediation of industrial conflict: A note on the literature." Journal of Conflict Resolution, 19(March).

ROSS, D.B. (1969). "The arbitration of public employee wage disputes." Industrial and Labor Relations Review, 23(October):3-14.

ROSS, M. (1976). "The local government budget crisis: Is bargaining to blame?" California Public Employee Relations. Berkeley, Calif.: Institute of Industrial Relations.

SALTZSTEIN, A. (1974). "City management and organized public employees: Sources of influence on city manager's attitudes." Midwest Review of Public Administration, 8(October):217.

SAYRE, W., and KAUFMAN, H. (1965). Governing New York City: Politics in the metropolis (2nd ed.). New York: W.W. Norton.

SELEKMAN, B.M., SELEKMAN, S.K., and FULLER, S.H. (1958). Problems in labor relations (2nd ed.). New York: McGraw-Hill.

SIMKIN, W.E. (1971). "Fact-finding: Its values and limitations." Arbitration and the Expanding Role of Neutrals. Proceedings of the Twenty-Third Annual Meeting of the National Academy of Arbitrators. Washington. D.C.: Bureau of National Affairs.

SIMON, H.A. (1957). Administrative behavior (2nd ed.). New York: Macmillan.

SPERO, S.D. (1948). Government as employer. New York: Remser.

STANLEY, D.T. (1972). Managing local government under union pressure. Washington, D.C.: The Brookings Institute.

STEVENS, C.M. (1963). Strategy and collective bargaining negotiations. New York: McGraw-Hill.

——— (1967). "Mediation and the role of the neutral." In J.T. Dunlop and N.W. Chamberlain (eds.), Frontiers of collective bargaining. New York: Harper and Row.

TAYLOR, G.W. (1957). "Wage determination processes." In G.W. Taylor and F.C. Pierson (eds.). New York: McGraw-Hill.

THOMPSON, J.D. (1967). Organizations in action. New York: McGraw-Hill.

THOMPSON, M., and CAIRNIE, J. (1973). "Compulsory arbitration: The case of British Columbia teachers." Industrial and Labor Relations Review, 27(October):3-7.

TURK, H., and LEFCOWITZ, M.J. (1962). "Towards a theory of representation between groups." Social Forces, 4(1962):337-341.

WALTON, R.E., and McKERSIE, R.B. (1965). A behavioral theory of labor negotiations. New York: McGraw-Hill.

WAMSLEY, G.L., and ZALD, M.N. (1973). The political economy of public organizations. Lexington, Mass.: Lexington Books.

WARREN, R.L. (1967). "The interorganizational field as a focus of investigation." Administrative Science Quarterly, 12(December):397.

——— (1972). "The concerting of decisions as a variable in organizational interaction." In M. Tuite, R. Chisolm, and M. Radnor (eds.), Interorganizational decision making. Chicago: Aldine.

WELLINGTON, H.H., and WINTER, R.K. (1971). The unions and the cities. Washington, D.C.: The Brookings Institute.

WORD, W.R. (1972). "Fact-finding in public employee negotiation." Monthly Labor Review, 95(February):60-64.

YAFFE, B., and GOLDBLATT, H. (1971). Fact-finding in public employment disputes in New York State: More promise than illusion. Ithaca, N.Y.: School of Industrial and Labor Relations, Cornell University.

YOUNG, O.R. (ed., 1975). Bargaining: Formal theories of negotiation. Urbana: University of Illinois Press.

ZACK, A.M. (1970). "Improving mediation and fact-finding in the public sector." Labor Law Journal, 30(1970):259-273.

9

Applying Behavioral Science to Urban Management

WILLIAM B. EDDY
THOMAS P. MURPHY

☐ ATTEMPTS BY URBAN MANAGERS to solve the problems of communities have been thwarted by a variety of contextual factors—including shortages of resources, political forces, and the sheer complexity and multiplicity of the problems. Over the past decade, however, there has been a growing consensus that traditional managerial approaches may not be adequate for effective problem solving and that the urban manager must reexamine and possibly revise his own practices.

In years prior to the "urban crisis," urban management was characterized by a "plant management" approach. Organizations were fairly static and were composed of functional units that handled the various specialized aspects of the city's business with relatively little interdependence. There was emphasis on efficiency, rational decision making, civil service systems to counter political influence, and a separation of policy and administration. "Delivery of services" was a mission more often referred to than "solving urban problems," "improving the quality of life," or "promoting equality of opportunity."

As attitudes about the proper role or urban government began to change, practitioners and scholars looked for new techniques to augment the managerial repertoire. At the time, several methodologies in the field of management were available for experimentation. Some were old, and others had been developed more recently by industry or by other levels of government.

For example, Leavitt (1965) proposed a model of organization change in which the change process may focus on one or more of the following four areas: *structure, technology, task,* and *people.* Efforts at improvement through structural change are well documented. They include reorganization, redefinition of roles and functions, and changes in procedures (Porter and Lawler, 1965). Technological changes have had major impact on almost all elements of urban government—from computerized accounting systems to fire fighting techniques. Task changes involve modifications in the kind of work that is done. Attempts to influence the performance of an organization through policy revisions may be thought of as task changes; i.e., they are intended to direct the efforts of people and units to assure that they will accomplish certain things.

Attempts to change and improve organizational functioning through interventions in the structure, technology, and task or policy dimensions are fairly standard practices. While the possibility of triggering organization improvement by applying a behavioral approach is not a new concept, it traditionally has been utilized less frequently. New developments in the applied behavioral sciences have enhanced the feasibility of working with employees, either individually or in task units, to develop viewpoints, understandings, attitudes, motives, skills, and norms which enhance individual and organizational effectiveness. Early efforts in this direction included human relations training, attitude surveys, management development programs, and personnel counseling. More recent applications are organization development, including such specific techniques as team building, action research or survey feedback, job enrichment, management by objectives (which may be but is not necessarily a behaviorally oriented approach to a task intervention), and various participative management arrangements.

These methods, individually or in combination, make up the current stock of techniques which are being attempted by administrators in efforts to enhance organizational effectiveness. They are sometimes described under the general rubric "organization development" (OD) but more correctly are referred to as *applied behavioral science methods,* since OD has for many a more specific set of meanings. They sometimes also have been referred to as "humanistic" approaches—more philosophical than scientific and more oriented toward satisfaction than productivity. OD proponents deny this, and there is evidence to support the use of OD by tough-minded managers (French and Bell, 1973).

Many observers have asserted that the Weberian bureaucratic model, while always an incomplete description of real organizations, is now outdated because of the many changes in the realities of contemporary urban systems. It is argued that the traditional view of the administrator's role does not provide adequate tools for handling real problems. A list of current issues forcing reevaluation of urban management approaches would include the following:

(a) *Instability.* Bureaucratic forms of organization were built to promote reliability in a relatively stable environment. They do not adapt readily enough to unstable and shifting conditions in urban communities.

(b) *Complexity.* The urban condition is characterized by a large number of interacting needs and problems. These problems often are not solvable if they are kept within the boundaries of traditional line departments. Solutions to a given problem may require the expertise of several functional units supplemented by resources from outside the city organization.

(c) *Open Systems.* Government organizations cannot think of themselves as closed systems with firm boundaries and limited avenues for access. Citizens, the press, political organizations, interest groups, contractors, other local governmental units, the state and federal governments, metropolitan councils of governments, and regional planning commissions all regularly enter the organization and attempt to influence it. This circumstance places new management demands on all levels of the system.

(d) *"Soft" Problems.* Thanks to technological progress, many of the physical problems of the city are under control—or would be if resources were available or if political obstacles were overcome. At least many of the answers are "known." This is not so in areas that have to do with mental health, family stability, crime, social incompetence, and poverty. Attempts are being made to shift more of the energy of local government organizations toward solution of socially oriented problems.

(e) *Employee Participation.* Significant internal changes, which are reflective of external phenomena, have occurred in the area of employee influence. Public employee unions, collective bargaining, equal employment opportunity programs, and a general societal thrust in the direction of more democratic institutions have created the need to reexamine the managerial style utilized to deal with employees.

In addition to all these systemic changes and pressures, there are even questions about basic strategies. Since many localities recently have moved from a pluralistic to a far more partisan political environment, former models for effective managerial leadership may be no longer viable. The "entrepreneurial" manager who formerly was praised for his ability to integrate diverse interests in a low cleavage pluralistic community may be ineffective and even dysfunctional in an ethnically or racially polarized community (Levine, 1974).

Effective response to the problem areas described above have forced organizations to adopt behaviors that deviate from traditional methods. There is more need for pooling information and expertise in problem-solving settings, more emphasis on intergroup collaboration, more need for responsibility, commitment, and initiative at lower levels in the organization, and better methods for coping with conflict. The applied behavioral sciences, with their emphasis on human interaction processes, have been sampled as possible solutions to contemporary organizational problems. Before undertaking an exploration of the application of behavioral science to urban institutions, we might first usefully describe more fully the origin and theoretical base of current behavioral science methods.

THE APPLIED BEHAVIORAL SCIENCE BACKDROP

It has long been acknowledged that the attitudes and feelings of employees affect organizational performance. The dilemma of the discrepancy between the needs of the individual and the demands of the organization is as old as history. Writers of every age from Plato to Machiavelli to Gulick have stressed the importance of *esprit de corps,* the consent of the governed, communication, and leadership skills. While the major thrust of management has always been control, one major problem always has been employee resistance. In the modern era, managers have been made fully aware of the significance of employee feelings and attitudes and are regularly exhorted to respond to them. It has been less than clear, however, why some managers create great employee loyalty and motivation while others are buried in personnel problems, and why some organizational units are invigorating, healthy, and creative places to work while others are psychological death traps. It has been even less clear how to develop and implement strategies for bringing about positive changes.

The frequently cited Hawthorne studies symbolized the first wave of new knowledge in current times about employee relations. They also triggered an action-oriented approach which went beyond traditional managerial strategies—the "human relations" movement. The findings of the various studies conducted in Western Electric's plants were significant. They documented the existence of informal organizational influences, the development of performance norms, the impact of group relationships, the character of basic employee sentiments, and a variety of other factors. Probably most important is the fact that the studies painted for the first time a picture of the sociological work environment of the factory employee. However, as often happens when a body of knowledge becomes popularized, the lessons of the Hawthorne studies and their successors were oversimplified by those who wanted to apply and profit from them.

Human relations training programs provided supervisors with training in understanding human needs, interviewing and listening, improving morale, and resolving human relationship problems. However, this "contented cow" approach to organizational improvement is now out of vogue. Its training programs often were based on the hypothesis that employee satisfaction leads to productivity—a belief that has not been confirmed in subsequent research. Although it opened up many organizations to training programs dealing with human behavior, it no longer can be considered an adequate application of behavioral science methods to organizations because of its incomplete view of employee effectiveness.

Current approaches to operationalizing social science concepts in organizations are only in part a second generation of the human relations movement.

Their origin relates more directly to the work of Kurt Lewin, early leader in American social psychology, and his students. The Lewinians were interested in the application of theory and did pioneering work in group dynamics, attitude change, organization change, and action-oriented research techniques. Their work led to the founding of the National Training Laboratories and to sensitivity training, which utilized intensive small group sessions (T-Groups) to foster understanding of group participation styles. Adaptations of the T-Group methodology currently are used in team development programs, group problem solving, and other group settings. (Marrow, 1969).

In a related area, Ronald Lippitt (Lippitt et al., 1958) conceptualized the role of the "change agent," a consultant whose expertise is not in the substantive area of the organization's functioning, but rather in helping the organization diagnose its own problems and implement self-directed change processes. Floyd Mann (1961) and others refined the applied research techniques into an approach labeled "survey-feedback." A similar approach has been followed by Neely Gardner (1972) in his Action Research.

Another line of thinking which has contributed to current applied behavioral science practice is that of humanistic psychology. Abraham Maslow (1971) and Carl Rogers (1961) are among those whose ideas have impacted contemporary thought. Douglas McGregor (1960), Chris Argyris (1970), and Warren Bennis (1966) have applied the ideas to organizations and the management process. They have advocated greater commitment to the development and growth of employees, more democratic and participative organizations, and more open and authentic styles of interpersonal communication. Organizational humanism is offered as a counter to autocratic management and technologically controlled work processes.

There are several other areas of thought which require at least brief mention as contributors to current practice in applied behavioral science. General systems theory has assisted in conceptualization of organizations as social systems or, more validly, as "sociotechnical" systems. Clinical and counseling psychology have provided tools in understanding communication processes and helping relationships. The considerable body of small group research has contributed perspectives on task performance, decision making and problem solving, leadership roles and other dimensions of group behavior. Other contributors include modern management theory, industrial psychology and sociology, and adult education.

Probably as significant as those theoretical developments is the fact that a growing number of social scientists, educators, and trainers have become interested in directly applying this knowledge in order to bring about organizational change. Labeling themselves applied behavioral scientists, change agents, or organization development consultants, they have sought to gain not only competence in the knowledge bases of the field but also mastery of intervention skills. Thus, applied behavioral scientists seek to use themselves as well as the literature of the field to facilitate change. Areas of skill include diagnosis of human problems in groups and organizations, assisting groups to become more aware and in control of their interaction processes, playing a facilitator or catalyst role in helping organizations plan for and carry out change, and teaching behavioral science techniques to managers (Schein, 1969).

Thus urban managers have available in the late 1970s an accumulation of social science theories about organizations and the people who work in them, a humanistic value system which proposes normative views about human relationships in organizations, an assortment of intervention technologies aimed at organization change, and a small but growing group of change agents.

ORGANIZATION DEVELOPMENT IN URBAN MANAGEMENT

Organization Development (OD), the application of applied behavioral science for the purpose of improving organizational functioning, is *one* of the applied behavioral science approaches with which urban governments are experimenting. OD has been variously defined and associated with an assortment of goals, which should be reviewed.

Several authors have attempted to evaluate the impact of Organization Development upon organizational performance (Friedlander and Brown, 1974; Kimberly and Nielsen, 1975; Kahn, 1974). Like other intervention approaches, such as management development and management by objectives, Organization Development is difficult to assess in terms of effectiveness. Confounding variables, lack of comparative output data, environmental changes and a variety of other factors contribute to the problem of evaluating OD in the public sector. In addition, and perhaps more importantly, applied behavioral science methods usually are not viewed as direct solutions to problems in organizational production systems. More frequently, such techniques are introduced in an attempt to iron out human problems, improve teamwork, facilitate communication, or better utilize human resources. That significant improvements in these areas are worthwhile in and of themselves and that they will ultimately have an impact on organizational performance must often be accepted as articles of faith. Most of the follow-up data are subjective and idiosyncratic. In efforts deemed successful, managers may say, "It was helpful," "We work together better," "Morale and involvement have improved," or "We can iron out more quickly the problems we cause each other."

DEFINITIONS

Organization Development is best described as part of the "humanistic" model. While the "positivist" model involves the application of a preexisting theory or a fixed set of techniques for the purpose of effecting predetermined changes, the thrust of OD is to identify problems and to discover a means for solving them as part of the process itself. The particular objects of change and the nature of the desired changes are not necessarily known in advance; this knowledge emerges from the diagnostic stages of the OD program.

OD, as an approach in improving urban management, includes therefore *both* a set of theories about groups and organizations and a set of developmental and educative intervention techniques. F. Gerald Brown (1974) made an exhaustive survey of scholars and practitioners who have attempted to define the OD process, particularly emphasizing the comments of those who have applied OD to the public sector. Some definitions of OD emphasize the technical aspects of the process, and others its goals and objectives. There is no similar agreement about the level at which the change process should be initiated; some argue that change must start at the top level of management, while others claim that it is possible or even desirable to start at other levels.

There is considerable agreement, however, about the *objects* of OD concern. The focus is always upon the "human systems" of the organization, including role conceptions; interrelationships at the dyadic, group, intergroup, and system levels; and the interface between interdependent systems. Further, there is general agreement that OD programs ought to be employed on an organization-wide basis, although many of the definitions and descriptions to date derive from application of the process only to subsections of organizations. Finally, most definitions stress its situational nature and the need to employ data and experiences unique to a particular organization.

There is also wide agreement that the object of OD at each system level to which it is applied is the "organizational culture and climate"—the attitudes, beliefs, values, and norms of the system and its subsystems, as revealed in observable behavior, patterns of relations, and organized structures. Concern with structure is derivative from and secondary to the concern with social processes within an organization—decision making, planning, communication, and other human interactions.

OD as applied to urban management, then, may be defined as an approach to maintaining, renewing, and changing the human system involved in local government operations. It entails a characteristic ideology, a program, and a technology rooted in recent behavioral science thought and findings. Perhaps the most significant assumption underlying the approach is that growth and development are natural and even inevitable and that individuals contribute most positively when their environment is supportive of their personal growth.

On the basis of this definition, several approaches often considered as OD do not qualify. For instance, OD cannot be considered as merely a new form of management development, because the focus of OD is organization-wide. Management development can be considered a valuable OD technique only when it addresses widely acknowledged organizational problems or when it contributes to improvements in the entire system. Further, OD must be considered as more than just "team building," even though this is one of its most useful tools. Again, team building, concerned with improving internal work functions of a group or its relationships with interfacing groups, is primarily ethnocentric in nature, whereas a fully developed OD program focuses upon the development of the organization as a complex set of interdependent systems. Finally, the application of OD techniques on a one-time or spot basis, usually motivated by limited and fairly precise objectives, does not itself demonstrate commitment to an ongoing organizational change process.

TECHNIQUES OF OD

By surveying the theoretical literature on the subject and by reviewing actual case studies, Brown (1974:42-46) constructed a five-category typology of the various techniques or "interventions" of OD that have been or could be applied to urban management systems. The first type consists of *particular training interventions*—conscious, structured efforts to change the knowledge, skills, behavior, attitudes, and degree of awareness of self and others on the part of individuals or classes of individuals. Such training techniques include sensitivity training as well as the more traditional management development training of the lecture and homework type, directed at either individuals or groups where a specific diagnosed need exists.

The second major type of intervention is *process consultation.* By employing a series of techniques, a consultant can help an organization solve its own problems by making it aware of organizational processes, their consequences, and the mechanisms by which they can be changed. In employing such techniques, the consultant ultimately aims to make the organization capable of doing for itself what he has done for it—conducting organizational self-evaluation and self-modification.

The trained observer focuses upon the social, behavioral process aspect of task performance within working groups or social systems. Process consultation can include team building, "real-time" interventions, and the use of paired relationships.

A third major class of OD intervention techniques can be labeled *"integrative interventions."* Closely related to process consultation, these techniques are designed primarily to promote the social integration of individuals or groups whose work is interdependent. Whereas process consultation was primarily concerned with promoting awareness of and improving decision making and procedures for accomplishing tasks, integrative interventions are designed to build patterns of communication and support. They are employed in situations where it is important to establish new connections or where dysfunctional conflict exists between persons and units.

Diagnostic data collection represents a fourth category of intervention tactics. Data collection provides one of the primary means for learning about the specific system in question—its boundaries, its principal components and their interrelationships, the nature of the work performed by the members overall and in their subgroups, the way in which members communicate with one another about their work, and, most important for the identification of problem areas, the discrepancies between stated goals or values and actual behavior or discrepancies among various individual or group perceptions of the same phenomenon. Further, the process of data collection itself actually can help to focus attention upon relational areas. Survey feedback, sensing, and mirroring are the primary data collection techniques.

The last major type of intervention tactic can be described as *"technostructural change."* It may be implemented to enhance the motivational character and problem-solving capabilities of the working environment. Structural alterations may be only temporary, such as the formation of task forces or other crosscutting groups to accomplish specific tasks, or they may be permanent as in the case of reorganized work units.

In any discussion of OD techniques, it is important to remember that OD is a *process* and *not a discrete event*. Thus, it is important that the particular tactics of the process be built into the system in question, and not merely applied to it. Accordingly, continuous feedback mechanisms must be developed and permanent follow-up support mechanisms for each of the primary tactics must be instituted.

IMPLICATIONS OF THE SPECIAL CHARACTERISTICS OF PUBLIC MANAGEMENT

Applied behavioral science methods, including Organization Development, came largely out of experiences in private industry. As with other management innovations which have been transplanted into local government, questions about their applicability naturally have arisen. It is commonly accepted among urban administrators that their organizations are different, though the real impact of the differences is poorly documented. There are no profit motive, more political pressure, and greater multiplicity of goals. The significant question, of course, is whether the special characteristics of urban government organizations preclude successful application of behavioral science methods (Rainey et al., 1976).

Several authors, including Golembiewski (1969), Eddy and Saunders (1972), and Culbert and Reisel (1971), have examined the fit between behavioral science and government agencies. It is asserted by these observers that there are a variety of factors unique to government which condition the applicability of OD techniques. These factors include the duration and succession of leadership, conflicts between political and administrative systems, favoring of strong control over participative and humanistic leadership, and a climate of distrust stemming from public suspicions. The cases discussed in the following sections of this article demonstrate the unique characteristics of urban government which limit the application of behavioral science methods. Effects take place at the levels of basic assumptions, operational practices, and interaction with the environment. However, the case examples also illustrate ways in which methods can be adapted to obtain at least a partial fit with urban organizations. Other cases not described here suggest additional differences (Garner, 1972; Rubin et al., 1974).

KANSAS CITY'S ORGANIZATION DEVELOPMENT (KCOD)

The Kansas City metropolitan area has been an experimental seedbed for organization development in local government and provides examples of the application of a variety of OD approaches. Since 1968, these approaches have involved the City of Kansas City, Jackson County, and a number of suburban jurisdictions acting collaboratively. They have dealt with department heads and city or county managers, as well as with elected officials and the relationship of those officials to professional employers and other elected officials.

THE KANSAS CITY CASE

The Kansas City program (which has been called KCOD) began somewhat by chance in the spring of 1968, when the city manager asked the University of Missouri-Kansas City for assistance in addressing problems within the organization. In the six years between 1968 and 1974, a variety of OD interventions were employed with top management, middle management, and the city council. These included sensitivity training, survey feedback, task force development, team building, confrontation meetings, process consultation, development of new line management roles, internal consultant development, and a variety of agenda building, problem solving, and management development oriented activities.

Brown's (1974:98-143) survey of the events in KCOD illustrates the precise application that an urban management system has made of the OD techniques. In 1968-1969, for instance, the survey feedback format was employed to interview members holding key administrative positions, including employees of the city manager's office, budget and systems analysts, and department and division directors. A confidential questionnaire was sent to each, and the data was summarized by the consultants. Interview data included assessments and critiques of leadership, structure, interdepartmental communication, staff development tactics, and staff meeting procedures. The results of the survey were discussed at a two-day feedback training session with the top management group. Small groups discussed the data and established six priority areas—management, internal communications, city goals, communication with the public, personnel services, and management development. Task forces were formed for all of these areas, and were instructed to organize and define an action plan.

Team building interventions also were employed, each preceded by an anonymous team building questionnaire seeking general information about such issues as perceived effectiveness, trust, commitment, and satisfaction with relationships with specifically named groups or individuals. Each session was tailored to the particular group and time in question. Meetings for organizational work evolved out of the team building approach. These were less intense and more wide-ranging in the nature of the subject matter covered. They focused sometimes upon problems of the "team," sometimes on broader internal organization questions, sometimes on current policy questions, and sometimes on interface issues. Over 30 meetings for organizational work have been held, and such meetings are the major surviving OD intervention format.

A means for interdepartmental work on organizational issues was provided by a series of interdepartmental confrontation meetings among all departments. Each department was paired with another department with which it needed to work. In the weeks preceding the training session, the departmental management groups met to develop lists of objectives or short statements describing how the department saw itself, how it thought the other department saw it, and the problems that the department believed it caused other departments. During the session, the departments completed mirror lists for their paired departments. After exchanging lists with paired departments and clarifying the meaning of critical terms, the departments withdrew to talk about their reactions to how the other department saw them and about what they could do to improve the situation. Then the paired departments reconvened for a joint discussion in which they attempted to formulate specific steps that could be taken to improve the relationship.

The KCOD experience affords a good example of the manner in which the forces which comprise the political arena, as previously described, may qualify the impact of OD in the public sector. The election of a new mayor and city council in April 1971 had a powerful impact on the city organization. The most notable change was in the alteration of political style from the patrician and solidly developmentally oriented approach of the outgoing mayor, Ilus Davis, to the press-conscious, self-proclaimed adversary approach of the incoming mayor, Charles Wheeler.

Although the OD program continued throughout this period and, in fact, provided forums in which the behavioral expectations of new assistant city manager roles were clarified, much of the momentum was lost. Attention shifted away from assessment of the overall organization and from creative problem solving to the more immediate concern with survival and reaction to the changes then being imposed upon the organization. By 1972-1973, the program had been reduced to one six-hour meeting per month with departmental directors, concerned exclusively with administrative operations and not with organizational planning.

Recent interviews have been conducted to assess the impact of KCOD. They reveal that the basic efficiency, adaptability, and humanistic goals were facilitated, at least to some extent, by the program. Two members of the city manager's office remarked that the monthly meetings were all that remained of the OD program. However, even though this meeting may be little more than an extended city manager's staff meeting, it is quite a significant event. Questionnaires indicate that it is highly valued by departmental directors, because it provides a regular face-to-face format for sharing information, solving problems, and working on issues that affect the overall city organization.

Probably as a consequence of the OD program, the department directors have developed the habit of occasionally polling themselves to see how things are going. They now also treat information about the relationships between departments and communication patterns as important inputs deserving of time in joint discussions. This is significant because there has been very little turnover in Kansas City department heads.

In general, the responsibilities of middle management have not been increased, nor have they been employed more as a resource for working on organizational issues. These factors have fostered some degree of frustration and cynicism. Nevertheless, middle and upper management workers probably have developed expanded interpersonal skills, problem-solving approaches, and some criteria for evaluating organizational patterns. Finally, one clear success of the program was the implementation of well over half of the task force recommendations.

One final way of viewing the outcomes of KCOD is to examine management's assessment of its impact. In 1968 and 1970, the top management group completed an "organizational status inventory" questionnaire which measured a number of variables considered important to the maintenance and renewal of organizations as responsive social systems. The surveys indicated that positive changes had occurred in the development and use of face-to-face groups, in the level of trust characterizing the environment, in the effectiveness of policy formulation and implementation, and in the recognition of talent and leadership capacity at all levels. Although it is possible that OD was not responsible for all of these improvements, clearly KCOD can claim some of the credit. However, the fact that the surveys indicated no notable improvement in the collaboration and communication among operating units only suggests that it is difficult to achieve interdepartmental integration and that, however useful, OD is no panacea. The impact of political problems and changes confirms the view that OD is more difficult to apply in the public than in the private sector.

THE SUBURBAN INTERGOVERNMENTAL NETWORK

One outgrowth of KCOD is the Suburban Intergovernmental Network for Management Development (SIGN), a consortium of the Missouri cities of Excelsior Springs, Gladstone, Grandview, Lee's Summit, and Liberty, in cooperation with the University of Missouri-Kansas City Center for Management Development (Heimovics, 1975; Brown and Saunders, 1973). The network was established in 1970 to provide training for all five cities and a program of OD for each. This approach to OD is both broader and more specific than KCOD. It is broader because of the networking aspects and because the total complement of administrative, professional, and technical manpower in each city is the target change group. It is more specific because experience has permitted advance definition of a number of specific training goals.

However, SIGN shares with KCOD the same fundamental process. None of the particular developments or outcomes were predictable in advance, and the overriding objective again was to increase internal organizational capability and to identify and examine administrative and policy issues important to the organizations' effectiveness. Training events and direct organizational interventions were selected and tailored to that specific goal.

For convenience, the SIGN effort can be described in two phases—management development followed by organizational development. First, a networking procedure was employed to select the participating cities and their representatives. About 12 individuals from each city were included as participants. The city managers and administrators were formed into one training group, and the remaining people were randomly assigned to two other groups.

The selected training groups then participated in management development, training in communication skills, role development, organizational perception, group behavior, problem solving, and decision making, primarily focusing upon the real issues of the five city organizations. The OD phase then represented a shift in format to single city groupings, but the network continued to provide support that could be summoned when needed.

In the first, or management development phase of the program, each of the three training groups worked simultaneously through a number of three-day intensive workshops, covering management styles, group behavior, and collaboration. A large variety of structured exercises, interspersed with lecturettes, were employed to encourage high levels of self-disclosure, as projective devices for surfacing data about the organizations and their role relationships, as formats for problem solving, and as tools for conceptual learning and communication skills development.

In the second or OD phase of the SIGN program, team-building tactics were employed. These were based upon the same goals and methods as those of KCOD, but there were significant differences. In SIGN team building, there was no prepolling of the group, and time was more strongly structured by the trainers. The focal issues varied from city to city, but two exercises were employed in each case. These were the use of organization model building as a framework for talking about roles and relationships and the use of a formal role negotiation in which each individual wrote and exchanged specific statements of "what I need from you in order to work effectively" and then discussed the requests received.

A number of outcomes of the SIGN effort indicate that it helped to promote basic efficiency and adaptability, as well as humanistic objectives. Probably the most favorable outcome of the SIGN program was the establishment of a network that still exists. A rather strong career support group has developed among the city managers. Since this position is so politically sensitive, so competitive, and so exposed to public criticism, it is quite an accomplishment that the five city managers have developed a high degree of trust and quite frequently meet to discuss sensitive organizational and policy issues. The network among other members of the training group is less intense but still functional. Employees of one city often contact employees of another city whom they now know in order to ask for advice, to request ordinances, or to obtain copies of items such as personnel manuals.

Other developments also have evolved from the SIGN effort. Four of the cities have instituted regular meetings between department heads and the city manager staff, as an outgrowth of the problem identification workshops; and two cities have established employee newsletters to help with general communication problems. During the program period, three of the cities set up elected "employee councils" to deal with general questions affecting employees and their working conditions. As a result of the ideas and collaborative skills developed during training sessions, two cities have been able to carry out a radically changed approach to making decisions about budgetary allocations and the uses of revenue sharing. Also, one of the cities developed a modified management-by-objectives process as an outgrowth of the feedback session. Like KCOD, then, the SIGN program indicates that there are measurable lasting effects from the application of consistent and carefully tailored applied behavioral science change efforts to urban management.

THE CHIEF URBAN ADMINISTRATOR AS CHANGE AGENT:
THE DAYTON TASK FORCE MANAGEMENT CASE

Most conceptions of applied behavioral science view the change agent as someone outside the line management system—a staff person or an internal or external consultant. It is useful, however, to consider the role of the chief administrator partly as that of a *change agent*—that is, one who brings about constructive change, not always through the use of direct control, but often through "facilitation."

In his study of cases of successful organizational change, Dalton (1970) found one of the necessary ingredients to be the "intervention of a prestigious influencing agent," "a respected and, ideally, a trusted source." In many cases, this agent was the formal head of the organizational unit involved. His behavior was instrumental in initiating changes and seeing them through to successful completion. In other situations, an outside agent facilitated the changes with the support of the unit head. At the level of social processes, behaviors which are effective for consultants are effective for managers.

We have suggested elsewhere that, in many important aspects of his position, the urban manager has relatively little formal "command power" in the traditional sense (Eddy, 1974); that is, in a variety of important areas, he cannot solve problems by ordering people to do something. For example, command is usually not feasible or effective under the following conditions:

1. Situations in which the line manager holds no authority over some or all of the personnel involved: for example, interdepartmental task forces, citizen groups, and interagency commissions.
2. Situations in which sophisticated or high-level behavior is required, but cannot be commanded: for example, complex decision making, problem analysis and solving, planning, and various professional activities.
3. Situations in which building collaborative problem-solving efforts is required: for example, situations in which people must work together smoothly, listen to and consider each other's ideas, and develop a spirit of teamwork.
4. Situations in which attempts are made to overcome resistance to change and to renew commitment and a sense of mission.

Thus, the urban manager often may find himself in a "facilitator" role. He brings people together to talk, he opens up communication, he works for objective problem solving, he tries to resolve conflict, he encourages acceptance of new ideas, he attempts to develop trust. These areas of social interaction process are the subject of applied behavioral science. The manager may increase his own effectiveness by borrowing from them.

A change agent-urban administrator can utilize a behavioral science approach by employing the technique of task force management training. A task can be distinguished from other kinds of committees in that it is a "temporary system." Many organizations, including local government units, find that not all necessary work can be done effectively within the framework of the formal structure. Some projects have short but intensive life spans; others require pooling of technical resources from several departments; still others are based on the collaboration of people with diverse interests. "Task Force" is the term commonly applied to a full- or part-time working group which is called together outside the formal structure to spend a finite amount of time on a defined task or project.

Task forces often experience difficulties, not because they have inadequate technical skills, resources, or motivation, but because of process problems: inability to clarify and operationalize the task to the satisfaction of all members, difficulties in resolving differences and reaching decisions, unclarified ground rules, leadership struggles, and covert competition. Task force management training techniques are intended to improve performance by helping members recognize and resolve process problems.

Kunde (1973) provided an example of a task force project in an urban community. As city manager of Dayton, Ohio, he was confronted with the need to keep its service delivery organizations relevant to the needs of the community. The city government was well organized as a large public service utility that delivered services such as street construction and maintenance, garbage collection, and water supply. But the main problems of the community were identified as race prejudice, crime, housing, and youth unrest. Further, the services relating to the aged, education, transportation and downtown development were not adequate. No department was equipped to deal in a total way with such problems. The problems were affected by the performance of all city departments as well as agencies outside of city government. Progress toward solutions required collaborative actions by people from a variety of professions and organizational affiliations.

Alternatives for responding to the situation were (1) to reject the problems as the responsibility of city government, (2) to reorganize the administrative system so that it more nearly fitted the array of problems, or (3) to attempt to create ad hoc devices which focused the attention of appropriate people on the problems without creating and breaking in a whole new organization. The latter alternative was selected as the least disruptive and expensive and the most politically feasible.

Sixty upper management personnel were assigned to seven task forces. They were advised that 50% of their salary evaluation would depend upon the success of their individual task force. Groups met for 2½ hours twice a week. They explored their assigned areas, conducted problem diagnoses, developed and recommended plans and policies, and participated in their implementation. According to Kunde, the task force program also promoted basic efficiency, adaptability, and humanistic goals. Spin-off benefits of the task force efforts included increased delegation of responsibilities by administrators participating in task forces, better focus on overall city problems and the roles of the various departments, and better recognition of the value of the planning process.

Outside consultants worked with the city to implement the program. Two off-site retreats were held to help participants understand the need for the new effort and to enable them to participate in the planning. An attempt was made to create an atmosphere in which concerns and reservations could be expressed and addressed. Preliminary training was provided in the leadership of task forces, and procedures were established for their conduct. After the task forces began functioning, the need for greater degrees of group skills was identified, and faculty members from a local university provided additional training in group communication and problem solving. To strengthen coordination, an upper level management person was assigned as coordinator. His role included regular meetings with the chairpersons to check on progress and iron out problems. The city manager assured that there was public recognition of the efforts of the task forces and provided continuing support and encouragement.

In the Dayton project, the consultants and city manager utilized several behavioral science concepts and practices. Questionnaire surveys and feedback of data provided a preliminary analysis of problems, concerns, and interests. An effort was made to have the changes be as internally generated as possible. Workshops and retreats provided an open climate for discussion and participation. Group dynamics techniques were used to help assure that the task forces would not flounder because of process problems, and consultants stayed in touch with the project as it evolved, offering suggestions for its administration. Thus, the consultants' role was facilitative rather than programmatic. They assisted in the organization's own diagnosis, problem solving, and implementation processes. They did not conduct their own study and write a manual prescribing structure, procedures, and objectives.

Evaluation of such a project is difficult because direct scientific comparisons with alternative approaches do not exist. Kunde deems the effort highly successful, and the basic effectiveness, responsiveness, and behavioral objectives previously cited do seem to have been facilitated by the program. A number of innovative programs and proposals were developed and implemented. Some task force members reported satisfaction, personal development, and greater organizational understanding. Others felt less positive, and few dropped out of the project. Some task forces were more successful than others. Kunde left the city of Dayton, and the new manager has taken the organization in some different directions. Many of the measures implemented by task forces still are operating, however, and it is reported that residuals of the task force effort still exist within the system. As presently utilized, task forces are more temporary and have distinct cut-off dates. Assignments are more specific—such as "development of a plan for reducing white flight." Increased emphasis has been given to putting noncity employees on the task forces, and, in some cases, city councilmen (commissioners) have served as members.

MANAGEMENT BY OBJECTIVES: THE CHESAPEAKE EXPERIMENT

Another management approach that demonstrates the effective use of behavioral science theory is the Management by Objectives (MBO) approach, originally designed for the private business sector but recently applied to the government sector at the federal, state, and local levels. MBO is a program by which organizational goals and objectives are set through a participatory process in terms of results expected. Subordinates participate in the goal setting, thereby increasing individual motivation toward organizational objectives. The approach is designed to promote interdependence between management and subordinates and to create greater organizational flexibility, greater efficiency, and greater job satisfaction for employees. Through the participative process, management hopes to increase organizational effectiveness by developing meaningful individual and organizational goals, clarifying responsibility, promoting supervisor-subordinate collaboration, and improving the administrative feedback process.

Clearly, MBO represents a potentially less instrumental, more humanistic approach to problems that interface between the individual and the complex organization. Superior and subordinate members of an organization define its common goals and each individual's major areas of responsibility in terms of the results expected of the individual. These measures then are used as a guide for operating the unit and for assessing the contribution of each member. Public administrators, therefore, are required to be more conscious of their role in clarifying organizational goals and operational objectives. Consequently, the use of MBO involves a large number of public administrators and extensive training within the organization. Implicit in MBO is the devolution of authority and power to successively lower levels of the hierarchy and the deemphasis of traditional hierarchal values. The emphasis is always upon self-management and the behavioral matters and values associated with contemporary organization theory.

MBO has established a mixed reputation in government, but it must be remembered that any kind of organizational change is hard to effect in public institutions. Changes face intense resistance, stimulated by deeply rooted suspicion and distrust within the bureaucracy. In order for MBO to work in government, an organization must be ready to move toward a more human and integrated system of management. If it is introduced into an agency where traditional bureaucratic values still prevail, it is likely to fail. The main problem in applying MBO to government agencies, then, has been the dominance of pyramidal values and centralized power and communication control.

The effective practice of MBO requires an organizational capacity to deal with humanistic goal-setting processes rather than to control procedural techniques. However, if administrators intend to introduce an MBO approach, they must be committed to several specific behavioral orientations, the most important of which are self-management and decentralization. By providing the individual with greater responsibility and accountability, management encourages individual self-control and self-evaluation, reducing tightly monitored supervision to a minimum.

MBO developed in many organizations as a reaction to the generally abortive attempts to introduce planning, programming, and budgeting systems (PPBS). The comprehensive PPBS approach proved to be far more demanding than many jurisdictions really could handle. Because in-depth analyses are necessary to use that system effectively and because there are relatively few qualified analysts to do the work, most such programs were doomed to failure before they started. Many managers were looking for a panacea, and PPBS provided only hard work and hard decisions. In many respects, MBO represented a face-saving device to lower the targets while still focusing on a few portions of the organization which needed extra management attention and evaluation. Also, MBO rests on a more participative management foundation than PPBS, which had more of a control than a behavior-modification orientation.

One city government, attracted by PPBS but unable to mount such an extensive effort, has experimented with an approach to MBO. Chesapeake, Virginia, with a population of 105,000, is a relatively new city, formed through the merger of South Norfolk and South Norfolk County in 1963. It has a stable city manager form of government with a substantial number of professional administrative employees. Utilizing 701 funding from the U.S. Department of Housing and Urban Development, Chesapeake undertook a study of its organizational and functional relationships for management and planning as a base for developing an MBO program.

One of the products of the study was a matrix of interdepartmental relationships. This revealed some surprises about organizational communications. For example, of the 23 departments reporting to the city manager, 11 indicated that they had to have contact with him more than once a day. However, an analysis of the functions involved indicated that there was no firm pattern in terms of function, degree of policy direction needed, or the sensitivity of the function. This is not surprising, since similar perceptions exist in many cities and relate in part to the level of security the department heads feel, the length of time that these persons have been departments heads, and the tenures of the city managers. Other findings from the study were that a number of departments and divisions perceived that they had a close functional relationship with other departments and divisions which did not express a reciprocal connection. This suggests that some functional interrelationships of the department were not mutually clear or significant (Murphy, 1974).

In addition, the elected officials—such as the city treasurer, the commissioner of revenue, the sheriff, the city clerk, and the assessor—perceived little relationship with the departments reporting to the city manager. This was a clear contradiction of the existing functional relationship. The survey also showed the several units such as the Bureau of Inspections, which was located in the Public Safety Department, did not perceive a close relationship between its functions and those of the rest of the department. The Department of Recreation did not indicate any functional relationship with the Social Services Department despite the fact that the Recreation Department had a senior citizens program, an arts and crafts program, and a cultural enrichment program which were directly related to social services.

The Inspections Bureau was not informing the Departments of Public Works, Public Utilities, and Planning, the Registrar, or the Real Estate Assessor about buildings that had been or were about to be demolished. Numerous types of information which were identified by the various departments in the survey were found to be of interest to other departments but not generally communicated to them. In turn, these other departments frequently were not even aware of the existence of the information which the city was expending money to collect.

The study concluded that the departments in the city of Chesapeake needed additional guidance in terms of their roles, their objectives, their relationships, and the measures of effectiveness which the manager would have to use in evaluating progress toward the city's goals. As a consequence of the study, the city manager embarked upon a limited Management By Objectives program. All 'the department heads were invited to a two-day training session in which they were asked to participate voluntarily in a follow-up interview process which helped to identify departmental objectives and priorities, to communicate needs, and to further define the perceived and required relationships with other departments. The willingness of a department to embark upon a detailed

Management By Objectives program also was determined, and a competition was held to select two pilot model units for the development of such a system, which would be geared to increasing productivity.

Ten of the 23 departments indicated a serious interest in proceeding with Management By Objectives, and, after further detailed interviews and considerations of all the factors involved, two departments finally were selected: one an operations unit and one a staff unit. The Purchasing Department was selected as the staff unit and proceeded to spell out a series of goals and objectives for a program beginning July 1, 1975. The operating unit considered most ready to begin the MBO program was the Sewer Divison. After six months of operation under the MBO system, both departments concluded that they had improved their management substantially and were ready to begin utilizing the system in preparing the Fiscal Year 1977 Budget (Murphy, 1975).

The significance of this is that these departments now had new data to justify budget requests and were in a better position than previously to indicate the actual effects of an increase or a decrease in their funding. This is not to suggest that the MBO process is completed. Some targets were missed, and some others proved to be unrealistic. These departments are currently improving upon their definitions of objectives and the measures by which to judge attainment. However, much progress has been made and the City Manager currently is considering an extension of the program to other departments in the city government.

The process also has revealed the need to maintain some central office capability in order to provide technical assistance to the departments undertaking such programs. In each department, the involvement of individuals in setting the department objectives and in helping to achieve them has been important. Initial suspicions of work-load counting and stop-watch efficiency experts have been dissipated, indicating that it is important for management not to use targets, even targets which employees have participated in setting, as means of demanding production beyond what is possible. Another department in the city government is experimenting with performance and productivity guidelines. The solid waste unit has assigned pick-up targets to a number of crews, which leaves them free to take off for the day as soon as their work is completed. So far, the change has encouraged the employees to perform more work and finish it earlier—with fewer complaints of refuse strewn about the neighborhood than under the previous system.

MBO, then, represents a highly specialized application of a behavioral science orientation and technology, designed to foster the basic organizational goals of efficiency, responsiveness, and employee development. Administrators who attempt to introduce MBO must regard it as an integrative approach to total management. Organizational goals must be integrated through employee participation at all levels. The attempt to integrate the major organizational functions to accomplish goals requires a systematic link between the budgetary process, manpower allocations, planning, training and development, research activities,

WILLIAM B. EDDY and THOMAS P. MURPHY [221]

and program evaluation. Also, the MBO administrator must be committed to a process of continual feedback between individual and organizational subunits in formulating and implementing goals and objectives. The organization must be engaged in constant research to develop meaningful goals and to determine indicators which may be useful in establishing priorities among various needs. Finally, the MBO manager must offer strong leadership support. In order to initiate MBO effectively, he must be committed to providing resources for training and to taking risks by experimenting with new management concepts. (For a recent, comprehensive discussion of MBO, see *Public Administration Review,* 1976.)

THE FUTURE OF BEHAVIORAL SCIENCE IN URBAN GOVERNANCE

Are behavioral science applications simply another set of acronyms in a long list of fads and nostrums, or will they survive as permanent tools in the repertoire of accepted practices? Most practitioners of OD are doing a brisk business, while critics of the field are already pronouncing its death knell. Far too few evaluation studies have been conducted for there to be any consistent scientific evidence. Some projects seem to succeed, others to fail—or at least to peter out. It is difficult to tell whether successes and failures are due to the methodologies themselves, the skills of the practitioners, variables within the organization, or the attainability of goals. Our predictions about the future of behavioral science applications must rely on assumptions about the fit between urban management needs and the power of the methods.

The following points bear consideration in assessing future trends:

1. The field of urban management—indeed, the total field of public administration—in the United States is in a state of flux. There is no accepted comprehensive theory or even agreement about whether a grand theory or paradigm is desirable or likely. Current practice is a patchwork of unrelated and sometimes conflicting approaches, heavily influenced by personal beliefs, intuitions, and styles of the managers. In such an arena, experimentation with new methods is likely to continue.

2. There appears to be an emerging viewpoint that solutions to urban problems will require broad involvement and commitment from citizens and groups on a metropolitan area basis. Better mechanisms for developing communication, consensus and collaborative problem solving are clearly needed. Behavioral science approaches for dealing with these process issues within organizations are being tested at the community level. If they prove successful, they will help meet an important need.

3. Organizational characteristics peculiar to public agencies will continue to provide resistances to behavioral science methods. The fact that merit boards and civil service systems originally were initiated by

reformers makes it ironic that these systems have become barriers to the adoption of progressive urban management.

4. Part of the fate of behavioral science methods in urban governance will be determined by trends in societal values. Some observers feel certain that they discern movement in the direction of more humanistic, open, and participative norms. Such a thrust would encourage further experimentation with managerial methods supportive of these norms. Others contend that economic pressures, technology, class conflict, and other forces will push values in more controlled, centralized, and mechanistic directions. Such a shift, one would predict, would provide a stimulus for managerial thought to veer away from humanistic stances and toward higher levels of control.

5. The more conscientious behavioral science practitioners point out that their techniques must be viewed as long-term approaches to change and that organizations which are seriously interested in development must commit themselves to a process which requires months and years. This stance does not square well with the short-time horizons of most governmental agencies or reform commissions, yet it is realistic for most substantial organization change processes.

The future, of course, will also be influenced by the general condition of the society in which the urban manager is operating. Like any other theoretical efforts, behavioral science approaches are limited by the political realities. Public sector unionization and racially based community conflict are two major barriers to the success of any change. However, even settling the conflicts generated by these conditions generally requires a strategy of confrontation and negotiation. OD is catching on in urban management but will not reach full bloom until urban managers have a clearer understanding of how OD can help meet strongly felt needs. The kinds of behavioral science approaches present in the OD movement, as well as those in the PPBS and MBO experiments, should equip urban managers with tools for making effective use of such strategies. It is important perhaps to bear in mind that the forces that made OD and the applied sciences work or fail are not unique to urban governance.

REFERENCES

ARGYRIS, C. (1970). Intervention theory and method: A behavioral science view. Reading, Mass.: Addison-Wesley.

BENNIS, W.G. (1966). Changing organizations. New York: McGraw-Hill.

BROWN, F.G. (1974). "Organization development in the city of Kansas City, Missouri: A case study." Ph.D. dissertation, University of Pittsburgh.

BROWN, F.G., and SAUNDERS, R.J. (1973). "Organization development in Kansas City KCOD and SIGN." In First tango in Boston. Washington: National Training and Development Service Press.

CULBERT, S.A., and REISEL, J. (1971). "Organization development: An applied philosophy for managers of public enterprise." Public Administration Review, 31(March/April): 159-169.

DALTON, G.W. (1970). "Influence and organizational change." Pp. 230-258 in G.W. Dalton et al. (eds.), Organization change and development. Homewood, Ill.: Irwin-Dorsey.

EDDY, W.B. (1974). "The management of change." Pp. 147-159 in S.S. Powers et al. (eds.), Developing the municipal organization. Washington, D.C.: International City Management Association.

EDDY, W.B., and SAUNDERS, R.J. (1972). "Applied behavioral science in urban administrative/political systems." Public Administration Review, 33(January/February): 11-16.

FRENCH, W.L., and BELL, C.H., Jr. (1973). Organization development. Englewood Cliffs, N.J.: Prentice-Hall.

FRIEDLANDER, F., and BROWN, L.D. (1974). "Organization development." Annual Review of Psychology, 25:313-341.

GARDNER, N. (1972). "Action training and research: Something old and something new." Public Administration Review, 34(July/August): 106-114.

GOLEMBIEWSKI, R.T. (1969). "Organization development in public agencies: Perspectives on theory and practice." Public Administration Review, 29(July/August): 367-377.

HEIMOVICS, R. (1975). "Training for organization change." Ph.D. dissertation, University of Kansas.

KAHN, R.L. (1974). "Organization development: Some problems and proposals." Journal of Applied Behavioral Science, 10:485-502.

KIMBERLY, J.R., and NIELSEN, W.R. (1975). "OD and organization performance." Administrative Science Quarterly, 70:191-206.

KUNDE, J. (1973). "Task force management." Nation's Cities, 11(October): 33-36.

LEAVITT, H.J. (1965). "Applied organizational change in industry: Structural, technological and humanistic approaches." Pp. 1144-1170 in J.G. March (ed.), Handbook of organizations, Skokie, Ill.: Rand McNally.

LEVINE, C.H. (1974). Racial conflict and the American mayor. Lexington, Mass.: D.C. Heath.

LIPPITT, R., et al. (1958). Dynamics of planned change. New York: Harcourt, Brace and World.

MANN, F.C. (1961). "Studying and creating change." Pp. 605-615 in W.B. Bennis et al. (eds.), The planning of change. New York: Holt.

MARROW, A.J. (1969). The practical theorist: The life and work of Kurt Lewin. New York: Basic Books.

MASLOW, A. (1971). Farther reaches of human nature. New York: Viking.

MCGREGOR, D. (1960). The human side of enterprise. New York: Viking.

MURPHY, T.P. (1974). Organizational and functional relationships for management and planning, Chesapeake, Virginia. Chesapeake, Va.: Office of the City Manager, March 29, pp. 108.

___ (1975). Pilot units project of the Chesapeake service evaluation system. Chesapeake, Va.: Office of the City Manager, May 9, pp. 62.

___ (1971). Metropolitics and the urban county. Washington, D.C.: National Association of Counties and Washington National Press.

Nation's Cities (1973). "Dayton's Jim Kunde: Part II task force management." October.

PORTER, L.W., and LAWLER, E.E. (1965). "Properties of organization structure in relation to job attitudes and job behavior." Psychological Bulletin, 65:23-51.

Public Administration Review (1976). "Symposium on Management by Objectives in the public sector." 37(January-February): 1-45.

RAINEY, H.G., BACKOFF, R.W., and LEVINE, C.H. (1976). "Comparing public and private organizations." Public Administration Review, 37(March-April):233-244.
ROGERS, C. (1961). On becoming a person. Boston: Houghton Mifflin.
RUBIN, I., PLOVNICK, M., and FRY, R. (1974). "Initiating planned change in health care systems." Journal of Applied Behavioral Science, 10(January-March):107-124.
SCHEIN, E.H. (1969). Process consultation. Reading, Mass.: Addison-Wesley.
WASHNIS, G.J. (1972). Municipal decentralization and neighborhood resources. New York: Praeger.

10

Issues and Problems in Applying Quantitative Analysis to Public Sector Human Resource Management

EUGENE B. McGREGOR, Jr.

□ TWO FACTS make human resources management a required topic in any discussion of urban management. One is the labor intensive nature of urban service delivery. The other is the unionization of the urban work force. The labor intensive nature of local government derives from its function as final deliverer of basic services to citizens. Thus, while state governments devote about 40% of total expenditures to payrolls, local governments spend, on average, 62% of expenditures on payrolls (Bahl et al., 1972:825). In the local government fuctions of "public welfare, police protection, fire protection, sewage, natural resources, housing and urban renewal, water transport and terminals, corrections and general control," the rate of labor intensity grew more than 1% per year from 1962 to 1970. In 1970, "police protection, fire protection, financial administration, general control, education, corrections, and sanitation (other than sewage)" headed the list of labor intensive services with ratios of payroll to direct general expenditures of .93, .90, .83, .80, .78, .77, and .69, respectively (Bahl et al., 1972:828).

The real impact of unions on local government is only now coming to be understood. Both argument (Warner, 1967) and evidence (Bahl et al., 1972:823-825; Stanley and Cooper, 1972:112-135) have been presented to suggest that levels of local government compensation—especially metropolitan government—will rise significantly as a result of union pressure. Thus, personnel compensation will put enormous pressure on the public financial system in terms of both the contentiousness of the budgetary process and the pressure to raise levies and cover personnel costs through bonding authorities.

Labor intensity and unionization combine to make human resource management one of the critical problems of local government administration. The most basic management problem is to find the right labor mix to provide a smooth flow of services while controlling the aggregate costs of the work force. The solution to the problem is found by avoiding feasts and famines of human resources which derive from imbalances between the work to be done, the people available, and the resources to pay the work force.

Implied in such a prescription are three intellectual problems. One is human resources *forecasting*: depicting over time likely configurations of supply and demand for labor. The second is *planning*: formulating and selecting alternative strategies to meet expected human resource needs. The third is *programming*: translating into reality the selected strategy. Table 1 describes the basic variables associated with forecasting, planning, and programming (each of which are discussed in succeeding sections of this paper).

On forecasting all else depends. In large organizations, forecasting is easiest when computerized data can be accessed and brought to bear on two fundamental problems: forecasting the organizational demand for human resources and projecting likely supplies of human resources. It is formulating and selecting alternative ways to equate the demand and supply of labor that constitutes planning and programming.

TABLE 1

HUMAN RESOURCE MANAGEMENT. VARIABLES ASSOCIATED WITH
FORECASTING, PLANNING, AND PROGRAMMING*

Problem	Short Range Time Variables	Long Range Variables
Forecasting demand for human resources	Productivity of labor, operating needs determined by budgets and plans, employment mix	Judgments about environmental and technological change affecting productivity
Forecasting internal supplies of human resources	Separation statistics, career mobility statistics, individual development plans, and stable trends	Judgments about changing character of employees and availability of human resources
Planning	Adjusting external and internal labor supplies to meet needs. Establishing a timetable of needs.	Minimizing likely discrepancies between demand and supply of human resources
Programming**	Implementation of plans through job design, recruitment and placement, manpower audits, appraisals, education/development, and motivation/compensation changes	

*SOURCE: Adapted from Walker (1969:156).
**SOURCE: Burack (1972:72).

For all three problems of forecasting, planning, and programming, time is taken into explicit account. The length of time involved depends upon the certainty with which forecasts can be made and plans formulated, which, in turn, depend upon the stability of the organizational environment and the adequacy of the data base and models to discern reliable trends. A common pattern (Walker, 1969) finds the "long range" horizon set at five years or more and the "short range" defined as some period less than five years.

The purpose of this paper is to discuss several kinds of tools available to public managers concerned with forecasting, planning, and programming. The aim is not to plumb the extent of the mathematical art but to explore the *applicability* of several quantitative modes of analysis to public sector urban systems, to discuss some assumptions which must be made in order to apply such tools, and to clarify several issues raised by the use of such tools. Emphasis is given to modes of analysis actually used or potentially useful in public agencies. For the sake of clarity, examples are used which should be helpful to readers without an elementary preparation in finite mathematics and statistics.

FORECASTING DEMAND

Forecasting public sector human resource needs is done frequently by working through the mechanics of a "manning table" which states the resource requirements of an organization for fixed periods of time. Manning tables vary in form depending on the variables important in producing the figures and the time periods for which needs are to be established. Major variables linked with short-run needs definition include work load or productivity measures, estimates of budgetary expansion (or contraction), and employment mix. In the long term, environmental and technological change must be taken into account. Schematically, Figure 1 shows the theoretical links among the needs-determining variables (Heneman and Seltzer, 1970:15; Burack, 1972:61-76).

Private enterprises typically tie forecasts of needs to production and sales of goods and services. Whether or not public sector enterprises, which do not engage in market transactions and whose outputs are subject to constant redefinition, can apply the same kinds of forecasting technologies is not clear.

The state of the art of needs determination now seems crude. Several obstacles may prevent the use of sophisticated forecasting methodologies. One is an organizational dilemma afflicting both private industry and government agencies: that "plans for acquisitions, mergers, conglomerates, and other organizational developments frequently fall into the category of privileged information" (Heneman and Seltzer, 1970:5). Thus, even if the forecasting tools were available, those responsible for human resource planning may lack the minimal information that they need in order to apply the tools available.

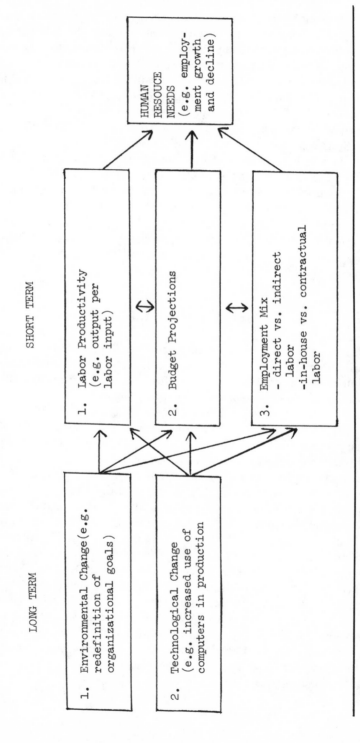

Figure 1. LONG AND SHORT TERM VARIABLES DETERMINING HUMAN RESOURCE NEEDS

Such secrecy may be especially prevalent when major technological or organizational change is being installed. Premature revelation of technological shifts can make basic change difficult. While there are instances recorded in which manpower forecasting and planning have been conducted in a reasonably open manner (Mosher, 1967:171-184), the extent to which sufficient openness can be maintained to use forecasting tools is not known.

A second problem endemic in public organizations is that human resource "needs" are determined by public policy. Defining certain city departments as "needy" can derive as much from political values that allow some functions to grow and others to decline (Thompson, 1975:33-43) as from any alternative, "objective" analysis.

Notwithstanding forecasting constraints, several variables seem immediately relevant to forecasting short-term human resource needs. The most important variable is the budget, in which the ancient public practice of making incremental adjustments in the input "base" is the dominant practice (Wildavsky, 1964). The standard practice finds urban policy makers using manning tables similar to the one found in Appendix A to make an across-the-board percentage increase or decrease in both dollars and the manpower "ceiling" authorized to the agency. The agency finds it expedient to regard the new ceiling as a target to be met in order to make the case for an expanded budget in the next fiscal year.

A hypothetical example of the kind of calculation that public agencies make is to define needs as the maximum number of people that the budget will allow. Thus the objective is to hire a mix of employees who will be paid salaries that the budget can sustain consistent with pay plans and negotiated wage and benefit packages. Thus, if the on-board number of employees is depicted by employment row vector e, and if the average salary that is paid each category in e is shown in column salary vector s, then the human resource budget totals $9.8 million, the product of e and s. The APT, SAM, and COMOT classifications are borrowed from the "Oliver Report" to the U.S. House of Representative's Committee on Post Office and Civil Service (1972, vol. 2).

$$
\begin{array}{cccccc}
 & \text{APT} & \text{SAM} & \text{COMOT} & \text{PF} & \text{LTC} \\
e = (& 200 & 100 & 400 & 100 & 100)
\end{array}
$$

APT = Administrative, professional, and technical employees

SAM = Supervisory and managerial

COMOT = Clerical, office machine operation, and technician

PF = Police and fire

LTC = Labor, trades, and crafts

$15,000	APT
$16,000	SAM

$$s = \quad \$ \ 8{,}000 \qquad \text{COMOT}$$
$$\$13{,}000 \qquad \text{PF}$$
$$\$ \ 7{,}000 \qquad \text{LTC}$$

If the agency budget for the next year is set at a 5% increase, a total of $10.29 million can be allocated in a variety of ways. For example, if there were no increase in the number of employees, then an across-the-board raise of 5% for all classes would produce a new salary vector, s' of:

$$s' = \quad
\begin{array}{ll}
\$15{,}750 & \text{APT} \\
\$16{,}800 & \text{SAM} \\
\$ \ 8{,}400 & \text{COMOT} \\
\$13{,}650 & \text{PF} \\
\$ \ 7{,}350 & \text{LTC}
\end{array}$$

An example of the extreme alternative is to grant no raises but use the increased appropriation to hire additional people. Such a policy results in a new employment vector e' of

$$e' = (210, 105, 420, 105, 105)$$

Clearly there is an infinite variety of ways in which managers can both assign salary raises and hire additional people to spend the budget authorization of $10.29 million.

The question for forecasting needs is whether statements of human resource requirements can have a basis other than mere determination to use up authorized budgets. Several attempts have attempted to break the traditional budget logic of making incremental adjustments in the resource base. Performance budgeting, program budgeting, planning budgeting (Schick, 1966), management-by-objectives systems (Newland, 1974; Brady, 1973), and zero-based budgeting (Pyhrr, 1977) are all strategies for tying budget allocations to statements of agency goals and measured outputs.

More sophisticated organizations link the determination of needs to the contributions which the various types of labor make to organizational productivity. For instance, one survey of private sector practice reports that the most sophisticated organizations divide manpower forecasting into projections for "direct" manpower (i.e. those directly involved in production) and "indirect" manpower (i.e. support personnel). The forecasts of "direct" needs were derived from such variables as production schedules, work loads, and estimates of "lost" time. "Indirect" labor needs were forecast by dividing the group into variable manpower (i.e., those whose work load varied with the size of the work force and the productivity of the plant) and fixed manpower (i.e., those whose work load is neither affected by variations in the workforce not the productivity of the plant). (See Heneman and Seltzer, 1970:17-18.)

While forecasting direct labor needs seems a straightforward process of linking needs to measured output and schedules, forecasting variable, indirect labor needs seems relevant to governmental situations when the "product" of an administrative unit such as a police department, public works department, or planning office is not analogous to factory production and when the contribution that employees make to the "product" is not easy to measure. In such situations one approach is to develop "conversion factors" (Heneman and Seltzer, 1970:19) which represent the estimated amount of time required either for completion of certain actions or for "production" of certain units of output. Thus, the conversion factor $_nC_i$ is the time t_i which n people require to produce N numbers of units of output i or complete N actions i, and it is expressed as

$$_nC_i = \frac{_nt_i \quad \text{(persons days, weeks, or months)}}{_nN_i \quad \text{(output, actions)}}$$

Thus, $_nC_i$ can be calculated either for one person (in which case $_1C_i$ will cover several actions or output units i such that 100% of a person's time is accounted for) or for a group of persons.

When the total projected load L_i for each action or unit of output i can be estimated (by means not to be discussed here) and multiplied times $_nC_i$, then the human resource requirements $_nR_i$, for a single person or a group of persons n, can be expressed as follows for each activity i:

$$_nR_i \text{ (man days, weeks, or months)} = {}_nC_i \text{ (time/output or actions)} \times {}_nL_i \text{ (outputs or actions)}$$

For example, a group of 10 welfare caseworkers may have their time apportioned among three activities: client office interviews and counseling, clerical work (e.g., processing forms), and client home visitation. In Table 2, if an average single caseworker spent 50% of his time on interviews and counseling, 30% on visitations, and 20% on processing forms, then the conversion factors could be calculated if the numbers of actions could be reliably estimated. In this example, we assume that 10 caseworkers handled 2,000 interviews, of which 1,600 resulted in processing forms and 500 involved visitations to the clients home.

If estimates for the following year are 3,000 interviews with proportional increases in visitations and forms processing, then the projected load $_nL_i$ allows us to estimate future resource requirements for each activity i. Adjusting $_{10}R_i$ to

TABLE 2
ESTIMATING CASEWORKER NEEDS IN MAN YEARS (n = 10)

Activity Type $_i i$	Percent of Time $_1 t_i$	Number of Actions/Years $_1 N_i$	Conversion Factor $_1 C_i$	Number of Projected Actions/Years $_n L_i$	Projected Required Person Years (CxLx10) $_n R_i$
Interviews	50	2,000	.00025	3,000	7.5
Processing forms	20	1,600	.000125	2,400	3.0
Visitations	30	500	.0006	750	4.5
Total	100				15.0

a base of 10 person years, the projected need is calculated at 15 man years, or an increase in five person years of effort.

Whether five new caseworkers should be added depends not so much on the simple arithmetic shown above but on the conclusions that analysts draw. For instance, managers might restructure work such that computers would be used to process forms and reduce the clerical load. Or work might be done on contract. Decisions can also be made to reduce the number of interviews subject to visitation from 750 to, say, 500, which would save 1.5 man years of effort. Such possibilities represent only a few strategies for eliminating the discrepancy between manpower needs and supplies.

In some public management situations, it is difficult to decompose people's time into discreet actions and to calculate conversion factors. An alternative strategy is to use surrogate measures of output which can be plotted against the numbers of on-board personnel for specific times. Should analysis reveal that the plots can be reasonably described by some simplifying statement such as regression line or a ratio, conversion factors can be generated easily.

A practical example of this was once demonstrated by a student researching the relationship between manpower ceilings and output measures of a procurement division of a federal agency. The example is transferable to an urban management setting in cases such as administrative staff offices in which the real functions can be only partially or indirectly measured. In this case, the output of a procurement division was defined as the number of procurement actions (i.e., purchases) made per year.

Such a definition of output is a radical simplification of the mission of a government procurement division which must balance the need to maintain an open marketplace with the need of operating divisions for timely deliveries of required and desired goods and services. Furthermore, a procurement "action" is often a highly variable output, embracing large, complicated, competitive procurements and relatively small, "sole source," purchases.

While the data gathered were limited to seven years, a simple analysis of the basic trends provided useful information to resource planners willing to accept the data limitations. Table 3 shows for a series of fiscal years both the numbers of professional personnel on board and the number of procurement actions done in each year (Rosenthal, 1973).

The simplest way to associate work force size with output is with a ratio of actions to work force size. The mean number of actions per person is shown to be 21 per person per year. Clearly, additional analysis (e.g., regression analysis) can be done to determine whether 21 actions per person is an appropriate work load figure from which a conversion figure of .048 (i.e., $1 \div 21$) can be derived. (Inasmuch as this procurement division managed large, complex procurements, a figure of 21 actions per person seems reasonable.) In regression analysis the slope of a "least squares" line, where the X axis denotes actions and the Y axis denotes personnel, becomes the conversion factor.

The measurement of outputs of urban government functions may be a more tractable problem than commonly believed, especially when there is a production process at least partially described by such measures as checks written, cases handled, tons of garbage collected, and acreage mowed. Productivity measurement is an elusive process, however. The federal finding that 67% of the civilian work force has been covered by some kind of productivity measure (Joint Financial Management Improvement Program, 1974) is important not because managers have completely described 67% of agency functions. The significance is that a tool for forecasting personnel needs is available which does not depend on strictly input-oriented budget strategies.

Notwithstanding productivity measurement advances, great problems remain in indexing urban service outputs (Wise and McGregor, 1976:5-13). Not only are there organizational pathologies which develop as a result of measurement programs (Blau, 1955; Ridgway, 1956), but there are also far-reaching policy

TABLE 3
NUMBERS OF PROCUREMENT DIVISION ACTIONS AND PERSONNEL FROM 1967 TO 1973

(1) Year	(2) Person Years	(3) Number of Procurement Actions	(4) Ratio: (3)/(2)
1967	40	790	19.8
1968	29	780	26.9
1969	31	716	23.1
1970	28	659	23.5
1971	32	493	15.4
1972	28	502	17.9
1973	24	475	19.8

mean = 21 actions/person

consequences attached to any set of measures. Police services provide a good example. Elaborate output statistics can be developed to measure the productivity of police departments; the number of tickets issued and crimes "cleared" (i.e., persons arrested and charged with crimes) are ways to measure police output (Hatry, 1972). Yet in communities in which the function of the police is to provide such services as smoothing community relations, offering lectures, organizing community projects, and counseling, an entirely different set of productivity measures becomes appropriate (Wilson, 1968). Thus, the key to successful needs forecasting depends on measures of output consistent with the political objectives of the organization and not on the quantitative manipulations which translate outputs to personnel requirements.

FORECASTING SUPPLIES

Forecasting human resource supplies depends on models of the career mobility of personnel rather than on measures of agency output. At the crudest level of analysis, the most elemental model is the attrition rate—usually expressed as the monthly or annual proportion of employees likely to depart from a given organization.

Single estimates of attrition can be misleading, however, since departures from an organization can vary over time. Unlike arrivals (i.e., hiring), which can be completely determined by the organization, departures (i.e., separations due to reductions in force, death, retirement, firing, or resignation) are best described by a random process. Since the size of any organizational population is a function of only two processes—arrivals and departures—one of which (i.e., departures) is partly beyond the control of management, population size can be defined as a stochastic departure process in which there is some probability that each of a number of possible departure levels will occur and there is certainty that one of them will occur (Katz, 1974:2). Clearly, a "best guess" about attrition needs to be developed with care since there is a likely chance that a single guess maybe wrong. For instance, to observe that in one month 10 people departed from category i initially populated by a work force of 50 would suggest that the probability of departure was .2. If at some later time the same category held 60 persons, and one wanted to guess the number of on-board persons one month later, a good "best guess" would be that 48 persons would remain. In reality, however, the chances are good that the real on-board labor force would, in one month, vary around the most likely estimate of 48. Thus, the best predictions of attrition will take into account the likelihood of ending up with 13 or 11, 14 or 10, or some other range expressing the most likely attrition. This range is important precisely because the departure process is stochastic: each person in an organization must, in the absence of a means of distinguishing among individuals, be regarded as having a probability of .2 of leaving and a probability of .8 of staying; the outcome of these independent "decisions" to depart or stay can be described as though a biased coin were being tossed for each individual as in a series of Bernoulli trials (Katz, 1974:10).

The outcome of the trials can be calculated using the formula:

$$f(n,x;p) = \binom{n}{x} p^x q^{n-x},$$

which gives the probability in an independent trials process of arriving at *exactly* x "successes" in n trials where the probability of leaving is arbitrarily defined as a success p and the probability of staying as a failure is expressed as q (i.e., l-p). (See Kemeny et al., 1957:148.)

Thus the probability of 12 people leaving an organization which originally had 60 people can be calculated as

$$f(60, 12; .2) = \binom{60}{12} (.2)^{12} (.8)^{60-12} = .128.$$

The probability of 13 people leaving (i.e., 47 remaining) is

$$f(60, 13; .2) = \binom{60}{13} (.2)^{13} (.8)^{60-13} = .118.$$

And the probability of 11 people leaving (i.e., 49 remaining) is

$$f(60, 11; .2) = \binom{60}{11} (.2)^{11} (.8)^{60-11} = .125.$$

The probabilities decline from .128 as the prospective number of employees leaving diverges from the "best guess" of 12 during the time period.

The management issue is to decide how to report the results in a useful manner. At least two basic formats are available depending on the managerial problems at hand. One problem common in many inner city urban service organizations is severe budget constraint which forces managers to operate under a "hard" personnel ceiling. In the case of stable or contracting organizations, the public manager's main problem is to maintain a work force which is at or just below the authorized personnel ceiling. Woe unto the manager who exceeds the ceiling!

A different reporting problem is encountered by some managers in smaller and medium-sized cities which have growing economic bases and expanding or stable agencies not beset by hard ceilings so much as by the need to maintain the on-board capacity to deliver services. This latter problem suggests a preoccupation with finding the numbers of people needed to accommodate an expanding demand for service and to replace those lost through attrition. Differently stated, the expanding agency finds its toughest problem is maintaining an adequate number of workers, not exceeding a ceiling, while the contracting organization's concern is to reduce the size of the workforce to a ceiling permitted by the budget.

In each case managers are best served when they have forecasts which advise them about the probability of either reaching a ceiling or confronting certain levels of attrition. Such forecasts leave managers free to impose their own attitudes toward risk on personnel policy decisions. Such forecasts may also encourage managers to consider a variety of programming options—including a revision of ceilings and authorizations—rather than rely solely on one "best guess" which can obscure the full range of options available.

THE CONTRACTING ORGANIZATION

For the manager faced with hard ceilings, the big problem involves assessing the chances of reaching a certain organizational size by a certain time and formulating plans to reach the required level in the event that "natural" attrition does not produce the desired size. One way to report the data is by calculating the effect of the probabilities generated under the Bernoulli trials on organizational size over time. This was done by one agency by incorporating the probabilities into a Markov chain transition matrix the end states of which were organizational size (Katz, 1974). A Markov chain is a stochastic process easily compatible with the attrition process discussed above, in which the probability that organization h will reach size n at time 2 depends only on its size n at time 1 and not on its individual history of past organizational sizes. The theory of Markov chains is explained in other sources (Kemeny, et al., 1957:217-232) and will not be discussed here. What will be explained is the applicability of Bernoulli trials and Markov chains to forecasting attrition in contracting organizations.

TABLE 4
ONE-MONTH TRANSITION MATRIX (M) FOR ORGANIZATION A
(Size Range 4 to 0 Persons)

		Number of Personnel j at time 2				
		4	3	2	1	0
	4	P_{44}	P_{43}	P_{42}	P_{41}	P_{40}
Number of Personnel	3	0	P_{33}	P_{32}	P_{31}	P_{30}
i at time 1	2	0	0	P_{22}	P_{21}	P_{20}
	1	0	0	0	P_{11}	P_{10}
	0	0	0	0	0	P_{00}

Table 4 depicts a Markov transition matrix for organization category h, defined to be a branch, division, directorate, or some other meaningful organizational unit. For the sake of simplicity we will consider only that organization A can vary in size from 0 to a total of 4 persons and that the transition period covers one month.

Each row of the matrix represents the probability P_{ij} of going from initial size i to size j where $0 \leqslant j \leqslant 4$ and P_{ij} lies within the range of $0 \leqslant P_{ij} \leqslant 1$. Because population size i must become some size j in one month.

$$\sum_{j=0}^{4} P_{ij} = 1$$

The factor which determines the size j in this model is the departure process established earlier as a stochastic process for which probabilities can be calculated using the binomial formula for the independent trials process. If we assume that the probability of departure is, like the earlier example, .2 and the probability of remaining is .8, then the binomial probabilities can be calculated for each row. Furthermore, since this matrix describes only a departure and not an arrival process, $P_{ij} = 0$ for all $j > i$ (Katz, 1974: 4).

The probabilities in the matrix are generated for each row. For instance, if the original size of A is 3 and we wish to know the probability of being a different size one month hence, clearly the probability is the same as a coin flipping experiment in which we wish to know the probabilities of one "success" (i.e., one departure) in 3 trials, two "successes," three "successes," and no successes (i.e., the organization remains the same size).

Thus we have:

$$P_{32} = f(3,1;.2) = .384$$
$$P_{31} = f(3,2;.2) = .096$$
$$P_{30} = f(3,3;.2) = .008$$

Since the probability of remaining the same size, P_{33}, is .512,

$$\sum_{J=0}^{3} P_{3j} = 1$$

The reader can satisfy himself that the derived probabilities are the same as respectively calling departure a "failure" and remaining a "success." In such cases:

$$P_{32} = f(3,2;.8) = f(3,1;.2)$$
$$P_{31} = f(3,1;.8) = f(3,2;.2)$$
$$P_{30} = f(3,0;.8) = f(3,3;.2)$$

From these probabilities two kinds of analyses can be done. The first is assessing the odds of ending up with a certain number of "on-board" personnel. In the example cited, the probability of three people in organization A becoming either one or no people in one month is .104, which suggests that a personnel ceiling of 1 can be met only by some kind of direct action (e.g., forced retirement or reduction in force) to meet the ceiling requirement.

A second kind of analysis involves extending the time horizon of the matrix to project what organizational size is likely to be on the basis of which planning and programming occur. Since the transition matrix is "memoryless" (i.e., transitions in month 6 have the same probabilities as in month 1), the transition matrix M can simply be raised to the power required for the forecasts. Thus, a six month-forecast can be derived from M^6 and an annual forecast from M^{12}, where M is a monthly transition matrix. For all matrices a probability density function can be computed for each row which shows the probability of moving from an original size i to some future size j.

Similar calculations can be performed for other organizational units B through Z of different initial sizes and departure probabilities. Aggregating probabilities to produce probability density functions for groups of organizational units over time (e.g., A-G) can be accomplished if the same probability density functions can be fitted to all units being combined (Wonnacott and Wonnacott, 1972: 71-79). A Poisson density function is a convenient distribution to approximate the transition probabilities. It provides the best fit for distributions in which the number of trials n is large (i.e., organization size is large) and the probability p of an event (e.g., departure) is small (Kurtz, 1963: 138-142). With all rows of the transition matrices fitted by a Poisson probability density function, the sums of independent probability density functions result in still another Poisson distributed density function (Katz, 1974: 8). Thus the use of transition matrices of the sort discussed have considerable flexibility and can be used to aggregate forecasts for several organizations of varying size.

One of the most useful reports which can be developed takes the form of a cumulative probability distribution function. For instance, Figure 2 shows a hypothetical Poisson cumulative distribution function after one month, where

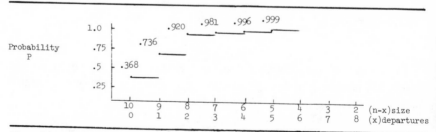

Figure 2. POISSON DISTRIBUTION FUNCTION AFTER ONE MONTH WHERE n=10 AND PROBABILITY OF "SUCCESS" P IS .1

an original organization size of 10 is subjected to a probability of departure of
.1. Even for such a small n and such a large value for p, the Poisson approximation provides a reasonable fit for the outcomes (Kurtz, 1963: 139). As n
becomes large and p becomes small such that m = np remains constant, then

$$\binom{n}{x} p^x (1\text{-}P)^{n\text{-}x}$$

approaches the Poisson approximation

$$\frac{e^{-m} m^x}{x!}$$

which is a vastly simpler computation. Figure 2 tells the manager that a staff of
10 has a .368 probability of remaining at the same size one month later, a .736
probability of becoming *nine or more*, and a .92 probability of remaining at
eight or more. Thus, if he is faced with a 30% reduction in staff in one month,
he can decide how his attitude toward risk will affect his need for a reduction in
force to reach the new ceiling of seven. Since the probabilities are .632 that *one
or more* persons will leave during the month and .264 that *two or more* people
will leave, he may well decide to reduce by one at the beginning of the month
and wait until the end of the month to see how much more he must reduce in
order to reach his ceiling.

For an organization of 10 people, Poisson distributions may not be necessary
to make personnel policy of the sort described. Line managers who know their
people can probably determine better than the probabilities suggest the likelihood of achieving a ceiling through attrition. For a large, complex organization
in which painful cutbacks must be allocated to a variety of divisions and
branches, however, such analysis may be a way to live with budget constraints
without larger than necessary cutbacks.

DEVELOPING TRANSITION MATRICES

There are two practical problems in developing transition matrices for forecasting organizational size. One problem is reducing the number of organizational states (i.e., the range of likely sizes) to a more manageable number and
thereby simplifying the calculations. The second problem is generating the best
departure probabilities for use in the binomial departure "experiments" just
discussed.

Because the problems of multiplying the matrices of exit states for large
organizations can be exceedingly burdensome (i.e., an organization of 1,000
people theoretically has $1,001^2 > 10^6$ discreet probabilities for 1,000,000 end
states), analysts will wish to calculate probabilities only for the most likely end

states. The most likely end states are those which fall above a lower bound below which the calculation of probabilities makes little sense. The lower bound is a function of the departure rate.

One estimate for the lower bound L^h for organizational category h is to use Chebyshev's inequality (Feller, 1968: 233-234). The inequality states the probability of obtaining deviations from a population mean, in this case the mean of monthly or annual departures. One version of the inequality is that the probability of the absolute value of a random variable (i.e., departure rate) falling more than K standard deviations from the mean is less than K^{-2}. Thus, if one desires L^h with a 99% lower bound, 10 standard deviations added to the departures mean, all of which is subtracted from the population size, will be the lower bound above which the population will fall 99 times out of 100.

For example, if the administrative, professional, and technical (APT) class of city department A has a population size of 300 with a mean annual departure rate of 50 and an annual departure standard deviation of 7.0, the lower bound L^A_{99} under Chebyshev's Inequality above which 99% of the possible organization sizes would fall is

$$L^A_{99} = \text{(total population)} - \text{(mean departures)} +$$
$$\text{(K standard deviations)} \times \text{(standard deviation)}$$
$$= 300 - (50 + (10)(7.0)) = 180 \text{ persons}$$

If one assumes a manpower ceiling of 300 persons as the upper bound for a contracting organization, then the forecasting matrix contains only 120 x 120 probabilities describing the end states of the APT subpopulation of department A.

For the purposes of data analysis, managers may wish to make forecasts for periods of less than one year (e.g., monthly departures). In such cases, the mean monthly departure rate is one-twelfth the mean annual departure rate. The relation between the monthly departures standard deviation and the annual departures standard deviation is (Katz, 1974:13)

Annual departures standard deviation = $12^{1/2}$ x
Monthly departures standard deviation

And the quarterly departures standard deviation is

Quarterly departures standard deviation = $3^{1/2}$ x
Monthly departures standard deviation

Other, more precise means of calculating the lower bounds of confidence intervals are available. While Chebyshev's inequality is a convenient tool, its practical utility for fixing confidence intervals must be discounted by its lack of accuracy (Kurtz, 1963: 114; Hays, 1963: 188-191). The inaccuracy of Chebyshev's inequality occurs because the inequality refers to any distribution. If one is willing to assume or can establish empirically that the distribution of departures is both unimodal and symmetrical, then more precise statements can be made about how large the number of departures from any organizational category is likely to be.

For instance, if the monthly departures for a period of two years (i.e., 24 observations) approximate a normal distribution, the 99% confidence interval for the mean number of departures is (Kurtz, 1963: 200)

$$\text{(Mean monthly departures)} \pm Z_{.01} \text{ (Monthly departures standard deviation)}$$

Since we desire the lower bound of organizational category i of initial size n, the maximum likely number of departures should be subtracted from n. Managers of contracting organizations will use a one-tail confidence interval, since they do not care about the fewest departures likely and therefore do not care about the lower confidence interval (although other analyses of attrition might desire both lower and upper bounds). Thus the lower bound $L_{99}h$ for organizational category h is determined as

$$L_{99}{}^h = n_h - (\text{(departure mean)} + Z_{.01} \text{ (departures standard deviation)})$$

where n_h is the population size of category h at some base date.

In the example involving 300 APT personnel in department A, the lower bound is established as follows:

$$L_{99}{}^A = 300 - ((50) + 2.33(7.0)) = 300 - 66.3 = 233.7 \text{ persons.}$$

Rounding partial departures off to an additional departure finds that our expectation of the lower bound of department A will be 233 or higher 99 times out of 100. Table 5 shows the comparison of the Chebyshev and the normal

TABLE 5
ORGANIZATION SIZE LOWER BOUND (99%)

Organization Code	Size N	Mean Departures $\overline{X}$	Departures Standard Deviation 6	Chebychev's at $L_{99}h$	Normal Approx $L_{99}h$
A	300	50	7.0	180	233
B	250	60	5.0	120	178
C	200	40	4.0	120	150
D	100	20	2.0	60	75

approximations of the lower bounds for 4 organizations at the 99% level of confidence. Clearly, the assumption of normality substantially reduces the number of transition matrix end states and thus simplifies calculation.

A final adjustment must be made in calculating the transition probabilities as a result of the lower bound, however. The binomial formula will not yield correct transition probabilities for those matrices for which $L^h > 0$, where the number of monthly departures takes on all values inclusively between L_h and n, the initial population size above which the "contracting" organization will not go. Thus, with $L^h > 0$, the transitional probability is conditional: the future size j of organization h, given the present size i, is premised on the future size being equal or greater than L^h. This is expressed by dividing $f(n,x;p^h)$ by the *sum of all the probabilities* for possible departure values x, where $x = L^h, \ldots, n$. Thus the values of the bounded transition matrix are normalized for the probabilities generated from the lower bound as follows (Katz, 1974:13):

$$f(n, x; p^h, L^h) = \frac{\binom{n}{\bar{x}} (p^h)^x (q)^{n-x}}{\sum_{x=L}^{n} \binom{n}{x} (p^h)^x (q)^{n-x}} = p^h_{n, n-x}$$

For $L^h > 0$, the above equation probabilities satisfy the stochastic requirement that

$$\sum_{x=L}^{n} h \, p^h_{n, n-x} = 1$$

If $L^h = 0$, however, then the usual binomial formula applies without normalizing the probabilities, since the denominator of the above equation = 1.

While the manager of the "contracting" organization seeks ways to reduce the number of cumbersome calculations in order to forecast probabilities for different organization sizes, the manager of an "expanding" organization is most concerned about sustaining unacceptable attrition. His problem is to calculate the odds of sustaining personnel losses which, when added to the personnel needed for expansion, provides the recruitment targets (i.e., people who must be placed in category i through such means as recruitment, promotion, or development) for meeting anticipated needs.

The two managers, thus, view departures in a different way. For the "contracting" organization, the probabilities of departure depend on a "best guess" on which a binomial experiment is based. Such probabilities are most easily generated from historical data as follows (Katz, 1974: 10):

Let D_i^h = number of departures from organizational category h during month i

i = 1, 2---n (months of fiscal year 1, 2. . k)

S_k^h = average monthly organizational category h population during FY k

p^h = probability of departure from organizational category h in any one month

Then the one month departure probabilities are

$$p^h = \sum_{i=1}^{k} D_i^h \bigg/ 12 \sum_{k=1}^{k} S_k^h.$$

For the expanding organization (and even organizations without computer support) the forecasting problem is more tractable. It is to estimate the most likely range of attrition which can be the basis of planning to meet anticipated needs. Estimates can be obtained in the form of confidence intervals of the sort discussed above. For instance, if a manager wanted to know the likely attrition for organization A in Table 5, he might wish to know what departures range would cover 75% of the recorded cases (the example does *not* involve a sampling problem since all the recorded cases are included in the analysis). Thus, if we find that the distribution of departures is approximately normal, the upper and lower limits of the two-tailed confidence interval for departure range r is expressed as:

$$r_{U,L} = \text{departures mean} \pm Z_{.125} (s)$$

In the case of organization A, the manager would find that $r_{U,L} = 50, \pm$ (1.15) (7.0) = (58, 42). Thus we would have a basis for planning to replace 42 to 58 persons who will leave organization A with a probability of .75. Clearly the forecasting problem is easier for managers of expanding organizations than for managers of contracting organizations. The latter group is responsible for making fine-tuned projections on which decisions concerning reduction in force can be made.

FORECASTING SUPPLIES OF SKILLS

Forecasting organizational size and departure rates is only one of the supply forecasting problems that a manager faces. The utility of such analysis is that

TABLE 6
ONE YEAR OCCUPATIONAL TRANSITION MATRIX (M)

		Time One Year Later				
		1	2	3	4	5
	1. SAM II	.9	0	.05	0	.05
Present time	2. SAM I	.1	.8	.05	0	.05
	3. APT II	0	.1	.7	0	.2
	4. APT I	0	0	.2	.6	.2
	5. Exit	0	0	0	0	1

aggregate supplies and costs can be established and compared with aggregate needs and overall budget ceilings. A second problem is to forecast the likely distribution of occupational skills *within* an organizational unit. To deal with this problem, the use of Markov chain analysis has also been well established (Burack, 1972: 144-150; Bartholomew, 1973; Young, 1971; White, 1970; Vroom and MacCrimman, 1968).

The standard format for such analysis involves the use of a transition matrix in which the end states are occupational groupings of the work force. Table 6 illustrates a hypothetical example of a matrix M where annual transition rates are depicted for two classes of supervisory and managerial personnel (SAM) and two classes of administrative, professional, and technical (SPT) personnel. In this example some mobility is contemplated between SAM classes and APT classes, as shown by P_{i3} and P_{23}. There are no demotions in the professional ranks as shown by P_{12} and P_{34}, which have values of 0.

A value for P_{55} of 1 indicates that persons who leave the system do not reenter; they are absorbed by the "exit" state.

If the number of full-time, on-board personnel at the beginning of 1976 is given in employment row vector e, then the number of on-board personnel for 1977 is row vector e' and is calculated as

$$e'_{1977} = e_{1976} \cdot M$$

The number of on-board personnel forecast for 1978 is denoted by e'' and is given by

$$e''_{1978} = e'_{1977} \cdot M = e_{1976} \cdot M^2$$

Personnel supplies can be projected as far into the future as needed by raising to the appropriate power the transition matrix multiplied by the initial row vector. For example, one could project personnel supplies by multiplying an employment row vector such as $e_{1976} = (50\ 200\ 100\ 200\ 0)$ times the transition matrix in Table 6. Vector e represents the fact that, in 1976, 50 persons were in SAM

II, 200 were in SAM I, 100 were in APT II, 200 were in APT I, and no people had yet left. The result of the multiplication is e'_{1977} = (65 170 122 120 72), the sum of which is slightly different from the sum of e_{1976} because of rounding. The results show that 72 people left the system during the year and may have to be replaced in the absence of declining demand for personnel, a constraining budget, or a planned increase in productivity by remaining personnel.

The forecasts are oblivious to the pre-1976 sequences of moves made by the individuals in e. All that is known are the aggregate probabilities of shifts between pairs of categories based on review of historical data. This, the SAM II category grows from 50 to 65 because 45 people (50 x .9) remained in the same category from 1976 to 1977 to which were added 20 persons (200 x .1) who were in SAM I in 1976 and promoted by SAM II by 1977. The numbers for other categories are derived in the same manner.

If the analyst believes that the transition matrix is a stable representation of movement among personnel categories, he may wish to project likely supplies further than one year. Thus, e''_{1978} = (76 148 122 72 133) where the elements in the set have been rounded to the nearest whole person. The results for 1978 are clearly that the organization will be even more shorthanded than in 1977, with 133 of an initial 550-person work force expected to leave by 1978. Should new recruits be added to the organization in 1977, the elements of employment vector e'' change. For instance, if the incoming "class of 1977" consisted of 72 people (15 SAMs and 57 APTs), then an additional employment row vector might be e^R_{1977} = (0 15 0 57 0), which shows that personnel recuited in 1977 were brought in at the lower grades from which they could be promoted. Applying the transition probabilities to e^R_{1977} produces an e^R_{1978}- = (2 12 12 34 12)which must be added to e''_{1978} to produce a total 1978 work force of (78 160 134 106 145).

PLANNING

Developing alternative strategies to meet personnel needs requires an appreciation of the alternative interpretations which can be given the forecasts from which plans are made. Transition matrices, for example, can be regarded in two ways: as *behavioral predictions* of likely mobility patterns or as *policy statements* describing mobility which must occur for an organization to be healthy and productive. Manpower planning varies with each interpretation.

In the behavioral view, the probabilities may be regarded either as precise statements of known and stable patterns of movement or merely as crude approximations of constantly shifting patterns of mobility. The former case would seem a priori to obtain in organizations in stable environments, using stable technologies, which are expanding or contracting at a slow to moderate

TABLE 7
HYPOTHETICAL DISTRIBUTION OF PERSONNEL SUPPLIES (e), PERSONNEL REQUIREMENTS (r), AND RECRUITMENT PLANS (p) FOR 1976, 1977, AND 1978

	(1)	(2)	(3)	(4)	(5)	(6)	(7)	(8)	(9)	(10)	(11)	(12)	(13)
Personnel Classes	1976 e	1976 r	1976 Discrepancies $(e-r)$	1977 e'	1977 r'	1977 Discrepancies $(e'-r')$	1977 Recruitment Plan p	1977 Discrepancies $(e'+p)-r$	1978 e''	1978 r''	1978 Discrepancies $(e''-r'')$	1978 Recruitment Plan p'	1978 Discrepancies $(e''+p')-r''$
SAM II	50	50	0	65	50	+15	0	+15	78	50	+28	0	+28
SAM I	200	200	0	170	200	-30	+15	-15	160	200	-40	+12	-28
APT II	100	100	0	122	100	+22	0	+22	134	100	+34	0	+34
APT I	200	200	0	120	200	-80	+58	-22	106	200	-94	+60	-34
Departure	0	---	---	72	---	---	---	---	145	---	---	---	---
TOTALS*	550	550	0	549	550	-73	+73	0	623	550	-72	+72	0

*Totals for e may not agree from year to year because of rounding.

rate. The latter case would seem to occur in more turbulent environments, during times of substantial technological change, and in rapidly expanding or contracting organizations.

The former "behavioral" case yields the most reliable predictions on which the most precise plans can be based. Both forecasts of needs and forecasts of supplies can be arrayed as in Table 7 with supplies based on the transition matrix M and requirements based on the hypothetical assumption that personnel needs remain unchanged from 1976 to 1978. Table 7 provides a format for relating the constant demand for labor r to the supplies of labor e. The discrepancies between e and r can be calculated and plans established for eliminating the discrepancies. The recruitment plans of Table 7 are based on the assumption that e and r should be equated. Thus, it would seem that programming is simply a matter of developing the recruitment program to supply 73 new people in 1977 and 72 people in 1978.

The planning and programming problem is not necessarily solved, however. Table 7 also projects an increasing surplus of high level managers (i.e., SAM II) and top-ranked APT personnel. From a "behavioral" perspective, such a pattern can be regarded as a natural and predictable phenomenon which may be incorporated as part of the manpower plan. Given a sufficient supply of salary money, the rising average grade in SAM and APT can simply be incorporated into budgets and personnel policy. This might be the case where APT II and SAM II not only cover executives and supervisors but denote a distinction between an "apprentice level" (i.e., SAM I and APT I) and a "journeyman level" (i.e., SAM II and APT II). Thus, the "excess" of SAM IIs and APT IIs could be viewed as a healthy development: more than minimally adequate supplies of personnel are available.

An entirely different view of Table 7 is obtained when the transition matrix becomes a statement of policy. The issue is not to discover so much what the "natural" pattern of movement is but to decide what mobility patterns are desirable and use the matrix to project the "quotas" which can be derived. An example is probability P_{21} from Table 6 which states that 10% of SAM I personnel will move up to SAM II in one year. Another example is P_{15} which can be read to mean that 5% of top ranked SAM IIs will be retired in each year. Should the matrix forecast unacceptable numbers of higher ranking and higher paid SAM and APT personnel, P_{15} and P_{35} can be raised with "outplacement" programs designed to achieve the desired probabilities. Such a strategy has been used in "up or out" personnel systems such as those in the military, foreign service, and universities which rely on a continued circulation of personnel for rejuvenation and to provide career opportunities for new recruits.

Still a third view of the same matrix recognizes both the ephemeral nature of the transition probabilities—especially in turbulent environments and changing organizations (Hoagland and Pazer, 1976)—and the planning problems of tying future recruitment efforts to past mobility patterns. In this view, forecasting is used by line managers for contingency planning. Recent discussions even suggest the use of computer simulations and mathematical programming to derive optimal transition probabilities (Glynn, 1976). In this kind of "what if?" planning mode, the real aim of the analysis is to project the results which will occur *if* the probabilities are either behaviorally correct or politically desirable and to explore alternative sets of probabilities and assumptions on which plans and programs can be based.

PROGRAMMING

Whether optimizing techniques will be widely used to formulate human resource programming by urban managers and to make real decisions about hiring, firing, career development, and other personnel actions is difficult to determine. Substantial work has been done in the area of goal programming applied to multilevel models (Charnes et al., 1972) in which resource planning and career planning are linked. Recent discussions focus on simulations of manpower planning in which optimal ways to establish service worker shift schedules are established (Raedels, 1976). The prospects seem good that, as cities acquire more analytic capacity, efforts will be made to find optimal solutions to manpower planning and programming problems.

The search for more rational management of the urban work force will force an integration of the advancing state of the quantitative art with the administrative analysis on which all models, forecasts, plans, and programs depend. Clearly one problem requiring immediate attention is that most forecasts now assume that people move through a static system of unchanging job classifications (Burack, 1972:144; Charnes et al., 1972, p. III-8). Yet the realities of urban service delivery suggest that changes in organizational function and in the design and classification of jobs do occur. For instance, to the extent that police service delivery can be organized according to either "team policing" or the traditional "beat" (Schwartz et al., 1975a) or can experimentally include different mixes of sworn and nonsworn personnel in administrative positions (Schwartz et al., 1975b) or can provide a widely varying range of services (Wilson, 1968), the categories used for manpower analysis ought to reflect the different organization designs. Similarly, when a city government or hospital facility takes on new functions such as the establishment of community mental health centers (Foley, 1975) or when health departments conduct environmental quality monitoring in addition to the usual health inspections or when administrators use computers to replace clerical workers, substantial shifts of occupational classification can be involved.

With the shifting structure of urban service delivery must occur the revision of personnel classification systems used for human resource forecasting, planning, and programming. Position classification in government has traditionally been used both as a means of guarding the merit system and of conducting job evaluation (McGregor, 1977). The classification tradition has established an elaborate technology designed to make precise distinctions among jobs (U.S. House of Representatives, 1972). Such distinctions are often too refined for forecasting and planning problems where aggregate numbers of people and salary dollars must be computed. Manpower analysis of the sort discussed in this paper must seek ways to collapse categories in order to derive statistically useful units of organizational analysis (Treires, 1971; Charnes et al., 1972:142). To have classification systems serve several administrative purposes suggests that considerable effort be given multilevel classification systems which can be aggregated and decomposed to serve the purposes of merit protection, job analysis, job evaluation, forecasting, planning, and programming.

The ultimate paradox is that human resources programming will probably render useless the forecasts on which the plans and programs are based. For instance, to project a shortfall in the numbers of APT and SAM personnel in Table 7 (columns 6 and 7) might well suggest the need to reorganize the structure of work for those persons who remain rather than automatically to recruit as indicated in columns 7 and 12. Such reorganization might well involve a new configuration of capital intensive production, contracting for services, and a redesign of remaining jobs. Clearly the implementation of such a program of organizational change can require enormous managerial skill of the sort not discussed in this paper but which is as important to human resource management as the quantitative concepts involved. This design, in turn, will require a reanalysis of personnel mobility through new organizational categories, a reassessment of output measures, and new forecasts and plans. Thus, the most useful tools of quantitative analysis in urban human resource management would appear to be those most adaptable to the constantly changing realities of urban service delivery.

APPENDIX A

MANNING TABLE FORM
Estimated Requirements, Fiscal Year 1 Through FY 3
Salaried Employees Only
Department___

	FY 1	FY 2	FY 3	Total FY 1 Through FY 3
I. Additional Hiring A. Supervisory and Managerial(SAM) B. Administrative, Professional and Technical(APT) C. Clerical, Office Machine, or Technician(COMOT) D. Police and Fire(PF) E. Labor, Trades and Crafts(LTC) Totals				
II. Turnover - Replacement Hiring A. SAM B. APT C. COMOT D. PF E. LTC Totals				
III. Total Hiring Effort for Department___ A. SAM B. APT C. COMOT D. PF E. LTC Totals				
IV. Total Salaried Employment for Department___ A. SAM B. APT C. COMOT D. PF E. LTC Totals				

REFERENCES

BAHL, R.W., GREYTAK, D., CAMPBELL, A.K., and WASYLENKO, M.J. (1972). "Intergovernmental and functional aspects of public employment trends in the United States." Public Administration Review, (November/December):815-832.

BARTHOLOMEW, D.J. (1973). Stochastic models for social processes (2nd ed.). New York: John Wiley.

BLAU, P.M. (1955). The dynamics of bureaucracy. Chicago: University of Chicago Press.

BRADY, R.H. (1973). "MBO goes to work in the public sector." Harvard Business Review, (March/April):65-74.

BURACK, E.H. (1972). Strategies for manpower planning and programming. Morristown, N.J.: General Learning Press.

CHARNES, A., COOPER, W.W., and NIEHAUS, R.J. (1972). Studies in manpower planning. Washington, D.C.: Office of Civilian Manpower Management, U.S. Department of the Navy.

FELLER, W. (1968). An introduction to probability theory and its applications (3rd ed.). New York: John Wiley.

FOLEY, H.A. (1975). Community mental health legislation: The formative process. Lexington, Mass.: Lexington.

GLYNN, J.G. (1976). "A simulation of the manpower process for a high talent management firm." Paper presented at the eighth annual conference of the American Institute for Decision Sciences, San Francisco, November 10-12.

HATRY, H.P. (1972). "Issues in productivity measurement for local governments." Public Administration Review, (November/December):776-784.

HAYS, W.L. (1963). Statistics for psychologists. New York: Holt, Rinehart and Winston.

HENEMAN, H.G., JR., and SELTZER, G. (1970). Employer manpower planning and forecasting. U.S. Department of Labor, Manpower research monograph no. 19.

HOAGLAND, J.S., and PAZER, H.L. (1976). "Manpower strategy in a pulsating environment." Paper presented at the eighth annual conference of the American Institute for Decision Sciences, San Francisco, November 10-12.

Joint Financial Management Improvement Program (1974). Report on federal productivity: Productivity trends FY 1967-1973 (Vol. 1). Washington, D.C.: U.S. General Accounting Office, Joint Financial Management Improvement Program.

KATZ, L.G. (1974). "A Markovian model for assessment of personnel hiring plans" (NASA Technical Note TN D-7640). Washington, D.C.: National Aeronautics and Space Administration.

KEMENY, J.G., SNELL, J.L., and THOMPSON, G.L. (1957). Introduction to finite mathematics. Englewood Cliffs, N.J.: Prentice-Hall.

KURTZ, T.E. (1963). Basic statistics. Englewood Cliffs, N.J.: Prentice-Hall.

McGREGOR, E.B., JR., (1977). "Future Challenges for Public Personnel Administration." In H. G. Frederickson and C. R. Wise (eds.), Public Administration and Public Policy. Lexington, Mass.: Lexington-Heath.

MOSHER, F.C. (1967). Governmental reorganizations: Cases and commentary. Indianapolis, Ind.: Bobbs-Merrill.

NEWLAND, C.A. (ed., 1974). "Management by Objectives in the federal government," Bureaucrat, (winter):349-426.

PYHRR, P.A. (1977). "The zero-base approach to government budgeting." Public Administration Review, (January/February):1-8.

RAEDELS, A.R. (1976). "Teller scheduling under uncertain customer arrivals." Paper presented at the eighth annual conference of the American Institute of Decision Sciences, San Francisco, November 10-12.

RIDGWAY, V.F. (1956). "Dysfunctional consequences of performance measurements." Administrative Science Quarterly, (September):240-247.

ROSENTHAL, E.A. (1973). "Productivity in procurement: Measures and results." Greenbelt, Md.: Goddard Space Flight Center, Summer Institute in Public Administration.

SCHICK, A. (1966). "The road to PPB: The stages of budget reform." Public Administration Review, (December):243-258.

SCHWARTZ, A.I., CLARREN, S.N., FISCHGRUND, T., HOLLINS, E.F., and NALLEY, P.G. (1975a). "Evaluation of Cincinnati's Community Sector Team Policing Program: A progress report" (Working paper 3006-18). Washington, D.C.: Urban Institute.

SCHWARTZ, A.I., VAUGH, A.M., WALLER, J.D., and WHOLEY, J.S. (1975b). Employing civilians for police work (paper 5012-03-1). Washington, D.C.: Urban Institute.

STANLEY, D.T., and COOPER, C.L. (1972). Managing local government under union pressure. Washington, D.C.: Brookings Institution.

THOMPSON, F.J. (1975). Personnel policy in the city. Berkeley: University of California Press.

TREIRES, J. (1971). "Counting heads in the federal service." Personnel Administration, (November-December).

U.S. House of Representatives. Committee on Post Office and Civil Service (1972). Report of the Job Evaluation and Pay Review Task Force to the United States Civil Service Commission. Washington, D.C.: Author.

VETTER, E.W. (1967). Manpower planning for high talent personnel. Ann Arbor, Mich.: Bureau of Industrial Relations.

VROOM, V.H., and MACCRIMMAN, K.R. (1968). "Toward a stochastic model of managerial careers." Administrative Science Quarterly, (June):26-46.

WALKER, J.W. (1969). "Forecasting manpower needs." Harvard Business Review, (March/April):152-164.

WARNER, K.O. (1967). "Financial implications of employee bargaining in the public service." Municipal Finance, 40(August):34-39.

WHITE, H.C. (1970). Chains of opportunity: System models of mobility in organizations. Cambridge, Mass.: Harvard University Press.

WILDAVSKY, A. (1964). The politics of the budgetary process. Boston: Little, Brown.

WILSON, J.Q., (1968). Varieties of police behavior. Cambridge, Mass.: Harvard University Press.

WISE, C.R., and MCGREGOR, E.B., JR. (1976). "Government productivity and program evaluation issues." Public Productivity Review, 1(3):5-19.

WONNACOTT, T.H., and WONNACOTT, R.J. (1972). Introductory statistics (2nd ed.). New York: John Wiley.

YOUNG, A. (1971). "Demographic and ecological models for manpower planning." Pp. 75-97 in D.J. Bartholomew and B.R. Morris (eds.), Aspects of manpower planning. London: English Universities Press.

11

Productivity and the
Measurement of Public Output

ROY W. BAHL
JESSE BURKHEAD

☐ THE STATE-LOCAL "FISCAL CRISIS" of 1975-1976 produced an unusual set of political and economic responses on the part of elected officials, politicians campaigning for office, citizen-taxpayers, and the media. Rhetoric was oversupplied; analysis was undersupplied. The result was an almost single-minded popular focus on the management problems of states and cities.

Such an emphasis vastly oversimplifies the problem and is not likely to result in appropriate remedial policy. On the other hand, the realities that face the 95th Congress, including militant public employee union demands, suggest that the heightened concern about the efficiency and productivity of government employees and the effectiveness of many government programs—federal, state, and local—is most appropriate. Citizen-taxpayers are certainly entitled to ask whether they are receiving a reasonable return on their tax dollar and whether public output is produced with increased efficiency. Unfortunately, these questions are easier to ask than to answer.

This essay is an overview of the problems and possibilities in measuring municipal output, with special emphasis on productivity. The following section

AUTHORS' NOTE: *This paper grows out of a larger Public Employment Project being carried out in the Metropolitan Studies Program, The Maxwell School, Syracuse University. The authors are indebted to the Ford Foundation for support of this project, although the foundation is not, of course, responsible for the views set forth here.*

[253]

explores the nature of public productivity and management issues in the setting of the current fiscal crisis. Attention is then directed to an examination of contemporary fiscal theory in an attempt to identify the conceptual issues that are relevant to public output and its measurement. The third section sets forth a framework for systematic thinking about output measurement. The final section reports on the efforts that economists and other social scientists have employed in their examination of the determinants of public sector output.

PRODUCTIVITY AND FINANCIAL MANAGEMENT

More than any other single factor, the widely publicized plight of New York City generated the impression that fiscal mismanagement is the root cause of big city financial difficulties. Credence was given to that position by President Gerald Ford's assertion that the city would have to learn to live within its means, a statement which implied a necessary change in the level, quality, and method of local public service delivery.

An alternative view of the New York City crisis was downplayed—by the media as well as the politicians. This view would hold that while fiscal misman- agement accentuated the problem, the root cause was the far-ranging set of social services which the city attempted to deliver to an enormous indigent population, complicated in recent years by inflation, recession, and the secular erosion of the city's economic base.[1] This view would lay New York City's problems on the absence of federal health and welfare reforms and on those policies which led to the sad state of the economy in 1973-1975.

Regardless of the correct assessment of New York's problems and those pending in other large cities, attention has continued to focus on the expendi- ture side of the fiscal equation, accompanied by rhetoric on the issues of efficiency and productivity. The key policy variable seems to have become expenditure control.

But expenditure control has come to mean many things—more work per- formed for the same compensation, the avoidance of long-term borrowing to meet short-term needs, cutting down an alleged oversupply of public services, increased use of modern capital equipment, a reduction in pensions and benefits, and the elimination of ordinary kinds of graft and corruption (Committee for Economic Development, 1976). All of these have and will be examined in the hope of reducing expenditure without lowering service levels. The best of studies seem to have had little impact (National Commission on Productivity and Work Quality, 1975). The less careful ones have generated a great deal of hortatory advice. In the meantime the concerned efforts to improve municipal manage- ment and enhance productivity, such as New York City-RAND, have been terminated by the current fiscal crisis.

The response of large-city mayors to the productivity concern has been very mixed. Some mayors, as in New York City, postponed confrontation with the financial crisis. Others, as in Detroit, declared a crisis and undertook immediate retrenchment. In both cases, the inevitable, eventual result is a cutback in public service levels. Politicians campaigning for office in 1976 frequently chose "big government" as the enemy, and the public servant, whether on the federal, state, or local government payroll, was depicted as lazy and unproductive, with unions that were too strong and pensions that were excessive. The media highlighted the crisis and overlooked the urban areas that were encountering little or no fiscal difficulty.

In the midst of this confusion it would seem useful to look once more at the general issue of public sector productivity, the reasons why so little is known about it and the roundabout approaches that must be taken to its measurement. The state of the theory is examined first.

FISCAL THEORY

The very great interest in the study of the public economy in the last 30 years is the obvious product of the growth of the public sector in the developed and less developed countries alike. The traditional field of public finance, which once emphasized taxation, with occasional reference to public debt and expenditures, has been greatly expanded in scope and with changed emphases. In particular, the analysis of public expenditures—their optimum level and composition—and techniques for judging the appropriateness of specific public outlay, such as cost-benefit analysis, are now a very significant part of the study of the public economy.

All of this has, of course, been accompanied by a good deal of fiscal theorizing, by both academics and practitioners. The purpose of fiscal theory or public sector theory varies greatly. For some, this is an effort to specify equilibrium conditions in the traditional framework of neoclassical economies in which the objective function is the maximization of consumer welfare. Other fiscal theorists are interested in the decision process—how, in the real world, decisions are made about the size and composition of the public sector. Others are interested in fiscal institutions, such as voting schemes in legislatures or intergovernmental relations and their impact on fiscal outcomes. These three approaches are usually described as "neoclassical", "incrementalist", and "public choice" (Burkhead and Miner, 1971:145-173).

For present purposes it is the neoclassical approach that provides the most useful insights into the nature of public output and the difficulties in its measurement. This "school" of fiscal theory, associated primarily with the work of Paul Samuelson and Richard Musgrave, emphasizes the peculiar nature of the demand for public goods (Samuelson, 1954, 1955; Musgrave, 1959:3-135). Such demands are characterized by nonexclusion and nonrivalry. This means that, in the so-called polar cases, such as national defense, no one is excluded from the benefits of the expenditure regardless of his or her willingness to pay for the particular public good. The national defense establishment that provides protection to A also provides protection to B. There are no price tags attached to units of output, and, indeed, none can be. The concept of nonrivalry, roughly equivalent to joint consumption, means that A and B will both consume (or enjoy) the same supply of public goods. The concepts of nonexclusion and nonrivalry, although they are conventionally defined in very much the same terms, are nevertheless distinct. It is possible to charge a fee for some public goods provision, as for a park, and thus practice exclusion. But once in the park the recreation services are provided on a nonrivalry basis.

Almost all public goods, except where they are provided on a fee or charge basis, share these characteristics to a greater or lesser degree.[2] The police department that protects one part of the city will protect, to a greater or lesser degree, other parts of the city, as will the fire department and the public health department. In short, there are no discrete units of public output. Therefore, public output cannot be measured in the same way that the output of steel, cement, or automobiles can be measured.

The foregoing is all familiar to first-year students of public economy, but the implications of the peculiar demand characteristics for public goods are by no means well understood, particularly by those who attempt program budgeting or productivity analysis. There seems to be an unerring tendency on the part of some academics and some practitioners to utilize private sector analogs for the public sector, even when they are inapplicable. It may, therefore, be useful to contrast some of the economic characteristics of the two sectors.

Some private sector outputs are highly susceptible to measurement. A barrel of cement of unvarying quality over time permits the measurement and pricing of discrete units of output. It is possible, thus, to associate a quantity of output with a production function showing the labor and capital employed for the output of quantities of cement. It is possible to measure productivity over time in terms of an input-output relationship. There are other commodities typically produced and priced in the private sector for which similar computations are possible—tons of steel, ingots of aluminum, barrels of petroleum.

In other cases of commodity production in the market sector such computations are more difficult because quality is not invariant over time. A 1977 model compact differs from a 1976 model; it may be "better" or "worse," but it is different, and therefore output and productivity measures are more suspect.

In the private service sector, the problems of output measurement begin to approach those in the public sector. The provision of recreation services, financial services, or personal services such as these provided by barbers and hairdressers perish in the act of performance; their quality varies over time and from establishment to establishment. But at least, unlike the public sector, there are some price tags attached to specific units of service. Some measures of private sector service output and even changes in productivity are thus possible, although they are admittedly crude.

In short, there is a spectrum of "measurability" in the private sector with some components susceptible to quantification and others much less susceptible. But in the public sector there is very little that is amenable to output measurement, and such measurements may be wholly misleading in terms of answering the question of whether the citizen is getting more or less for a tax dollar.

It is not without interest that some noneconomists who have analyzed organizational objectives in the context of organizational theory and behavior have arrived at a similar conclusion by a very different route. Peter Drucker, in a recent article (1976), wrote:

> *Objectives* in public service agencies are ambiguous, ambivalent, and multiple. [italics in original; p. 13]
>
> When the police department tries to make operational the vague term "maintenance of law and order" it will immediately find there is a multiplicity of possible objectives—each of them ambiguous. "Prevention of crime" sounds very specific. But what does it really mean, assuming that anyone knows how to do it? Is it, as many police departments have traditionally asserted, the enforcement of all the laws on the statute book? Or, is it the protection of the innocent law abiding citizen, with respect both to his person and his property? Is it safety on the streets or safety in the home, or is it both? [p. 14]
>
> In attempting to reduce pious intentions to genuine objectives, the administrator will invariably find that equally valid objectives are mutually incompatible or, at least, quite inconsistent. [p. 14]

Drucker's last point, which has been noted by other students of organizational behavior, is significant (Gross, 1964:167-537). Almost all public sector activities are characterized by multiple and conflicting outputs or objectives. Should the high school mathematics teacher, whose classroom time is limited, attempt to teach the "average" student, raise the lowest quartile somewhat closer to "average" performance, or concentrate attention on the most able students? If a teacher's output or productivity is to be judged on the basis of an improvement in students' mathematical ability, what improvements are to be counted? Similar examples could be elaborated in almost every area of government service delivery.

Finally, there is the troublesome matter of interdependencies, or reciprocal externalities. The police department's effectiveness depends on the condition of the streets, which is determined by the effectiveness of the street maintenance department. The fire department's ability to respond to alarms depends on the sanitation department's effectiveness in keeping the streets clean of trash and garbage. Similar examples of interdependencies that will affect productivity measurement at the municipal level could be multiplied almost indefinitely.

It must be concluded that for the public sector any attempt to establish a simplistic relationship between inputs and outputs is conceptually flawed. And this, of course, is why the strenuous efforts of recent years to apply systems analysis to governmental activities have encountered such difficulties and why so many have been tried and abandoned. This is the case with Planning-Programming-Budgeting (PPB) and with many of the efforts to measure government productivity. This may also be the fate of Management By Objective (MBO).[3]

A CONCEPTUAL FRAMEWORK

In 1969 Bradford, Malt, and Oates introduced an important distinction between outputs and consequences in public sector service delivery. They pointed out that it is possible to conceptualize a vector of inputs in the production of public goods (I). This maps through a production function into a vector (D) of directly produced outputs. In police service, for example, the components of I would consist of patrolmen, police cars, communications equipment, uniforms, and the like. The D vector would consist of numbers of city blocks patrolled, numbers of intersections provided with traffic control, etc. Bradford, Malt, and Oates pointed out, however, that the D vector is not what the citizen is primarily interested in. He is interested in the consequences of D, which they described as the C vector. The citizen, then, is concerned with the degree of safety from criminal activity or with the smooth flow of traffic through the intersection. The local government purchases I to produce D, but citizen is interested in C (Ross and Burkhead, 1974:51-54).

The Bradford-Malt-Oates framework can be expanded to embrace some additional vectors. This expansion will, hopefully, indicate with somewhat more precision the potentialities and limitations in a systems analysis of local government service delivery. For these purposes a municipal department of sanitation will serve as an appropriate example. Waste disposal will be assumed to be conducted by a separate department.

The sanitation department will be conceptualized as a "system" consisting of five vectors:

1. Environment (E)
2. Input (I)
3. Activities (A)
4. Output (O)
5. Consequences (C)

Each of these, and their interrelationships, will be examined.

The environmental vector is intended to describe neighborhood conditions that generate the necessity for garbage and trash removal. These are the "needs" of the citizenry and will be conditioned by the character of the neighborhood—

family income and tradition, the type of dwelling, and the types of commercial or industrial activity that occur within the neighborhood. Neighborhoods within a city could be indexed with respect to the tonnage and type of garbage and trash that they generate.

The input vector consists of the labor services and capital equipment that are purchased by the sanitation department. Presumably the department manager has some concept of appropriate factor combinations and some understanding of available technology. And, presumably, the input vector, even as in the private sector, reflects least-cost combinations of capital and labor.

Since most municipal services are labor intensive, the department manager must necessarily be concerned with the manpower capacity that he can purchase as inputs. This extends to an examination of the skills and quality of his labor force, which in turn will structure the kinds of deployment or activities that he can undertake (see below).

As in the private sector, the ability of the manager to effect a least-cost combination (equate the marginal productivities of labor and capital) is structured by budget constraints, by custom and tradition, and possibly by union rules and regulations.

The activities vector is similarly rather close to the private sector analog. The manager will deploy the sanitation crews on an efficient routing pattern with a knowledge of the environmental conditions that will be encountered and the type of garbage or trash that will be picked up. However, the manager (or the mayor or the city council or the voter) must make a crucial decision—how much in private costs shall be imposed to conduct the public function? Shall garbage and trash be picked up at curbside or at the back door? Must commercial and industrial firms dispose of their own trash, or is this a public responsibility? If it is a public responsibility, must garbage and trash be compacted or baled? The manager's deployment of resources will thus depend on decisions that are made as to the quantity and quality of sanitation service to be provided. These kinds of decisions about the activities vector, will, of course, determine the kinds of inputs that the manager will purchase (Greytak et al., 1975:3-18).

Apart from the conventional problems in the attribution of overhead costs to inputs or activities, as defined here, the measurement possibilities for both the input vector and the activities vector should be stressed. A conventional line-item budget will, of course, provide adequate information on input cases. Activities can be measured with or without dollar signs attached. In sanitation it is possible to measure deployment patterns in terms of numbers of blocks and frequency of garbage and trash collection or in terms of numbers of crews on the street or numbers of hours for sanitation runs. The transition from the activity vector to the output vector does, however, require that dollar costs be computed.

The output vector, given the decisions that have been reflected in the activities vector, is now relatively straightforward and measurable in dollar terms. It consists of carts of tons of garbage collected and tons of trash collected. Useful ratios can be computed from the output vector that will give some insight into productivity. These could include costs of tons per crew, tons per day, or tons per man. Such ratios do not resolve all measurement problems in this vector, since the composition of the trash or garbage may change over time, or the garbage may be compacted or noncompacted, but, nonetheless, some measures of output and productivity are possible at this point.

The final vector—the consequences vector—is not a scalar multiple of the output vector. There may be an improvement over time in the output vector measured in terms of technological efficiency, but the streets may not be cleaner than they were before. The environmental vector—the starting point—may intervene to generate a larger volume of trash and garbage. The consequence may be a larger pickup and dirtier streets. The productivity of the sanitation service may have improved but its effectiveness may have declined (Hatry and Fisk, 1971).

The five-vector systems analysis approach to local government services delivery may at least clarify the complex patterns of interrelationship. But even these vectors do not embrace the full range of considerations that should be taken into account in judging the adequacy of local government services. The ultimate objective function—that which is to be maximized—should be described as a social state: are citizens better or worse off as a result of a particular government service delivery?

A social states analysis would require two additional considerations. The first of these, which is important for every local government service, is the spatial distribution of that service. The distribution of real income in a municipality is not affected alone by governmental actions with respect to taxes and transfers. It is also affected by service delivery. One neighborhood may have better sanitation service than another, or better policing or better fire protection. Neighborhood welfare is higher or lower in consequence.

A second element that should be embraced in an analysis of changes in social states is a noneconomic value—neighborhood participation. In many communities there are neighborhood associations that have an informed interest in the quantity and quality of local government services delivery and assume responsibility for participating in decisions that affect the quality of their environment. Neighborhood participation in the decision process, for some governmental programs, may be as important as the service provision itself.

Sanitation, the example chosen here for illustrative purposes, stands relatively high on the measurability spectrum for both activities and outputs. Water supply stands even higher, since there are relatively few output characteristics that are important—pressure, taste, smell, and purity—and evaluation standards can be established. Fire protection becomes more complicated, since a municipal fire department may or may not emphasize inspections and may be more concerned with the protection of property than with the prevention of deaths (Wallace, 1975).

Police protection is probably at the low end of the measurability spectrum. The final output is a social state measured perceptually by citizen attitudes as to whether streets are safe. But policing involves many activities including but not limited to patrols, traffic regulation, apprehension of criminals, and appearances at judicial proceedings. The productivity of the police force is heavily dependent, in any city, on the general effectiveness of the criminal justice system and is, of course, influenced by environmental factors.

A different set of output measurement problems is encountered in any attempt to systems-analyze central staff functions. There are no useful measures extant of the productivity of a personnel office, a budget office, or a purchasing department.

The foregoing conceptual framework is not intended to overcomplicate the reality of local government services delivery. It is intended to demonstrate its complexity and to illustrate that those things that can be measured—usually in the activity vector or in the output vector—may be very far from that which *should* be measured but which cannot—the social state of the citizenry.

There is a basic conflict here. The department manager, in his concern for inputs, activities, and direct outputs operates in a technological capacity, which may, of course, extend to concerns for the morale and organizational behavior of his staff. But the citizen views the department's efforts in perceptual terms, and for him the important issue is whether he perceives the streets to be cleaner this month than last month. For some local government services technological efficiency may be quite irrelevant to citizen perceptions. The police department may be most inefficient in its deployment of forces, but, if the citizen demands more police protection, he does not ask for more technological efficiency—he asks for more police protection.

This conceptual framework also suggests that the fundamental *economic* reason why program budgeting and productivity analysis usually ends up as activity measurement, not output measurement, is because there are relatively few government programs that have a measurable output. This conclusion does not suggest that activity measurement is useless. To quote (Drucker, 1976:13):

"Saving souls" as the mission of a church is totally intangible. At least, the bookkeeping is not of this world. However, the goal of bringing at least two-thirds of the younger people of the church into the congregation and its activities is easily measurable.

On a more secular note it may well be that improving the technological efficiency of government activities and occasionally finding some measurable direct outputs is the best that can be done. If this route is pursued with diligence, there is a high probability that citizens will get more for their tax dollars. Whether their social state (social welfare) will be improved cannot be determined.

EXPENDITURE DETERMINANTS ANALYSIS

In the face of the almost insurmountable conceptual and measurement barriers to quantifying public output, it is not surprising that major efforts have been directed to empirical studies of the level of state and local government *expenditures* rather than to empirical studies of the level of government output. The chance of meaningful return on a reasonable research effort in the area of public-sector productivity analysis is slim at best. On the other hand, available data and easily used statistical procedures almost guarantee a scientific-looking empirical expenditure analysis (Bahl, 1970). But perhaps most important, even though expenditure analysis cannot deal directly with the efficiency of government operations, it contributes some understanding of why state and local government expenditures increase. Such results may provide insights into how expenditures may be predicted and possibly controlled and, in that sense, overlap the productivity concern.

The lack of a generally accepted theory of the determination of local government expenditures has given rise to numerous attempts to statistically identify the determinants of spending behavior. The rationale for such studies is straightforward: how communities actually reveal preferences for public goods in a context of varying or changing community characteristics is essentially an empirical question. Accordingly, these statistical analyses usually have involved estimating a relationship between per capita local government expenditures and varying sets of socioeconomic and demographic variables (e.g., racial mix, population density) and elements of the budget constraint (e.g., per capita state and federal aid) and identifying those variables with statistically significant coefficients as the "determinants."[4] The statistical results derived in the more careful of these studies have tended to square with economic theory, although the basic assumptions of the model—particularly the output proxy choice—and the estimation problems and data comparabilities are common criticisms (Morss, 1966).

Unfortunately, most expenditure analyses are subject to the same basic shortcoming as productivity studies—the absence of a usable measure or proxy for public output. As a result, these studies identify factors which underlie higher or lower expenditure levels—the latter due to some combination of quality and scope of service, factor cost, productivity, and needs differences. The determinants literature has not yet disentangled these separate effects and hence can add little to knowledge about the relationship between productivity and expenditure level. Indeed, these studies use a variety of conceptual and statistical approaches to finesse the output measurement issue.

GOALS AND GENERAL APPROACH

As noted above, expenditure determinants studies do not address the productivity question. Rather, they do analyze many of the issues that have come to be associated with concerns about the cost of government operation:

- What is the expenditure effect of various population characteristics which would seem to indicate a higher level of public service needs?
- Do cost determinants (e.g., unionization) exert a stronger effect on spending levels than "needs" characteristics of the population;
- Are there economies of scale in the provision of state-local government services?
- Does the form of local government (e.g., strong mayor, city manager) make any difference in expenditure level?
- What is the fiscal response of state and local governments to an increased flow of federal aid?
- What explains the wide variation among similar governments in the level of public employment?
- What is the "independent" effect of strong unions on public sector wage rates and hence on expenditures?

On the surface, it would appear that analyses of such questions would provide useful insights into the possibilities for controlling the level of state and local government expenditures. This is unfortunately not the case. A set of conceptual as well as empirical limitations, described below, almost fatally restricts the policy usefulness of most of the determinants literature.

There are three basic approaches to empirical analysis of the determinants of state-local government expenditure level. By far the most common, and probably the best developed, is the *constrained maximization approach.* This assumes the existence of a community preference function and uses the traditional utility maximization approach to deduce the expected effects of various socioeconomic factors on the level of expenditures. *A bureaucratic approach* to explaining expenditure level would attribute expenditure level to the budget maximization objectives of bureaucratic decision makers. An *organization theory approach* explains public expenditure levels primarily on the basis of institutional constraints and a rigid set of decision rules (Johnson, 1976). Because of the concern here with indirect treatment of output, the constrained maximization approach will be examined in relatively more detail.

CONSTRAINED MAXIMIZATION APPROACH

Although there are many varieties of constrained maximization approaches, they fall into two camps in terms of how they finesse the output measurement problem. First, there is an approach which, by various justifications and with varying theoretical models, uses per capita expenditures as a proxy for output. Second, a more recent and growing literature uses public employment as a proxy for output. Both are obviously weak proxies, but they do suggest differing underlying assumptions which must be understood if policy implications are to be drawn from the empirical results.

Expenditure Approaches

Although conceptual justifications may differ widely, the expenditure approach has in common its treatment of per capita expenditures—either directly or by implication—as a proxy for public sector output. The more purely statistical analyses simply treat per capita expenditures as the dependent variable in a regression, making no assumption about the form of the utility function or the production function facing the government decision maker. These regressions are typically on a cross-section of data; hence the independent variables will explain why per capita government expenditures differ among states (cities, countries, etc.).

The treatment of output in such cases may be handled in one of two ways. First, independent variables may be introduced to account for variations in the "quality" of expenditures, that is, differences in the scope of services provided, the amount paid for factor inputs, and the level of a service provided. Largely because these indicators are difficult to measure, there has been little work of this type. Notable exceptions are Borcherding and Deacon (1972), who attempted to deflate expenditures for interstate variations in the wage and salary level, and Schmandt and Stephens (1960), who attempted to measure variations in the number of subfunctions provided.

The second possibility for acknowledging the output issue is to simply assume that the dependent variable—usually per capita expenditure—measures variations in both the quantity and quality of output. The implications of this assumption do not square with interpretation of results usually given because of the interactions among the vectors discussed above. For example, consider the meaning of the (consistent) finding of a positive relationship between per capita city government police expenditures and poverty population, if the dependent expenditure variable is assumed to reflect both quantity and quality. A larger poverty population requires a greater range of police services and more policemen; a larger poverty population calls forth higher quality police inputs. Since any, or some combination, of these three are likely explanations of the statistical relationship, the findings do not help measurably in understanding expenditure variations or in formulating control measures.

There is a third possibility—that the scope and quality of services is constant when considered on average across government units, e.g., across all state and local governments in each state. Even if this assumption were valid (and a cursory examination of the range of subfunctions provided in Mississippi and New York should reveal that it is not) there would remain the problem of separating expenditure differences due to factor price differentials from those attributable to differences in units of output or the range of activities.

There is an additional conceptual problem in the nature of the choice model implied when public expenditures are assumed to enter household utility functions. In these circumstances a household may derive satisfaction from either an increase in public goods consumed or from an increase in the price of those goods. Put another way, this assumes that a household would be willing to trade away units of private goods for public good price increases.

A final view of the implications of using the expenditure proxy for public output is the assumption of a constant public good price of unity, which in the context of cross-section analysis is tantamount to the assumption that the price of public goods does not vary across the government units studied. This very restrictive and unrealistic underlying theoretical model would limit the explanatory power of determinants analyses, almost fatally.

Some research using the expenditure proxy has specified a production function, i.e., a specific functional relationship between inputs and outputs. With a Cobb-Douglas form for the function (constant returns to scale) and assuming constant factor prices, it is possible to deduce an optimum and test it without ever measuring output. The key assumption is a known relationship between labor and nonlabor expenditures. While informative in that all relationships are more precisely drawn out, these studies still employ per capita expenditures as the output proxy.

Public Employment Approaches

An alternative to the expenditures proxy for output is the use of numbers of public employees (Bahl, Gustely, and Wasylenko, 1976). The argument is essentially that since the public sector is so labor intensive, public employment is roughly proportional to public output. This suggests that employment rather than expenditures enters community preference functions.

Certainly, citizen-voters are likely to attach more value to an increase in employment than to an increase in the public sector wage rate. (In the expenditure approach, the same dollar increase in either component would produce the same increment in utility.) In the employment approach, the output proxy can be given a price whereas in the expenditure approach it cannot.

In addition to the conceptual advantages, this approach may yield more informative policy direction. Since the expenditure form considers only the reduced form expenditure equation, it masks important relationships among the components of the dependent variable—per capita expenditures—thereby making it difficult to explain the *process* by which any particular determinant affects expenditures. For example, state and local government expenditures increase primarily because of increases in employment and/or in wage rates, but the traditional expenditure determinants literature provides little indication about how the underlying explanatory variables should or actually do affect expenditure levels through differential effects on wage rates and employment. As a result, it is not possible from such research to determine whether the usual finding of a significant, positive coefficient for the per capita income variable explains demand influences on the level of expenditures. Income may reflect a supply influence—higher wage requirements for public employees. More generally, it is not possible to say *how* any particular determinant affects expenditures.

The employment approach is not without drawbacks. First, it assumes, in a cross-section analysis, that the quality of a unit of employment is everywhere the same, i.e., that the amount of citizen satisfaction produced by a fireman in Omaha is equivalent to that produced by a fireman in Los Angeles. In a time series, the analogous assumption is that public employee productivity does not increase over time. Second, the proxy measure of output is estimated without regard for the amount of capital or nonlabor input used. There is with the employment approach, as with the expenditure approach, a production function form which is consistent: a Leontief fixed factor relationship where nonlabor expenditure is constant per employee. That is, a unit of police output would require a policeman plus $X for his uniform, gun, club, etc. No capital-labor substitution is possible. Hence, the employment approach also finesses the measurement of output by assuming only one productive factor in the public sector.

The expenditure determinants extensions of the public employment approach are limited, but a related literature has been developing in labor economics. This work centers on the demand for public employee (Ehrenberg, 1973a, 1973b; Ashenfelter, 1970) and the determinants of their compensation (Schmenner, 1973) and contains many useful insights for expenditure determinants. Where estimation of a demand equation is involved, these studies have followed the convention of assuming employment and output proportional.

BUREAUCRATIC APPROACH

An alternative to the consumer maximizing model—which assumes that the decision maker reads community preference for public goods—is that of a bureaucrat who simply strives to maximize his own budget (Niskanen, 1971). Hence, the utility function to be maximized belongs to the bureaucrat, and the arguments in the function are factors such as prestige and power. The major exponent of this theory, William Niskanen, contends that such factors are directly related to the size of the budget.

This theory of expenditure determination has very little directly to do with an optimum level of societal output, although the function requires a constraint that society values output at least above total cost. This approach does have a great deal to do with efficiency considerations in the operation of government in that, if accepted, it all but guarantees an overproduction of public goods. Moreover, to the extent that such an explanation is realistic, there is little point in using sophisticated techniques for government decision making or worrying about output measurement and productivity.

ORGANIZATION THEORY APPROACH[5]

An organization theory approach to government expenditure determination does not require a utility function for the decision maker to maximize, rather just a yearly incremental budget decision. This incremental decision is made under a set of political and fiscal constraints (balanced budget) and is shaped by a yearly set of "emergencies." Accordingly, the primary determinant of this year's spending is last year's, with marginal adjustments to reflect current needs and situations—for example, available matching aid or a particular service level deficiency.

The measurement of output is not directly a part of such an approach. The primary determinants of marginal adjustments in spending are institutional and political.

CONCLUSIONS: STATE OF THE ART

During the years from 1965 to about 1972 a great many states and local governments experienced a dramatic increase in public expenditures. As national economic growth receded in 1973 and continued through 1975 there were new manifestations of the state-local fiscal crisis, particularly in cities and states in the Northeast and Middle West. This, in turn, gave rise to demands for expenditure control. Many of these demands have centered on the alleged lack of productivity growth in the state-local public sector, and productivity improvement has become as fashionable a remedy for state-local fiscal ills as was program budgeting 10 years ago.

But productivity analysis in the public sector has not proved to be easy, and the reason for this lies in the intractable nature of the measurement difficulties that are encountered. Public output, characterized to a greater or lesser degree by nonexclusion and nonrivalry, cannot be quantified in discrete units. In consequence, those who have worked in this vineyard have very often emerged with a mishmash of "productivity indicators" that cannot be compared, aggregated, or utilized effectively for expenditure control.

This essay has attempted two things. First, it has sought to untangle the conceptual morass by examining local government services delivery in terms of a series of vectors: environmental, input, activities, output, and consequences. Second, it has reported on the efforts of public finance economists to analyze expenditures as a proxy for the analysis of output. This latter approach is usually described as "determinants studies."

The determinants literature has not produced an analytic way around the output measurement problem. Hence, it contributes little to an understanding of the relationship between government spending and government productivity. Constrained maximization approaches finesse the problem with assumptions and sleight-of-hand. Organizational theory and bureaucratic approaches are not helpful because knowledge of output/productivity effects are immaterial to the expenditure determination process.

Where then does this leave the field? The empirical expenditure analyses do reveal much about why expenditures vary—*notwithstanding* the ambiguity in interpreting many of the results. Moreover, research on the public sector labor market will likely carry us further toward an understanding of why expenditures grow, whether they should be controlled and how they may be controlled.

Econometric research of the determinants variety, however, will never give more than very general results based on aggregate analysis. It will not help the engineer or systems analyst in local government who works at such seemingly mundane tasks as an improved routing system for sanitation trucks, nor will it help the personnel manager who works out an incentive system that raises activity levels, increases wage rates, and enhances job satisfaction. Since productivity analysis cannot be conceptually rigorous without measures of public output and since such measures, except in isolated instances, are not available, the public administrator is probably best advised to concentrate his efforts on the technological improvement of activities.

NOTES

1. If New York City employment had grown at the national average between 1965 and 1974 the city would have added 1.03 million jobs, that would have produced an estimated $800 million in additional revenue in 1975. (See Bahl, Jump, and Puryear, 1976.)

2. The conventional exception is merit goods, such as school lunches or public housing, where non-exclusion and non-rivalry do not obtain.

3. Some are more optimistic. See *Public Administration Review* (1976).

4. Reviews of the determinants literature may be found in Bahl (1968) and Wilensky (1970).

5. The classic contribution is Lindblom (1961). See also Crecine (1969).

REFERENCES

ASHENFELTER, O. (1970). "The effect of unionization on wages in the public sector: The case of firefighters." Industrial and Labor Relations Review, (January):191-202.

BAHL, R. (1968). "Studies on determinants of public expenditures: A review." Pp. 184-207 in S.J. Mushkin and J.F. Cotton (eds.), Functional federalism: Grants-in-aid and PPB systems. Washington, D.C.: State-Local Finances Project, George Washington University.

——— (1970). "Quantitative public expenditure analysis and public policy." Pp. 547-569 in Proceedings of the National Tax Association, 1969. Columbus, Ohio: National Tax Association.

BAHL, R., GUSTELY, R., and WASYLENKO, M. (1976). "The determinants of local government police expenditure: A public employment approach." Unpublished paper, Metropolitan Studies Program, Syracuse University.

BAHL, R., JUMP, B., and PURYEAR (1976). "The outlook for state and local government fiscal performance." Testimony, U.S. Congress, Joint Economic Committee, January 22.

BORCHERDING, T., and DEACON, R. (1972). "The demand for services of nonfederal governments." American Economic Review, (December):891-902.

BRADFORD, D., MALT, R., and OATES, W. (1969). "The rising cost of local public services: Some evidence and reflections." National Tax Journal, (June):185-202.

BURKHEAD, J., and MINER, J. (1971). Public expenditure. Chicago: Aldine-Atherton.

Committee for Economic Development (1976). Improving productivity in state and local government. New York: Author.

CRECINE, J.P. (1969). Government problem solving: Computer simulation of municipal budgeting. Chicago: Rand McNally.

DRUCKER, P. (1976). "What results should you expect? A users' guide to MBO." Public Administration Review, (January/February):12-19.

EHRENBERG‘ R. (1973a). "The demand for state and local government employees." American Economic Review, (June):366-379.

——— (1973b). "Municipal government structure, unionization and the wages of fire-fighters." Industrial and Labor Relations Review, (October):36-48.

GREYTAK, D., PHARES, D., and MORLEY, E. (1975). Municipal output and performance in New York City. Lexington, Mass.: D.C. Heath-Lexington.

GROSS, B. (1964). The managing of organizations. New York: Free Press.

HATRY, H., and FISK, D. (1971). Improving productivity and productivity measurement in local government. Washington, D.C.: Urban Institute.

JOHNSON, M. (1976). Two essays on the modelling of state and local government fiscal behavior. Unpublished Ph.D. dissertation, Syracuse University.

LINDBLOM, C.E. (1961). "Decision making in taxation and expenditure." Pp. 296-336 in Public finances: Needs, sources and utilization. Princeton, N.J.: Princeton University Press for the National Bureau of Economic Research.

MORSS, R. (1966). "Some thoughts on the determinants of state and local expenditures." National Tax Journal, (March):95-103.

MUSGRAVE, R. (1959). The theory of public finance. New York: McGraw-Hill.

National Commission on Productivity and Work Quality (1975). Employee incentives to improve state and local government productivity. Washington, D.C.: U.S. Government Printing Office.

NISKANEN, W., Jr. (1971). Bureaucracy and representative government. Chicago: Aldine-Atherton.

Public Administration Review (1976). "Symposium on Management by Objectives in the public sector." 37(January-February):1-45.

ROSS, J., and BURKHEAD, J. (1974). Productivity in the local government sector. Lexington, Mass.: Lexington.

SAMUELSON, P. (1954). "The pure theory of public expenditures." Review of Economics and Statistics, (November):387-89.

——— (1955). "Diagrammatic exposition of a theory of public expenditures." Review of Economics and Statistics, (November):350-356.

SCHMANDT, H., and STEPHENS, R. (1960). "Measuring municipal output." National Tax Journal, (December):369-75.

SCHMENNER, R. (1973). "The determination of municipal employee wages." Review of Economics and Statistics, (February):83-90.

WALLACE, R. (1975). Productivity and the fire service (Occasional Paper No. 26). Syracuse: Metropolitan Studies Program, Syracuse University.

WILENSKY, G. (1970). "Determinants of local government expenditures." Pp. 197-218 in J.P. Crecine (ed.), Financing the metropolis: Public policy in urban economies (Urban affairs annual reviews, vol. 4). Beverly Hills, Calif.: Sage.

12

National Politics, Local Politics, and Public Service Employment

RANDALL B. RIPLEY
DONALD C. BAUMER

AN OVERVIEW OF PUBLIC SERVICE EMPLOYMENT

□ EVEN BEFORE JIMMY CARTER BECAME PRESIDENT on January 20, 1977, he laid out an economic program in early January that had as one of its central features the expansion of public jobs programs in at least three different guises. His proposals (including some public works) were estimated to cost about $5 to $8 billion for fiscal year 1978. Naturally, it remains to be seen what emerges from Congress in response to these proposals, but they dramatize the centrality of public jobs in the economic strategy of the new President.

At the same time, when they are compared to competing proposals, Carter's proposals also dramatize the degree of disagreement between various major actors along the political spectrum. Shortly after Carter revealed the outline of his proposals in early January, President Gerald Ford sent his budget for Fiscal 1978 to Congress. It contrasted sharply with the Carter proposals and placed

AUTHORS' NOTE: *The research on which this article is based has been supported by the U.S. Employment and Training Administration and by the Mershon Center at Ohio State University. Intensive fieldwork was done in all CETA prime sponsorships in Ohio (there were 17) between mid-1974 and mid-1976. Additional fieldwork was done in 15 prime sponsorships spread throughout the country in 1976-1977. The article does not necessarily reflect the views of the Employment and Training Administration, which follows the admirable practice of encouraging its grantees to express their own opinions based on their research.*

[271]

little stress on the combination of public works and public jobs. In direct reaction to the Carter proposals, the AFL-CIO, on the other hand, called for a much larger effort. Table 1 compares the three sets of proposals and also includes gross figures on the programs in place as calendar year 1977 began. The differences are significant. Also significant is the fact that a conservative Republican President preferred to eliminate all major federally supported public service jobs in 1978 (except for a 50,000 job program under Title II of the Comprehensive Employment and Training Act, or CETA), whereas a moderate Democratic President proposed a substantial increase in all categories, and the major spokesman for organized labor proposed a very large increase. Given a heavily Democratic Congress, substantial increases seem the most likely outcome.

Although there are different types of public service employment, the political debate that has taken place in the United States since the 1930s is usually couched in sweeping terms that ignore differences between programs. Perhaps it is true that, on most issues in American politics, party and ideological lines are so blurred as to be almost unrecognizeable. In the case of public service employment, however, such is not the case. Put in oversimple terms "liberals" on domestic economic questions (primarily Democrats) tend to be supportive of public service employment programs, and "conservatives" (primarily Republi-

TABLE 1

MAJOR PUBLIC SERVICE EMPLOYMENT PROGRAMS AND PROPOSALS, 1977-1978

Program Area	Program in Existence for FY 1977 (as of Jan. 1977)	Ford Proposal for 1977-1978 (Jan. 1977)	Carter Proposal for 1977-1978 (Jan. 1977)	AFL-CIO Proposal for 1977-1978 (Jan. 1977)
Public service employment at state and local level (Titles II and VI of Comprehensive Employment and Training Act)	310,000 jobs funded (50,000 in II and 260,000 in VI)	Existing level through FY 1977; phase out Title VI in FY 1978; retain Title II at 50,000 in FY 1978	Expand to 725,000 jobs in VI by end of FY 1978; some expansion of II	Expand to more than 1 million jobs in FY 1978 (add $8 billion)
Public works at state and local level	$2 billion	No additional funding	Addition of $2 billion; possible addition of $2 billion more	Addition of $10 billion
Countercyclical revenue sharing	$1.25 billion for five quarters (7/1/76- 9/30/77)	No additional funding	Addition of $1 billion	Addition of $2 billion

SOURCES: *Employment and Training Reporter,* January 19, 1977, p. 236; *National Journal,* December 25, 1976, pp. 1794-1798; *Congressional Quarterly Weekly Report,* January 22, 1977, pp. 120, 122, 123, 126, 129.

cans) tend to be against or at least skeptical of them (see Davidson, 1972; Van Horn, 1976; Shields, 1976). The disagreement between Ford and Carter underscores the point.

The core of the political debate is how far, if at all, the federal government should go in providing money for additional public jobs beyond those "normally" funded by all levels of government. The central rationale for supporting or opposing such job creation is, of course, economic in language, although the content of the debate leads to the reasonable conclusion that liberal Democrats simply believe in the effectiveness of public service job creation and that conservative Republicans simply do not share that belief. Like most national political debates on complicated issues, fundamental gut feelings on an all-encompassing question are dominant.

When the debate does become more sophisticated—usually in pages produced by academic economists—no clear conclusion emerges based on widespread agreement. Seemingly plausible arguments are made on both sides about the effectiveness of various kinds of public service employment—either tied to public works projects or not. (For good examples of differing views taken by economists see Congressional Budget Office, 1975; Fechter, 1975; Levitan and Taggart, 1974, 1976; Solow, 1976; Ulman, 1976; Wiseman, 1975.) It seems reasonable to suspect that at least some of the economists also operate partially on the basis of ideology when arriving at their respective conclusions.

There are at least four major types of public service employment (beyond "normal" employment of government employees) that need to be distinguished. The first type involves the large funding of public projects of one kind or another. The intent of such programs in basically countercyclical—that is, to offset downturns in the business cycle and to help stimulate the next upturn. The large public jobs programs created in the 1930s as part of the New Deal's efforts to fight the Depression (the Civilian Conservation Corps and Work Projects Administration) fit in this category. At present, Title I of the Public Works Employment Act of 1976 also falls in this category. It contains $2 billion for public works projects typically involving "construction and renovation of municipal offices, courthouses, libraries, schools, police and fire stations, detention facilities, recreational facilities, civic and convention centers and museums" (Singer, 1976:1797). About a quarter of the money is likely to be used for wages.

Second, there are limited, special programs aimed at opening public careers to the disadvantaged. Thus their intent is to deal with structural unemployment and to attack various forms of discrimination. In the 1960s a variety of programs such as Operation Mainstream, Public Service Careers, and New Careers all had this focus. Title II of CETA had this as its original purpose, although—as will be explained below—this original purpose has become blurred by subsequent events. Some localities are using some Title I money from CETA for small programs of this character, usually direct successors to Mainstream or New Careers.

Third, there are countercyclical programs the main focus of which is on helping local units of government sustain vital services and at least theoretically expand services through providing money for hiring employees that can be used for any local purpose. The Emergency Employment Act of 1971 had this purpose (see Levitan and Taggart, 1974) as does Title VI of CETA, added in 1974. In practice, Title II of CETA is also treated as if this were its purpose. Some CETA Title I money is also being used for this purpose. Title VI was complicated by new provisions added in the autumn of 1976 that potentially tied some expenditures to projects and stressed hiring of low-income, long-term unemployed.

Fourth, there is general countercyclical revenue sharing, which provides additional money to local units of government in areas with substantial unemployment for the purpose of maintaining services by retaining public employees faced with layoff or rehiring those already laid off. Title II of the Public Works Employment Act of 1976 provides $1.25 billion for this purpose for the 15 months from July 1, 1976 through September 30, 1977.

When these programs are considered together it becomes apparent that by far the largest program aimed primarily at presumably new jobs and services is Title VI of CETA (supplemented by Title II as it has come to be understood, which is at substantial variance from congressional intent in late 1973). Thus a focus on CETA PSE (Public Service Employment—that is, II and VI together) is reasonable in trying to understand the overall politics and program impact of existing public service jobs programs.

Furthermore, given that the structure of CETA is such that most important decisions are made at the local level (with some steering from the national and regional offices of the U.S. Department of Labor), it seems sensible to move from a national overview to specific empirical attention to local decisions—their politics and impact—and to the federal "steering" role, such as it is.

The remainder of this article focuses on Titles II and VI (PSE) of CETA, using three different levels of data: data from all localities with CETA funds in the United States aggregated; data from all localities in the State of Ohio aggregated; and data from three Ohio cities as examples of what is happening with PSE in a disaggregated sense. The following sections (1) describe the CETA PSE program, (2) comment on the role of the U.S. Department of Labor, especially the regional offices, and (3) analyze the local implementation of CETA PSE. Four questions guide the analysis of local implementation: (a) what people get the jobs? (b) what organizations get the employees? (c) what kinds of jobs are created? and (d) what happens to the employees in employment terms after they start the program? A very brief concluding section summarizes some of the most important live issues for both scholars and public policy makers and administrators to address.

CETA PUBLIC SERVICE EMPLOYMENT: A DESCRIPTION

CETA PSE programs are implemented through the basic CETA delivery system, which includes two major components: (1) local units of government called prime sponsorships and (2) program agents. To qualify as a prime sponsorship, a jurisdiction must have a population of at least 100,000. Combinations of jurisdictions that include at least one jurisdiction with more than 100,000 people can form prime sponsorships called consortia, and all local units not included in any other prime sponsorship form a balance of state prime sponsorship. (In 1976-1977 there were 445 prime sponsorships of all kinds.) The chief elected officials (for example, mayors and county commissioners) of the eligible jurisdictions are designated as the prime sponsors. The prime sponsors are responsible to the U.S. Department of Labor for all CETA programs operated within their prime sponsorship.

The program agent designation was created specifically for PSE programs and applies to the chief elected officials of any jurisdiction with a population of at least 50,000. Program agents are given the "administrative responsibility for developing, funding, overseeing and monitoring" local PSE programs. Thus, Title II and VI funds are earmarked for program agents, and prime sponsors serve as little more than a pass-through for these funds even though they are formally responsible to the Department of Labor for making sure that they are spent properly. The relationship between prime sponsors and program agents varies, but in most cases the CETA Title I programs are operated through the prime sponsor administrative machinery, and PSE funds are passed on to the eligible program agents, which operate the programs using their own city or county staffs.

The purpose of Title II is summarized in the act: "to provide unemployed and underemployed persons with transitional employment in jobs providing needed public services in areas qualifying for assistance and, wherever feasible related training and manpower services to enable such persons to move into employment or training not supported by this Title."

Most observers considered Title II, which was part of the original CETA package passed in late 1973, not to be primarily countercyclical in intent, but, rather, aimed at persons with structural employment problems. Title VI, which was passed in December of 1974 as a response to rapidly rising unemployment, was regarded from the beginning as a countercyclical program. For Title II, emphasis was placed on serving "the most severely disadvantaged in terms of the length of time they have been unemployed and their prospects for finding employment." In Title VI the emphasis changed to favor victims of the recent economic downturn. Title VI states that "special attention" should be given "to serving those who have exhausted or are not eligible for unemployment benefits."

This apparent difference in intent, however, had little effect on most local implementors. Soon after the passage of Title VI, both programs were viewed by most local implementors as being essentially the same, which meant that, for the most part, they were implemented as countercyclical programs.

Originally, there were a number of other differences in the Title II and VI legislation and the regulations. Some of these differences remain, others were gradually eliminated during fiscal years 1975 and 1976. One basic difference that remains is the way Title II and VI funds are allocated. Title II funds go only to areas experiencing a rate of unemployment of at least 6.5% for three consecutive months during the year in which the funds are allocated. Title VI funds are allocated through a complicated formula that rewards areas of high unemployment, but every area in the country receives some Title VI money.

Eligibility requirements for Title II and VI also differ. To be eligible to participate in a Title II program, a person must reside in an area with a rate of unemployment of 6.5% and be unemployed for 30 days or underemployed. Anyone within the jurisdiction of a program agent was eligible to participate in the original Title VI program as long as he or she was unemployed for 30 days or underemployed. Title VI also included a provision that allowed program agents with rates of unemployment above 7% to reduce the length of unemployment necessary to qualify for the program to 15 days.

The first set of Title II regulations included a nonmandatory goal of placing 50% of the Title II participants served each year into unsubsidized employment. (This rate of placement was required under the Emergency Employment Act of 1971, the immediate predecessor to CETA PSE.) When Title VI was passed, the placement goal for both Title II and Title VI was modified with the inclusion of new provisions emphasizing that the placement goal was not a requirement and establishing a process by which prime sponsors could waive the goal altogether.

Some important features are common to both Titles II and VI as originally passed: public service jobs can be provided by either public or private nonprofit agencies; participants are to receive the same wages, fringe benefits, promotional opportunities, and working conditions as other persons employed in similar jobs by the same employer, as long as the wages do not exceed $10,000 annually; 90% of all PSE money spent at the local level has to be spent on participant wages or fringe benefits, thus setting a 10% ceiling on the amount that can be spent for program administration; and all PSE jobs are supposed to be net additions to those which would have otherwise been available, which has meant that no existing worker can be displaced, nor can PSE jobs be substituted for existing jobs.

The last aspect of PSE policy mentioned above concerns what is called "maintenance of effort." As CETA was being implemented in fiscal years 1975 and 1976, Congress and the Labor Department devoted a great deal of attention to clarifying what constituted a maintenance of effort violation, because some program agents had used substantial portions of their PSE funds to rehire laid-off municipal employees. This development also attracted the attention of the American Federation of State, County, and Municipal Employees (AFSCME).

The practice of rehiring laid-off regular employees using CETA funds was never prohibited, but a series of restrictions was placed on program agents who decided to spend their PSE funds in this way. First, they had to show that the layoffs were "bona fide," which meant that they were based on a legitimate shortage of local revenue. Second, no PSE participant could be hired in the "same or substantially equivalent job" in a department in which regular employees had been laid off unless all the regular employees were rehired first (this was AFSCME's main concern). In April of 1976 a new provision was added that stated that the number of rehires participating in a PSE program should be consistent with the relative percentage of the total unemployed population that such rehires constituted. If this percentage was less than 10%, then a maximum of 10% of the PSE participants could be rehires. However, the regulations issued in June of 1976 allowed for a waiver of this 10% restriction, and efforts to reduce the number of rehires in some of the large cities (which were the most visible violators of the 10% restriction) have not been very successful.

Table 2 shows the national allocations for Titles II and VI in fiscal years 1974, 1975, and 1976. Also shown are the dates in which the allocations were announced. By linking the dates with the allocations, much of the history of the implementation of Titles II and VI of CETA can be told.

In the first and second quarter of fiscal year 1975 only Title II was operative. Because of the late announcement of the fiscal year 1974 allocation both the fiscal year 1974 and 1975 Title II funds were available to be spent in fiscal year 1975. In addition, the Labor Department allocated $240,000,000 to phase out the previous PSE program under the 1971 Emergency Employment Act. Despite the availability of over $700,000,000, the implementation of Title II proceeded slowly in the first half of fiscal year 1975. By December of 1974 only 55,950 Title II participants had been hired.

The pace of implementation stepped up dramatically in January, February, and March of 1975 after the announcement of the first Title VI allocation. The combination of the Title VI, Title II, and EEA allocations represented a PSE

TABLE 2

NATIONAL ALLOCATIONS FOR TITLES II AND VI

Allocation	Title II	Title VI
FY 1974	$365,000,000[a]	
	(June 1974)	
FY 1975	$350,000,000	$850,000,000
	(September 1974)	(January 1975)
FY 1976	$400,000,000	$1,625,000,000
	(June 1975)	(June 1975)
FY 1976	$1,200,000,000	
(supplemental)	(April 1976)	

a. The Title II base allocation was announced in June 1974 and was $3,000,000. In September of 1974, when the FY 1975 Title II allocation was announced, a $65,000,000 discretionary allocation was added to the FY 1974 base. This made the total FY 1974 Title II allocation $365,000,000.

SOURCE: Manpower Information Service Reference File (1976: 71:1035-71:1080).

effort of just under $2 billion for fiscal year 1975. By June of 1975 there were 247,572 participants in Title II and VI programs.

By June of 1975 the pace of hiring began to slacken because Congress had not yet appropriated any new PSE money for fiscal year 1976. Just before the beginning of the new fiscal year the new allocations were announced. These allocations again amounted to roughly $2 billion for fiscal year 1976. Some new hiring did take place after the fiscal year 1976 allocations were announced, but for some program agents there were only enough funds to maintain the existing number of participants. By September of 1975 Title II and VI enrollments were up to approximately 300,000 and remained at that level through fiscal year 1976.

In late fiscal year 1976 the uncertain future of PSE funding caused prime sponsors and program agents another period of anxiety, wondering whether more PSE money would be allocated before they were forced to begin laying off the participants. With the Title VI extension stalled, the Congress passed an Emergency Supplemental Appropriation that added $1.2 billion to Title II for the purpose of keeping all CETA PSE participants working through December of 1976. Shortly thereafter both legislative and administrative actions were taken that would allow a continuation of PSE at then-current levels (about 310,000 jobs) through fiscal year 1977 (until October 1, 1977).

ROLE OF THE U.S. DEPARTMENT OF LABOR IN CETA PSE

Once Congress passed the CETA PSE legislation, the U.S. Department of Labor assumed a major role in making sure that the programs were administered in a way that reflected congressional intent. Its national office's primary responsibilities in overseeing the implementation of CETA PSE have been to write the regulations for Titles II and VI, to allocate PSE resources (both those allocated by formula and discretionary funds), and to communicate CETA policy to local implementors through its regional offices.

The regional offices have been given primary responsibility for communicating national policy to the prime sponsors, monitoring local operations, providing technical assistance, reviewing prime sponsor plans, receiving reports, and advising prime sponsors on programmatic decisions. Much of this burden rests on the shoulders of the federal representative assigned to each prime sponsorship.

Most regional offices and federal representatives have adopted a policy of not interfering with the programmatic decisions of CETA prime sponsors unless clear violations of federal regulations are involved. Therefore, federal representatives have not offered prime sponsors and program agents a great deal of guidance in developing the design of local PSE programs, unless an alternative being considered constitutes a violation of the regulations. They occasionally have responded to national priorities by taking on a more active role in influencing local decisions on certain matters (the most notable was their vigorous encouragement of prime sponsors to increase PSE enrollment levels in the winter of 1975), but, in general, they have left most programmatic decisions to the discretion of local implementors. Consequently, federal representatives are perceived to have had little impact on the shape of most PSE programs (see Ripley and et al., 1977).

Although the federal representatives and the regional offices in general were not perceived as having great influence in local CETA decisions, national policy and resource allocation as communicated by the regional offices did have a major impact on local implementation activities. The early period (June to December of 1974) of slow PSE implementation was due, in large part, to the fact that most federal representatives were portraying Title II to the prime sponsors as a very restrictive program. They were insisting that PSE hiring be carefully monitored to insure that only people from eligible census tracts were hired. They were also communicating the 50% transition goal as a requirement and asking prime sponsors to include the means by which this goal would be met in their Title II plans.

In January of 1975 the character of regional office communication changed noticeably. The regional offices began a new campaign of encouraging prime sponsors to hire people and spend PSE money quickly. Labor Department communications even contained the threat that Title II money not spent by the end of the fiscal year would be reallocated. The dramatic rise in PSE enrollments between December of 1974 and June of 1975 was a reflection of the impact that this new regional Labor Department posture had on the pace of local implementation.

By May of 1975 the messages coming from Labor Department offices changed again. Prime sponsors were told to plan to carry over unspent Title II funds into fiscal year 1976. The emphasis on spending and hiring had changed to one of maintaining existing levels of enrollment. This same emphasis continued throughout fiscal year 1976.

The inconsistent nature of Labor Department communication was a source of constant complaints among the prime sponsors and program agents during fiscal years 1975 and 1976. Even though many of the inconsistencies were due to changes taking place in Congress or the national office of the Department of Labor, the local implementors usually blamed the regional offices. They also objected to other regional office policies. Staff people at the local level frequently complained about the limited amount of time that they were given to plan PSE programs. Final plans were usually due at the regional offices less than a month after each new PSE allocation was announced. Local staff members also claimed that they received little or no useful technical assistance from regional office officials in designing their local programs. Many staff people indicated that the regional office could not legitimately complain about the poor design of PSE programs unless they were willing to offer prime sponsors more useful expertise and give them more time to plan the programs.

LOCAL IMPLEMENTATION OF CETA PSE

The legislation and the regulations impose some restrictions on the options available to local decision makers, but CETA prime sponsors and program agents still have a great deal of latitude in deciding how to spend Title II and VI funds. The existence of this local discretion has created a number of interesting implementation issues. The first issue concerns the types of clients who are served through CETA PSE programs. The second issue involves the distribution of jobs among the potential employing agencies. The third question focuses on the kinds of jobs created in CETA PSE programs. A fourth area of interest concerns the length of time that PSE participants remain in the program and the rate at which they are placed into unsubsidized employment.

In the sections that follow we will examine each of these issues in detail by presenting and discussing PSE data from the nation as a whole, the state of Ohio, and three major cities in the state of Ohio. Because national and state data are not always available in a form that allows for intensive investigation of some of the important questions under each topic, the three cities will serve as the primary units of analysis in some sections. The cities—labeled A, B, and C to preserve the anonymity guaranteed in exchange for complete access to data—are between about 110,000 and 280,000 in population.

WHAT PEOPLE GET THE JOBS?

The data shown in Table 3 demonstrate that the typical PSE client nationally is a white male between the ages of 22 and 44 with a high school diploma. This closely parallels Levitan and Taggart's (1974:18) findings regarding the public jobs program under the Emergency Employment Act. Ohio totals resemble those for the nation, but the state tends to serve more blacks, fewer women, and fewer disadvantaged participants.

TABLE 3
CLIENT CHARACTERISTICS FOR TITLES II AND VI, MARCH 1976

| Characteristic | Title II | | | | | Title VI | | | | |
	U.S.	Ohio	City A	City B	City C	U.S.	Ohio	City A	City B	City C
22-44 years of age	64%	66%	58%	67%	64%	65%	64%	62%	73%	71%
With 12 or more years of formal education	75	74	85	89	74	77	77	83	87	88
Black	25	35	51	47	67	23	25	43	39	39
Economically disadvantaged	45	40	74	12	36	44	39	53	14	32
Female	37	31	24	40	21	35	34	34	25	31

SOURCES: Quarterly reports submitted by prime sponsors and program agents.

National client service patterns have not changed a great deal over time. Since December 1974, aggregate client characteristics for Title II show small increases in the tendency to serve people between the ages of 22 and 44 and those with high school diplomas. Service to women has also increased slightly, while service to the disadvantaged has declined. For Title VI the only changes have been increases in service to those with high school diplomas and to females (see Ripley et al., 1977:76-85).

The data for the three cities illustrate how client service patterns can vary. They also suggest that local implementors have a great deal of discretion in determining the types of clients that will be served in their programs. In general, the three cities served a higher percentage of blacks and people with high school diplomas and a lower percentage of females than the national averages. Service to the economically disadvantaged varied. City A had a very high rate of service to this group, while in City B service to the disadvantaged was very low. City C served a percentage of economically disadvantaged participants just below the national average.

While the demographic makeup of an area has an important effect on the characteristics of the CETA service population, demography alone does not explain some of the variations shown in Table 3. For example, of the three cities studied, City B had the highest percentage of blacks in its population, but the lowest rate of CETA service to blacks. City C had the lowest percentage of blacks in its population, but had the highest percentage of blacks served in Title II. City A had the highest rate of service to blacks in Title VI. City B also had the highest percentage of its residents below the poverty level, yet service to the economically disadvantaged in City B was the lowest of the three cities. City A served a percentage of economically disadvantaged participants in both Titles that was much higher than either City B or City C, even though the percentage of residents below the poverty level in City A was approximately the same as City C and lower than City B. Thus, it seems that a number of factors besides demography play a role in determining client service patterns for CETA PSE.

The client service data presented in Table 3 suggest that many of the CETA PSE clients would not appear to fit the description of "the most severely disadvantaged." A number of factors help to explain why this has occurred. The most often cited factor is a practice known as "creaming." Most PSE employing agencies are given a choice from a number of referrals in selecting a person to fill a PSE opening. What usually occurs is that they choose the most experienced and highly qualified candidate referred to them. The incentives to do this are obvious, and the net result is that older, more experienced, better educated applicants are systematically selected.

Another important influence on the types of clients served in CETA PSE programs is the guidance offered by the Department of Labor. At first the department's position seems to have been that PSE was designed to serve those with long term employment problems. When unemployment rose and Title VI was passed, the department changed its position and stressed the countercyclical side of PSE. Emphasis was placed on the rapid hiring and rapid expenditure of PSE funds. In attempting to comply with the department's demands, most program agents abandoned any attempt to employ special need criteria in selecting PSE participants. The most disadvantaged tended to get lost in the shuffle when large numbers of recently unemployed and highly qualified workers began to apply for CETA jobs.

The goals of local implementors were also an important influence on the nature of client service patterns. If the local PSE staff was committed to the goal of serving a disadvantaged clientele, they could in many cases overcome the tendencies working to serve those less disadvantaged. City A achieved some success of this kind.

The role of the elected officials in CETA decision making is often a critical factor in determining the level of service to the most disadvantaged. In most cases that we have observed, direct and extensive involvement by elected officials in CETA matters has led to low service to the disadvantaged. This seems to be related to their perception that relatively less disadvantaged people represent a more reliable source of potential political support than the severely disadvantaged. An example of this pattern is provided in the three cities being analyzed. In City B, where the percentage of economically disadvantaged PSE participants was very low, the mayor participated in all aspects of CETA decision making and personally selected all PSE participants. In City A, where the rate of service to the disadvantaged was very high, the mayor paid practically no attention to CETA.

A final set of factors that have a direct effect on the types of participants served in a PSE program are the other programmatic decisions made at the local level. In particular, decisions regarding the agencies and/or departments that will receive PSE job slots and regarding the types of jobs that are created have a profound impact on the characteristics of the clients served. These two types of programmatic decisions are discussed in the sections that follow.

WHAT ORGANIZATIONS GET THE EMPLOYEES?

One decision that each program agent must make is how to distribute PSE jobs among the public and private nonprofit agencies in their jurisdiction that are requesting PSE positions. In this analysis the potential employing agencies will be grouped in three categories. The first category will include city or county departments within the program agent's unit of government. The second covers all other public agencies (for example, boards of education, state agencies, and public housing authorities). The third category is reserved for private nonprofit agencies (for example, community-based organizations and community action agencies).

The distribution of PSE positions among the three types of employing agencies for the state of Ohio and cities A, B, and C is shown in Table 4. For the state as a whole there was very little difference in the percentage of jobs going to

TABLE 4

THE DISTRIBUTION OF PSE JOBS AMONG EMPLOYING AGENCIES BY TITLE, 1976

	Title II			Title VI		
	% Program Agent's Unit of Gov't.	% Other Public Agency	% Private Nonprofit Agency	% Program Agent's Unit of Gov't.	% Other Public Agency	% Private Nonprofit Agency
State total	67	24	9	67	23	11
City A	100	0	0	60	29	11
City B	100	0	0	63	21	16
City C	100	0	0	47	25	28

SOURCES: Program agent records from the spring of 1976.

each category of employing agency between Title II and Title VI. For the three cities, however, there was a significant difference in the distribution of PSE jobs between the Titles. All three cities kept all the Title II jobs within city departments. Actually, this distribution does not diverge from statewide trends as much as it first appears. Many of the large cities in the state kept a very high percentage of the Title II positions available within city departments. The counties generally gave a much higher percentage of their Title II jobs to outside agencies. Title II is a permanent part of CETA, while Title VI was originally enacted to operate for only one year; thus the decision to retain all Title II jobs in city departments is a reflection of the desire of program agents to concentrate the PSE jobs with the longest life expectancy in their own unit of government. This reduced, to some extent, their uncertainty concerning the number of federally subsidized employees whom they would have available in the following year.

For Title VI, cities A and B were fairly close to the statewide pattern of job distribution, though city A gave more jobs to other public agencies and retained fewer in city departments than the state averages. City C, however, diverged noticeably from the state trends by giving an unusually high percentage of jobs to nonprofit agencies and retaining a low percentage within city departments.

There were a number of factors considered by local decision makers in deciding how to distribute PSE jobs among employing agencies. Elected officials were encouraged to locate jobs in city departments because of the prospect that PSE employees would reduce the burden on regular employees and allow the city to expand the services it offered. This possibility was particularly attractive to cities facing fiscal crises, which had caused city officials either to impose a freeze on hiring more regular employees or to lay off some employees. The disadvantage of hiring a large percentage of PSE employees in city or county departments was the possibility that future PSE funding would be decreased or eliminated, which would force program agents either to absorb PSE people or to lay them off. Few elected officials wanted to commit themselves to hiring a large

TABLE 5

CHARACTERISTICS OF PSE PARTICIPANTS IN THREE OHIO CITIES
BY TYPE OF EMPLOYING AGENCY, 1976

	Program Agent's Unit of Gov't.			Other Public Agency			Private Nonprofit		
	% B	% F	% E.D.	% B	% F	% E.D.	% B	% F	% E.D.
City A	46	16	52	49	53	71	30	56	43
City B	36	28	15	50	38	16	63	44	15
City C	46	12	28	33	45	26	67	53	40

Key: B = Black; F = Female; E.D. = Economically Disadvantaged

SOURCES: Program agent records from the spring of 1976.

number of PSE employees on their regular work force, and the political reper-
cussions that might accompany a large layoff were very unappealing.

The fiscal condition of the city or county government was the most important factor in determining the distribution of slots. Where local revenue was
short, there was usually a high concentration of PSE employees within the
program agent's unit of government. This pattern is reflected in all three cities.
Both cities A and B were experiencing fiscal difficulties and had imposed freezes
on the hiring of new regular employees. Cities A and B also retained a much
higher percentage of PSE employees in city departments than City C, which
enjoyed relatively good fiscal health through fiscal years 1975 and 1976.

Local PSE staff members cited other reasons for favoring certain types of
agencies. In general, the nonprofit agencies and some of the other public
agencies were considered to have a better record of hiring women and minorities
than the city or county departments. However, the quality of some of the jobs
created in the nonprofit agencies was suspect, and the possibility for permanent
absorption was not considered to be good. The public agencies were seen as
providing jobs that included more worthwhile training and work experience and
a better possibility of absorption.

The data presented in Table 5 on the three cities shed some light on the
accuracy of staff assessments of the client service records of the different types
of employing agencies. The one pattern that was clearly supported in these three
cases was that city departments gave very low rates of service to women. Women
received the most jobs from the private nonprofit agencies but also received
relatively good service from the other public agencies. The private nonprofit
agencies in cities B and C served a higher percentage of blacks, thus con-
firming staff impressions, but in city A blacks received a higher rate of service from the public agencies that had the highest rate of service to the
economically disadvantaged.

WHAT KINDS OF JOBS ARE CREATED?

A second type of PSE program design decision that all local implementors
must face concerns the kinds of jobs that will be created. The theory behind

TABLE 6

THE DISTRIBUTION OF PSE JOB CATEGORIES BY TYPE OF EMPLOYING AGENCY

Job Category	City A			City B			City C		
	City	OP	PNP	City	OP	PNP	City	OP	PNP
% Professional	2	7	11	4	5	23	3	5	4
% Paraprofessional	6	18	13	15	38	47	9	29	43
% Service worker	19	39	51	25	18	3	32	29	20
% Skilled labor	8	3	2	4	0	0	10	10	2
% Clerical worker	`17	26	18	11	21	23	8	26	28
% Laborer	48	8	4	41	13	3	38	0	2

Key: OP = Other public agencies; PNP = Private nonprofit organizations.

SOURCES: Program agent records and reports.

CETA PSE is that local programs should provide unemployed people with temporary subsidized employment in jobs that provide a meaningful working experience so they can develop to the point where they can compete in the regular labor market. Therefore, it is of critical importance to a program of this type that the jobs created provide the kind of working experience that will enhance the future employability of the participants. One of the standard criticisms made by the opponents of PSE is that the programs simply create a large number of "dead end jobs."

As is the case with most PSE decisions, there are a number of factors that local implementors must take into account in trying to decide what types of jobs should be created in the program. Most local implementors recognized the desirability of creating challenging jobs that allow participants to develop good working habits and learn skills that can help them in finding future employment, rather than simply creating a lot of labor and clerical positions. The problem was that the disadvantaged rarely had the education or the skill to perform these more demanding jobs adequately. Most PSE staff members said that basic, low-skill positions were the only jobs in which the disadvantaged could be hired and continue working. Furthermore, many of the elected officials expressed a preference for these types of jobs because they were expendable if PSE funding was cut off.

Most of the jobs funded in city or county departments tend to be traditional low-skill positions (laborers, service workers, and clerical workers). The private nonprofit agencies and the other public agencies often provide more innovative community-service-oriented positions, but, in many cases, these jobs may not be helping the participants to develop marketable skills.

In order to examine the patterns of PSE job creation in greater detail, we have classified the jobs created in cities A, B, and C into six categories according to the job title associated with each position as shown in Table 6.[1] In table 7 these same jobs have been categorized according to wage rates. The data presented in Table 6 display a number of patterns. In all three cases a high percentage of the jobs located in city departments were concentrated in the laborer, clerical, and service worker categories. The private nonprofit and the other public agencies also hired a number of clerical and service workers but had many more paraprofessionals and fewer laborers than the city departments. None of the employing agencies hired a high percentage of skilled laborers, but the city departments were slightly better in this area than the other agencies. The percentage of professionals also tended to be low for all types of employing agencies, but the private nonprofits in cities A and B had the highest percentage of jobs in this category.

Table 7, which shows the distribution of wage rates for the different employing agencies, also reveals some consistent patterns. Most of the participants employed in city departments fall into the middle wage categories. This reflects the heavy emphasis given to hiring laborers and service workers within city departments. These types of jobs usually included a wage between $3.50 and $4.50 per hour. The other employing agencies both tended to hire more participants in the lowest wage categories. (This was especially pronounced in city A). In many cases the private nonprofits and the other public agencies also had a fairly high percentage of jobs in the highest wage category. This was related to their practice of creating a substantial number of professional and paraprofessional positions, which usually included high wages. The exception to this pattern was the other public agencies in City B, which had wage schedules that looked more like the city departments.

Most of the PSE staff people interviewed have suggested that there is a very strong relationship between the types of jobs that are created and the characteristics of the people who fill those jobs. Thus women fill most of the clerical positions and are excluded from those involving physical labor. Blacks and the disadvantaged tend to be served in the greatest numbers in the low skill jobs that usually fall in the laborer, service worker, or clerical categories. White males are said to predominate in the professional and skilled labor categories.

The data presented in Table 8 allow one to explore the validity of these impressions. Overall, it appears that the jobs funded in city departments conform most closely to the patterns suggested by PSE staff members. All three city

TABLE 7
THE DISTRIBUTION OF WAGE RATES FOR PSE JOBS
BY TYPE OF EMPLOYING AGENCY

	City A			City B			City C		
Wage Rate	City	OP	PNP	City	OP	PNP	City	OP	PNP
% $2.01-$3.00	12	31	61	4	3	17	7	43	30
% $3.01-$3.50	6	27	20	22	27	21	10	4	22
% $3.51-$4.00	35	10	18	40	30	7	39	9	13
% $4.01-$4.50	27	16	0	21	22	14	36	2	15
% $4.51+	20	16	0	13	19	41	7	33	20

Key: OP = Other public agencies; PNP = Private nonprofit organizations.

SOURCES: Program agent records and reports

TABLE 8

THE DISTRIBUTION OF PSE JOB CATEGORIES BY CHARACTERISTICS OF PARTICIPANTS
AND TYPES OF EMPLOYING AGENCY

		Professional			Paraprofessional			Service Worker			Skilled Labor			Clerical Worker			Laborer		
		%B	%F	%ED	%B	%F	%ED	%B	%F	%ED	%B	%F	%ED	%B	%F	%ED	%B	%F	%ED
City A	City	33	17	50	20	33	47	30	4	62	0	0	46	49	80	46	54	4	50
	OP	75	50	88	50	75	70	38	36	80	33	0	33	75	93	100	13	0	50
	PNP	20	80	20	33	50	17	32	48	45	0	0	100	29	75	43	50	50	50
City B	City	0	0	0	29	0	0	34	37	14	14	0	43	27	73	13	46	2	18
	OP	0	40	0	50	40	10	56	33	22	0	0	0	75	100	25	20	0	20
	PNP	17	66	17	71	38	7	100	0	0	0	0	0	83	100	17	100	0	10
City C	City	50	50	0	50	30	20	31	3	3	36	0	36	44	88	44	49	0	29
	OP	0	67	0	43	50	14	60	0	20	0	0	0	25	67	33	0	0	0
	PNP	50	100	50	88	47	53	45	27	9	0	25	75	75	100	58	100	0	0

Key: B = Black
 F = Female
 ED = Economically Disadvantaged
 OP = Other public agencies
 PNP = Private nonprofit organizations

SOURCES: Program agent records and reports.

departments served the greatest number of blacks in the laborer category, but blacks also tended to be well represented among clerical and service workers. Service rates for blacks were usually lower for professionals, paraprofessionals, and skilled laborers, but these types of jobs were not funded by the cities in great numbers. City C diverged from this pattern by having a relatively high percentage of blacks served in the professional, paraprofessional, and skilled labor categories.

Service patterns for women, within the city departments, were very clear in certain job categories. They served in great numbers among the clerical workers, while they were virtually excluded from the labor and skilled labor jobs. In the other categories women received varying rates of service. They were not well represented among professionals in cities A and B but received 50% of the professional jobs in city C. In the paraprofessional and service worker categories, rates of service to women tended to be low (around 30%) but were very low in some instances (service workers in city A and paraprofessionals in city C).

Service to the economically disadvantaged was not as patterned as expected. In cities B and C most of the disadvantaged participants were either laborers or clerical or service workers, but in city A service to the economically disadvantaged was relatively constant across job categories.

The service patterns for the private nonprofit agencies and the other public agencies in some cases mirrored those of city departments, but there was more variation present. The one pattern that held across agency types was that women were served at very high rates in the clerical category, but very few of the laborers or skilled laborers were women. Overall, the opportunities for women in the professional, paraprofessional and service worker categories appeared to be better in these agencies than within city departments, but there were examples of very low rates of service to women in these types of jobs within private nonprofit and other public agencies. The private nonprofit agencies in cities A and B had the highest rates of service to blacks and the economically disadvantaged in most of the job categories. In city A the other public agencies had the highest rates of service for these two groups.

The data presented in Table 9 provide some insights into the relationship between the characteristics of PSE participants and the wage they receive. In general, the data reinforce the patterns suggested above regarding the relationship of job type and participant characteristics. Among city departments, service rates for women are clearly highest in the lowest wage categories and are very low in the highest wage categories. This pattern also holds for the private nonprofit and other public agencies, but the drop-off in service in the higher wage categories is not as severe as in the city departments.

Service to blacks also tended to fall off as wage rates increased, but not as sharply as was the case for women. This pattern was more visible for the city departments than for the other employing agencies. Many of the private nonprofit and other public agencies displayed a pattern of gradually declining service

TABLE 9
THE DISTRIBUTION OF PSE WAGE RATES BY CHARACTERISTICS OF EMPLOYEE
AND TYPES OF EMPLOYING AGENCY

		$2.01-$3.00			$3.01-$3.50			$3.51-$4.00			$4.01-$4.50			$4.51+		
		% B	% F	% ED	% B	% F	% ED	% B	% F	% ED	% B	% F	% ED	% B	% F	% ED
City A	City	62	86	59	53	53	67	54	4	50	30	3	54	35	8	56
	OP	67	100	83	53	63	77	56	11	78	13	6	69	47	79	26
	PNP	29	62	46	33	56	22	22	20	22	0	0	0	0	0	0
City B	City	20	100	0	45	77	13	48	6	19	33	0	15	19	0	19
	OP	100	50	0	67	100	33	42	17	17	17	33	17	40	0	0
	PNP	100	75	25	100	60	20	33	67	33	50	50	0	50	42	8
City C	City	50	75	13	60	50	20	38	0	36	30	0	24	40	20	0
	OP	39	56	22	25	75	0	0	67	33	0	50	0	18	18	35
	PNP	56	69	50	67	78	22	100	60	80	57	29	29	88	12	25

Key: B = Black
F = Female
ED = Economically Disadvantaged
OP = Other public agencies
PNP = Private nonprofit organizations
SOURCES: Program agent records and reports.

to blacks in higher wage categories. This pattern was reversed in the highest wage category, which often included a substantial proportion of blacks.

Patterns of service to the disadvantaged did not appear to follow a clea pattern. There seemed to be some tendency for the highest rates of service to the economically disadvantaged to be located in the middle wage categories, bu there were a number of exceptions to this trend.

In summary, the data presented on the patterns of job creation for PSI programs support a number of observations. Jobs funded in city department tend to be heavily concentrated in the service worker and laborer categories which typically involve wages between $3.50 and $4.50 per hour. The privat nonprofit and other public agencies usually offer more professional, paraprofes sional and clerical jobs than the city departments. The wage schedules for th outside agencies are more skewed than in city departments. Thus, these agencie often fund a high percentage of jobs in the lowest wage categories but also offe a number of positions which fall in the highest wage category (above $4.50 pe hour).

Blacks and economically disadvantaged participants are most often foun among the service workers, clerical workers, and laborers. Women are highl represented among clerical workers but are virtually excluded from the labore and skilled labor jobs. Women are better represented in the professional an paraprofessional jobs offered by private nonprofit and other public agencies tha they are in these same jobs in city departments. Sex discrimination with regar to wages (women tend to receive the lowest paying jobs) is quite clear regardles of the type of employing agency, but it is slightly more pronounced in cit departments than in the other types of employing agencies. There is also a les pronounced pattern of hiring black participants in lower paying positions i other agencies compared to city departments.

WHAT HAPPENS TO THE EMPLOYEES?

Everyone connected with PSE programs recognized that, at some point, th participants should move out of the program and into an unsubsidized jot either within the agency in which they were employed as a PSE participant c with some outside employer. There are, however, obvious incentives for progra agents and employing agencies to ignore this aspect of the program. Like an other employer, they prefer to have a stable work force of experienced em ployees who do not require constant training and supervision. Therefore, whe they employ a PSE person who performs well on the job, they usually try t keep this person on the program for as long as possible. This strategy has tw advantages from the employers point of view. It allows them to augment the work force with a capable employee, and the wages for this employee con from federal revenue rather than their own resources.

These sorts of considerations led Congress and the Department of Labor t impose placement requirements for EEA and placement goals for CETA. Ho ever, the CETA transition goals were effectively abandoned by the Department of Labor in January 1975 because of the rapidly rising rate of unemployment. With unemployment at such high levels it was thought unrealistic to expect that many PSE participants could be placed in unsubsidized jobs. The decision to relax the placement goals was also seen as a way of promoting rapid hiring in the program so that congressional demands for increased PSE enrollments could be met. Up to that point many prime sponsors were having trouble meeting their plans for PSE enrollments because employing agencies were wary of making commitments to absorb CETA people on a regular schedule.

As a result of this Labor Department policy, most program agents did not place a great deal of emphasis on moving PSE participants into unsubsidized employment during fiscal years 1975 and 1976. The general pattern was that once participants began working in a PSE job they remained in that position as long as CETA funds continued to be available. Many of the heads of employing agencies indicated in interviews that they planned to absorb many of their PSE employees, but they would not do so until they were no longer able to get CETA funds to pay the wages of these people. It was not uncommon to find a number of CETA participants, who were originally hired under EEA, in a PSE program. Some of these people had held federally subsidized jobs for four or even five years. There is nothing in the law or the regulations that limits the length of time that a person can be employed in a PSE job.

Local PSE staff people cited a number of reasons for not insisting on better placement rates in their programs. First, they pointed out that high rates of unemployment made placement difficult. Second, they claimed that the absence of federal requirements made local efforts to encourage placement nearly impossible, especially if the local elected officials were not willing to make it a matter of official policy. Finally, they said that there were good reasons not to emphasize placement. Employing agencies tended to react negatively to placement quotas, and the pace of implementation slowed when there was high turnover in PSE jobs. In addition, performance assessments of the regional offices of the Labor Department had been based mainly on the prime sponsor's ability to spend PSE money within the planned time frame, which meant that

TABLE 10
PSE PARTICIPANT ENTERED EMPLOYMENT RATES[a] THROUGH JUNE 1976

	Title II	Title VI
National totals	11%	16%
City A	27	11
City B	7	18
City C	8	8

a. The entered employment rate represents the percentage of the cumulative total of participants served in each title who have entered unsubsidized employment since receiving CETA services.

SOURCES: Prime sponsor and program agents quarterly reports.

slow PSE implementation caused by emphasis on placement could lead to a poor performance assessments. In short, it was much easier to operate a PSE program when placement was not emphasized.

The placement figures shown in Table 10 demonstrate the effect of the incentives working against high rates of placement in CETA PSE programs. By the end of fiscal year 1976, 11% of those enrolled in Title II and 16% of those enrolled in Title VI nationally had been placed into or found unsubsidized employment. As might be expected, placement rates varied among program agents, as the figures shown for the three cities suggest. While local economic factors, especially the fiscal soundness of local governments, are certainly important influences on the ability of program agents to place PSE participants, there were at least two strategies that some program agents employed to achieve better than average placement rates. One strategy was for PSE staff and elected officials to pressure employing agencies to make absorption pledges and then force them to live up to those pledges. This was done in the Title II program in City A and resulted in a better than average placement rate of 27%. The same emphasis was not present for the Title VI program, and the placement rate of only 11%. The other strategy was to hire nondisadvantaged recently-laid-off workers on the program. In time many of them would find their own unsubsidized jobs or be recalled to their old job. City B followed this strategy for its Title VI program and achieved a placement rate of 18%, which was based on a large number of self-placements. City C did not follow either of these two strategies. The staff did not stress placement, and they hired many disadvantaged participants, which resulted in low placement rates in both titles.

GENERAL POLITICAL AND PROGRAMMATIC ISSUES

The foregoing discussion suggests that there are a number of open and important questions about the noneconomic dimensions of PSE. (And recall that the economic issues are far from settled.) Ultimately, the noneconomic questions are both normative and political. Who should benefit from the programs both in terms of individuals and in terms of agencies? What kinds of jobs are appropriate for these programs? Do job holders move on to permanent jobs? Does employment in PSE enhance the employability of the participants after they leave PSE slots? Under what political conditions can both elected officials and professional staff members resist the seemingly strong pressures to use PSE money primarily to guard city or county payrolls against shrinkage by employing mostly white males in the better jobs and females and minority persons and economically disadvantaged persons in seemingly "dead end" jobs?

The data that we have presented illustrate the problems with CETA PSE if one is committed to trying to use the program to enhance the employability of the disadvantaged, to serve high proportions of women and minorities with good jobs, and to stress transition to permanent unsubsidized employment. A number f the individual patterns are suggestive both of additional research that needs to e done on a much larger scale and of steps that Congress and the Department of abor might take to alter the situation. The individual patterns also support the eneral proposition that individual prime sponsors do have latitude to behave ifferently. The forces pushing for the general patterns noted above are strong, ut with resolution individual prime sponsors can also succeed fairly well in inning against national or state trends. At minimum it would seem wise for ocial scientists and government officials alike to continue exploring how to reate the conditions under which the chances of such success are increased.

NOTES

1. These categories are loosely based on U.S. Bureau of the Census occupational issifications. The professional category follows the guidelines of the Bureau of the Census issifications for professional occupations. The paraprofessional category is basically used r all nonprofessional, nonmanual labor, and nonclerical positions. The category of service ain follows the Bureau of Census guidelines and includes mostly janitors and maintenance orkers, protective service workers, and health and recreational aides. The skilled labor tegory is slightly more broadly defined than the Bureau of Census listing of craft cupations. It includes all the crafts and some noncraft occupations requiring formal skills. ie clerical category includes secretaries, clerks, typists, etc. and follows the Bureau of ensus guidelines. The laborer category is used only for jobs with titles that were specifi- lly designated "laborers" or "semiskilled laborers."

REFERENCES

ongressional Budget Office (1975). Temporary measures to stimulate unemployment: An evaluation of some alternatives. Washington, D.C.: Author.
AVIDSON, R.H. (1972). The politics of comprehensive manpower legislation. Baltimore: Johns Hopkins University Press.
ECHTER, A. (1975). Public employment programs. Washington, D.C.: American Enter- prise Institute for Public Policy Research.
EVITAN, S.A., and TAGGART, R. (1974). Emergency employment act: The PEP genera- tion. Salt Lake City, Utah: Olympus.
-- (1976). The promise of greatness. Cambridge, Mass.: Harvard University Press.
IPLEY, R.B., et al. (1977). The implementation of CETA in Ohio. Washington, D.C.: U.S. Government Printing Office.
HIELDS, P.E. (1976). "Subgovernments; An application to the Comprehensive Employ- ment and Training Act of 1973." Unpublished paper.
INGER, J.W. (1976). "Carter's first big job—Getting people back to work." National Journal, 8(51-52):1794-98.
OLOW, R.M. (1976). "Macro-policy and full employment." Pp. 37-58 in E. Ginzberg (ed.), Jobs for Americans. Englewood Cliffs, N.J.: Prentice-Hall.
LMAN, L. (1976). "Manpower policies and demand management." Pp. 85-119 in E. Ginzberg (ed.), Jobs for Americans. Englewood Cliffs, N.J.: Prentice-Hall.
an HORN, C.E. (1976). "Intergovernmental policy implementation: The Comprehensive Employment and Training Act." Unpublished Ph.D. dissertation, Ohio State University.
ISEMAN, M. (1975). "On giving a job: The implementation and allocation of public service employment." Washington, D.C.: U.S. Congress, Joint Economic Committee.

13

Public Employment Policy-In-Use: The Atlanta Experience with CETA

CHARLES W. WASHINGTON

☐ THE COMPREHENSIVE EMPLOYMENT AND TRAINING ACT (CETA) of 1973, as amended, is designed to achieve objectives and goals which reflect different political preferences at both the national and local levels. CETA represents the maximum reduction of tensions between competing interests within the Congress, between the Congress and the President, between the Congress and manpower client advocates and traditional service providers, and between potentially new (prime sponsor) service providers and old manpower service deliverers.

This article presents an overview of a simplified model of the current manpower public policy system[1] and reports the findings of an on-sight study of the Atlanta, Georgia, public employment policy-in-use under CETA.

The focus is on Public Service Employment (PSE) under Title II of CETA. The outcomes of PSE are highlighted by such questions as these: Were the jobs made available under PSE in addition to existing jobs in the local jurisdiction? Did the PSE jobs that were created help to provide otherwise unprovided public services? Were PSE positions integrated into the city's governmental structure in such a manner as to assure continued delivery of city services? Were PSE clients used to replace city employees, thus becoming substitutive in nature with the salary of such employees comprising, in effect, a federal subsidy to local government budgets? Did the prime sponsor maintain its local effort in level of government service and employment when PSE positions under CETA were authorized? Were jobs created quickly? Were PSE positions created "loosely and handed out with no consideration of an applicant's ability or potenti ability to do the job? Were the jobs that were created meaningful jobs? Wa transition into permanent jobs with city or nonprofit private or publi agencies likely? Did PSE jobs help the economically disadvantaged as well a the unemployed?

These are difficult questions to answer even with the best of data an unlimited time and research resources. Certainly, they could not be adequatel answered in a case study in which there had been limited experience with a PS program and in which the state of analysis was during the first year of operatio However, the results produced by the end of the first fiscal year in Atlant provide some insight into several of these questions. Many of the first (trans tional) year outputs are intertwined with administrative and procedural concern which can influence or alter their relative significance. For this reason, substar tial procedural and administrative concerns are reflected in this paper.

THE MANPOWER POLICY SYSTEM

As enacted and amended, CETA became the "national espoused publi policy" (NEPP) for manpower. It prescribed the national preferences for mar power training and development opportunities, established the relevance an preferred use of public employment, outlined the framework for the design o artificial organizational structures and administrative processes by state and loca prime sponsors, and stated the types of anticipated outcomes. While the mar power NEPP was being formulated, the philosophical concept of "new federal ism" (Nixon, 1971), under which CETA was proposed, underwent modifica tion—if not, in fact, redefinition. Contrary to the intent to leave all manpowe policy and program priority considerations to state and local discretion, nationa goals, purposes and preferences were either explicitly stated or implied in th manpower NEPP.[2]

The CETA NEPP may be appropriately viewed as a theory for achievin national manpower goals.[3] It includes a number of assumptions about a pre sumed causal relationship between fiscal resources utilization, programs, activi ties, administrative arrangements, procedures and guidelines, and desired man power outcomes.

The statements and propositions that comprise titles of the policy ar primarily prescriptive and control in nature rather than descriptive, explanatory and solely permissive or discretionary. Virtually every section of the policy (th act itself) includes an assumption that if "X" is done, then "Y," some outcome is realized or is likely to be realized (Suchman, 1967). For example, there is th assumption that, if federal revenues are shared with state and local governmen prime sponsors, rather than with community action agencies, then greate coordination of manpower programs will result when one or a number o lternative activities is conducted. Or there is the assumption that if funds are rovided for public service employment (PSE), then transitional employment to meaningful jobs will result.

OMPONENTS OF THE MANPOWER POLICY SYSTEM

The national espoused public policy (NEPP) is only one aspect of the total manpower policy system, and CETA must be viewed in that context. Other lements of the public policy system, whether for manpower or otherwise, under ne "new federalism" block-grant approach include public policy rule modified PPRM), public policy discretion modified (PPDM), local espoused public policy LEPP), and—what we see when we review a given policy system in operation— ublic policy-in-use (PPIU).[4]

An understanding of the relationship and linkage between these system lements is critical to any policy analysis—i.e., essential to making any judgment bout the extent to which observed outcomes of a policy at the recipient or lient level are congruent with national policy intent and purposes, given esources and costs.[5] This is true of any substantive policy, but it is especially true manpower and other block-grant revenue-sharing policies which recognize rime sponsors' rights to formulate policy plans to achieve a set of desired utcomes.[6]

It is a necessity (and often a practice) to look at the relationship between the ational espoused public policy and the public policy-in-use when one conducts n evaluation of a particular policy. Between the NEPP and the PPIU (the set of olicy related activities put into effect to achieve local or state preferences) and esultant outcomes are several changes which occur in the original national spoused policy. These changes are produced by federal rules and regulations nd the exercise of discretion by decision makers in the PPRM and PPDM stages r by others permitted to do so by provisions resulting from these stages. One ffect of these changes is that the original intent and purpose of the NEPP ecomes obscured, if not significantly modified, to the point that the measures f outcomes (usually quantifiable outputs) should be made against a different et of goals and objectives. This different set of goals and objectives would nclude changes resulting from public policy rule (PPRM) and discretion modifi- ation (PPDM), taking into consideration the nature of the implementation rocess in the PPIU stage.

HE SIMPLIFIED POLICY MODEL

Public policy rule modified (PPRM) can and does affect the form of policy utputs as well as the types of decisions made at the discretion of federal, state, nd local program administrators in the public policy discretion modified stage. simplified version of the relationship among the stages and components of the ublic policy model is presented in Figure 1.

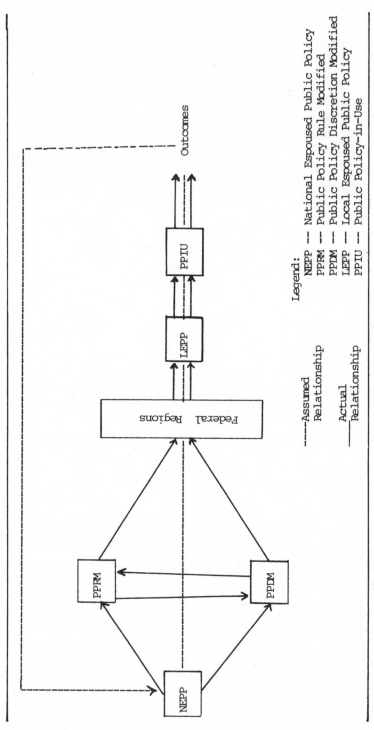

Figure 1.   SIMPLIFIED MODEL OF PUBLIC POLICY SYSTEM EVALUATION LOOP

The evaluation loop in the model is often seen as the way to determine the effectiveness and fate of a given policy without consideration of the modifications in the NEPP which are produced by the PPRM, PPDM, and the LEPP stages and processes. And too often at the urban policy level, the complexities of the implementation system, the PPIU stage, are simply glossed over. The public policy rule modification stage produces changes in the national espoused public policy by promulgation of federal rules and regulations, which may also influence or expand the degree of discretionary powers of certain decision makers in the PPDM stage. Public policy is further potentially modified by the decisions of individual federal representatives in the 10 federal regions (as indicated by the vertical rectangle between the NEPP, PPRM, and PPDM on the one hand and the LEPP on the other.[7]

The conventional assumptions about the public policy system includes the assumption that there is congruence between what is espoused nationally (the NEPP), what is espoused or planned at the state or local level (the LEPP or SEPP), and what is actually implemented (the PPIU).[8] Each of these stages involves a set of actors—decision makers, service providers, clients, interest groups—and an environment. All but one stage—the PPRM—also includes fiscal resources. The set of actors in each stage differs from those of other stages but also includes some common linkage with the NEPP and the other components—a perceived program objective. It is quite difficult then, even with the most sophisticated of quantitative methodology, to make accurate judgments about the success or effectiveness of a given policy without an understanding of the influence of these intervening stages and associated factors on the shape of the original policy. This is especially true when a congruence between local and national preference (LEPP and NEPP) is desired but not mandated by the public policy as codified in law. This is the case with the block grant for manpower, whereby local prime sponsors (including Indian tribes and designated Concentrated Employment Programs) may develop their own preferences for resource use within defined constraints. These constraints are specified by administrative rules and regulations (PPRM) and discretionary decisions (PPDM) made by individual decision makers at the federal and regional levels as permitted by explicit statements or implied provisions in the NEPP.

CETA AS A NATIONAL MANPOWER POLICY-IN-USE

NATIONAL ESPOUSED PUBLIC POLICY INTENT AND PURPOSE

The legislative intent of CETA is (1) to provide a flexible and decentralized manpower planning and administrative system to state and local governments' political decision makers—governors, mayors, county commission chairmen, and chief elected officials—and (2) to assure that manpower services, not just training but also job opportunities, will be made available to those most in need of them including low-income persons of limited English-speaking ability (CETA, 197: sec. 2).

Under CETA, a strong federal role was retained in the establishment (national objectives, priorities and performance standards, the review and ap proval of prime sponsors' and eligible applicants' plans and plans modification the assessment of manpower performance, and the provision of technica assistance.

The purpose of CETA is to provide job training and employment opportun ties for the unemployed, the underemployed, and the economically disadvan taged. The purpose is also to assure that training and other manpower service will lead to maximum employment opportunities and enhanced self-sufficiency These purposes are to be achieved through four types of programs authorized b CETA, as amended in 1974: comprehensive manpower services, public servic employment, special federal programs, and emergency jobs.[9]

CONTEXT OF ATLANTA'S PUBLIC SERVICE EMPLOYMENT UNDER CETA

Atlanta is the capital of the state of Georgia and the county seat and centra city of Fulton County in a 15-county Standard Metropolitan Statistical Are (SMSA). The estimated population for the city of Atlanta in midyear 1975 wa 474,600, for the 15-county SMSA 1,728,400.

The Atlanta area growth is typical of patterns in SMSAs across the country Suburban growth has been accompanied by declining central city population The city's population increased steadily from 1900, with black population lagging behind white population. After 1960 the black population began t move toward becoming the majority population, which was achieved in 197(when the city's total population reached its peak. Black population continued t increase marginally, but total city population, caused by white flight, began t decline. Although the black population has continued to increase in the city since 1970, it is becoming increasingly a smaller proportion in the city sinc 1970, it is becoming increasingly a smaller proportion of the total metropolita area population, dropping from 34.5% of SMSA population in 1970 to 29.0% i 1975.

Employment. The industrial mix in Atlanta has remained over the pas decade rather stable, with annual employment growth occurring more in ser vices, government, retail trade, and construction. The percentage employmen growth between 1960 and 1972 for each of these industries was respectively 110.5%, 104.4%, 92.1%, and 80.9%.

The unemployment rate for the city in each month during 1974 exceede(that of any of the surrounding jurisdictions qualified to be prime sponsors unde CETA. Unemployment ranged from a low of 5.2% in May to 9.0% in Decembe 1974, compared to a low of 2.9% in DeKalb County in May (the lowest of al surrounding jurisdictions) to 7.7% in December in Cobb County, the highest, i the same year. The average unemployment in the city for the first five months o

1975 was 13.2% compared to 10.0% unemployment in Cobb County for the same period of time, the highest unemployment within any of the surrounding jurisdictions.

Income. In 1960 the median income for the Atlanta metropolitan area was only 53.8% of what it was in 1970 and was $408 less than the 1960 national median income for all urban areas. Median family income in 1960 in the Atlanta SMSA was $5,758 compared to $10,695 in 1970. In the city of Atlanta the median income was $5,009 in 1960 and $8,399 in 1970 (Research Atlanta, 1973). Over the past decade, Atlanta has lagged behind other jurisdictions in the SMSA in the improvement in median family income. DeKalb, Cobb, and Fulton counties lead in higher family income levels.

Governmental Structure. CETA was implemented in Atlanta shortly after a significant governmental reorganization, effective January 7, 1974. With the reorganization, political power changed from predominantly white control to control by a racially sensitive black mayor and a racially balanced city council comprised of 50% black and 50% white members, with a white city council president. A strong mayor-council form of government replaced a weak mayor-council form of government. Executive powers were vested in the mayor, elected at-large to service a four-year term. All legislative powers were vested in a 19-member city council, 12 of whom are elected from single-member districts and 7 of whom, including the council president, are elected at large and may serve for four-year recurrent terms. The mayor was also authorized to appoint a professional staff in the office of the mayor to effectively carry out the responsibilities of government under the new charter.

Prior Manpower Experience. Prior to CETA, categorical manpower programs were operated by state and local public and private nonprofit agencies in the Atlanta SMSA. These agencies established various degrees of community-based confidence and demonstrated effectiveness. They constituted the "parties at interest".[10] Included among such agencies and programs were the Atlanta Board of Education, the State Department of Vocational Rehabilitation, the Work Incentive Program, the Neighborhood Youth Corps, the Atlanta Residential Manpower Center, the Atlanta Area Technical School, the Georgia Training Employment Service, the Atlanta Urban League, the Atlanta Skills Center, the Organizations Industrialization Incorporated, and the Atlanta Business League.

The major effort to coordinate manpower planning before CETA was made through the Atlanta Manpower Area Planning Council (MAPC), comprised of Atlanta and the counties of Fulton, DeKalb, Clayton, Cobb, Gwinnett, Rockdale, and Douglas. The MAPC was the metropolitan area's version of the 1971 restructuring of the earlier Cooperative Area Manpower Planning System (CAMPS) that developed in 67 labor market areas across the country between 1968 and 1971 (Levitan and Zickler, 1975). The MAPC was a regional manpower effort comprised of chief elected officials, clients, and business and labor representatives from within the metropolitan area.

The MAPC plan for fiscal year 1974 urged the planning, coordination, organization, and implementation of a comprehensive manpower program affecting the Atlanta multicounty area and included recommended levels of funding for individual programs (Massell, 1973:1-8).

The experience of local governments cooperating before the passage of CETA, plus the consortium provision and accompanying incentives under CETA, were insufficient to encourage an Atlanta metropolitan area consortium rather than single prime sponsor jurisdictional manpower programs. Atlanta is not a member of a consortium comprised of itself and surrounding jurisdictions. There are, instead, five prime sponsors for CETA programs in the Atlanta metropolitan area: DeKalb, Fulton, Cobb, and Clayton counties, and the city of Atlanta.

PEP Program. A modest amount of transitional public service employment was retained by Atlanta from the Public Employment Program (PEP), authorized under the Emergency Employment Act of 1971, and extended through the end of the first fiscal year of CETA. Between September 8, 1971, and June 30, 1975, the city received $3,186,602 of EEA funds and spent $2,170,585 of that amount for public employment programs. Of the 2,690 participants enrolled during the grant period, only 113 (or 4.2%) were transitioned into unsubsidized employment.

ATLANTA'S LOCAL ESPOUSED PUBLIC POLICY PRIORITIES
FOR PUBLIC SERVICE EMPLOYMENT CLIENTS

Priority for Title II enrollment under CETA from the significant segments of the population (areas within the city with an unemployment rate of 6.5% or more) was given first to the unemployed[11] then to minorities, females, special veterans, and recipients of public assistance.

The local espoused public policy intent was to allocate work experience job slots under Titles II and VI to various departments of city government.[12] As positions in these departments became vacant through normal attrition, CETA-funded employees would be transitioned onto city payrolls as regualr unsubsidized employees.

Under Title II, the public service employment provision of CETA, the Atlanta LEPP anticipated filling 1,168 PSE slots by the end of the first fiscal year and transitioning 25 of these onto the public payroll. This anticipated outcome carried with it an expenditure of $1.9 million, 98% of which would be expended for client wages. The balance was to be used for services to clients. Title II outcomes were to take the form of work experience.

Under Title VI, the emergency jobs program, the focus was on work experience rather than PSE. The anticipated outcomes included the provision of 505 new temporary jobs to be distributed among the significant segment in the following preference order: unemployed 15 weeks or more, 380; minority persons, 330; females, 200; and Vietnam veterans, 150. An anticipated expenditure of $1.489 million was to produce this outcome, with 87% of this amount going to temporary work experience and 13% to 195 PSE jobs.

PUBLIC SERVICE EMPLOYMENT: POLICY-IN-USE

Program Design. Atlanta utilized a concept of "dual service" in the design and implementation of its public employment efforts. The concept means that the employed CETA client would be placed in a PSE job to provide a service to some other needy segment of the population or would be placed in jobs in which additional services to other public goods recipients were warranted.

Jobs were designed to meet both the needs of the economically disadvantaged and the demand for additional services made upon public and private nonprofit agencies. To achieve this "dual service" objective, the Atlanta PSE program entered into contracts with public and private nonprofit agencies for placing PSE clients. Job positions, called slots, were negotiated with the intent that the cooperating agency would work toward transitioning the employee onto its payroll once CETA funds had been exhausted. Other PSE positions were created within city government departments through the city personnel system. In each instance, the upper limit on the salary of CETA PSE clients was the federally imposed $10,000 maximum. A public service job under Title II in the Atlanta Manpower Program had a specific title, job description, and salary based on an established pay schedule. A work experience job (for which Title VI was most exclusively used and for which a substantial part of Title II was used) was defined as a job with one of three general titles—Work Experience Aide I, Work Experience Aide II, or Work Experience Aide III.[13] All clients within each of these job titles received the same salary irrespective of the work assignment.

Administrative Design and Initial Operation. The PSE program got off to a slow start in Atlanta. It was not until November 1974 that the Title II office was fully staffed and program planning and implementation were underway. The PSE program was administered by a special projects coordinator on loan from the U.S. Department of Labor under the Intergovernmental Personnel Act of 1970. The staff of the Title II office in fiscal 1975 consisted of the Title II administrator, two manpower coordinators, two manpower counselors, and a stenographer. This staff is accountable to the deputy director of the Atlanta Manpower Program, who reports directly to the director. The director and deputy director work cooperatively with the various committees of the Employment and Training Advisory Council, the full council, the commissioner of the Department of Community and Human Resources, and various city departments affected by the PSE program.

The initial identification of community needs for Title II was more a function of administrative preference, advisory council agreement, and external (noncity) agency agreement than of any determination based on "objective" criteria of need. Except for administrative sensitivity, there existed no criteria against which to determine whether the selected priorities were more important than some others.

The first concern for PSE usage focused on inadequate building and grounds maintenance of public housing structures. The second concern was the problem of excessive delays and administrative backlog in rendering timely service to persons qualified to apply for food stamps and unemployment insurance. The third concern was the rising incidence of battered children thought to be related to the unemployment condition of persons and families not able to sustain them economically.

Contracting. By the end of fiscal 1975 contracts had been negotiated and made operational between the city and a number of community-based public and private nonprofit agencies. These included the Atlanta Housing Authority, the Georgia Department of Labor, the Fulton County Department of Family and Children Services, Economic Opportunity Atlanta, Inc., and The Community Coordinated Child Care Program. The Neighborhood Arts Center and the United Way of Metropolitan Atlanta were later added.

The 40 positions negotiated with the Atlanta Housing Authority consisted of 35 maintenance, clerical, and social-service type positions. Twenty positions with the Fulton County Department of Children and Family Services were earmarked for the Children Protection Services Division to help handle the rise in battered children cases. Some were assigned to the food-stamp-processing operation to help reduce the delay in processing food stamps applications.

The 13 positions negotiated with the Georgia Department of Labor were designed to expand the Unemployment Insurance Division to improve the service delivery of insurance claims to eligible recipients.

Under the EOA contract PSE persons were assigned to the Georgia Citizens Coalition on Hunger to do outreach activity to identify citizens eligible to receive food stamps. So, on the one hand, PSE enrollees were hired to generate demand for a service by those eligible to receive it and on the other hand to supplement the staff of agencies rendering such services. Other EOA slots were designed for different purposes. The 289 work experience positions (some funded under Title VI) were designed to render services related to handling federal offenders, rehabilitating the emotionally disabled, rehabilitating of drug and alcoholic clients, processing Title I clients, registering Title VI clients in the emergency jobs program, and taking job orders at the Georgia Department of Labor.

Contracts with EOA, United Way, and other private nonprofit agencies were undertaken pursuant to a public policy rule modification made by the U.S. Department of Labor in an interpretation of the NEPP. Prime sponsors were authorized to expand public service jobs into noncity government areas. A regional interpretation of the regulations permitted Atlanta to create work experience positions (typically found in Title I type programs) under Titles II and VI.

The contract with the United Way of Atlanta earmarked funds to expand the staff of the Atlanta Legal Aid Society, a community service legal agency, through PSE to improve the agency's ability to provide legal services and consultation to CETA clients. Under this contract, the first work experience positions were made available to offender clients from state and county work release centers and to probationers, parolees, and ex-offenders. These clients were enrolled by the United Way and detailed to its member agencies within the community for work assignments at $3 per hour, plus fringe benefits, on the agency's payroll. Each agency made its own selection of clients from a list of eligibles certified by the Title II office.

The large number of positions negotiated with the EOA was a result of a decision by the regional office of the U.S. Department of Labor that funds which the city had earmarked in its LEPP plan from Title II for a summer youth program in 1975 had to be moved directly into the economy for PSE before or by June 30, 1975, the end of the fiscal year. This was an effort to boost the use of Title II funds as a countercyclical device. The effect of this decision was an increase in the number of positions with EOA (some of which were temporary work experience) and a negotiated contract with the Neighborhood Arts Center. Twelve positions were established in the Neighborhood Arts Center, a nonprofit organization of local artists. The artists were required to do a minimum of 20 hours of instructional tutoring per week in and around the Neighborhood Arts Center and to devote 20 hours to the Center's planned activity in the artist's medium for his own benefit.

The Community Coordinated Child Care Program, a contractor under the city's Title I program component, as well as under Title II for fiscal years 1975 and 1976, provides another example of the "dual service" concept in operation. Its program under Title II was designed to use PSE positions—child-care trainees—to establish home day-care centers within the community and in the homes of mothers or female heads of households who could not leave the home because of the presence of either preschool children or invalid relatives.

A child-care trainee was placed in the home of a mother or female head of household interested in learning how to run a day-care service. The mother or head of household learned about child-care services from the PSE client and earned income from parents who left their children for child-care services in order to pursue for themselves employment or training.

Noncity agencies contracted to hire persons in PSE positions were required to use their respective personnel systems for recruitment according to job description and job title within their organization. The Title II office verified the eligibility of the applicant according to federal rules and regulations, especially the residency and unemployment requirements. Once hired, the client became a part of the Title II central management system but was to be treated by the agency the same way that other agency employees were treated.

City PSE Positions. The Atlanta Manpower Program (through its Title II office) and the Personnel Department established several PSE positions within city government as part of the classified service. These were referred to as CETA II positions. To avoid charges that less trained individuals were employed with CETA moneys to replace regular employees, Atlanta established a set of procedures for creating CETA II positions. The city did not during the first fiscal year of CETA lay off any of its employees because of either budget constraints or the receipt of CETA funds. A hiring freeze was in effect in the city when CETA funds were received.

No positions with salaries at or over $10,000 were created within city government. This kept the city from having to supplement salaries. Each classified PSE position was to be a new position created over and above the existing staffing levels. To create a CETA II PSE position in city government, the particular department head was required to submit to the Atlanta Manpower Office the position title, pay grade (less than $10,000), and justification for the position. The manpower office submitted the request to the Personnel Department for review and approval prior to being placed on the Civil Service Board's agenda. If the Civil Service Board approved the position, the Finance Department then had to review the request and justification and decide whether the need for the position was over and above the positions budgeted to the department and whether the need was sufficient to warrant funding. If the Finance Department approved the position as additive and justifiable, an ordinance establishing the position was drafted and submitted to the finance committee of the city council. The finance committee and the full city council had to approve the ordinance if the position was to be established. If the position was established, the department head originating the request had to submit a second request to the Personnel Department to fill the position through the regular civil service procedures. The Personnel Department then advertised the vacancy and established an eligible register for the CETA position.

This process accounted for much of the delay in creating classified PSE positions in the city, but it also was designed to assure that PSE positions were not "loosely" created and that such positions were in addition to existing positions. The 100 classified Title II positions filled within city government as of June 30, 1975, were not the only public employment positions within the affected department. Other positions were filled through work experience or classified Title VI positions. Thirty-two of these PSE positions were located in the Public Safety Department, 18 in Environment and Streets, 9 in Administrative Services, and the remainder distributed nearly equally among seven other departments. These 100 positions comprised 2% of the city's work force. However, when all 1,840 CETA employees are taken into account (administrative positions, PSE, and work experience), they comprised 18.7% of the 9,855 total city employees in the 1975 calendar fiscal year for Atlanta. Similarly, the total actual budget expenditure for CETA in the calendar fiscal year 1975 comprised 18.9% of current operating and capital outlays and 22.2% of current expenses.

Expenditures. Based on the cumulative planned and actual expenditures of the Title II PSE program for Atlanta, a final allocation of $2.1 million was received by the manpower program for Title II operations for fiscal 1975. Just under 75% of the cumulative planned expenditure was spent. The initial plans did not include expenditures for work experience positions. Less than one-half of all dollars spent under Title II was spent for public service employment. On the enrollment side, while 74.8% of the planned funds was spent, 60% of planned enrollment was achieved (PSE and work experience combined).[14]

Client Characteristics. The performance gap between what was planned (the local espoused public policy) and what was achieved (the PPIU) under Title II in Atlanta varied by objectives, as is shown in Table 1. The planned objective to serve 1,168 unemployed persons was 60% achieved. The special veterans goal was 82% achieved. Only the goal for serving public assistance recipients was exceeded. The poorest performance was in achieving the goal set for serving females. In the case of offenders and vocationally handicapped persons, there was no initial planned goal, but those placed in PSE jobs of these two categories comprised 29% of PSE jobholders.

Table 2 compares Title II client characteristics in Atlanta with the results of the NEPP in aggregate at the end of the first fiscal year. The clients nationally, and in Atlanta, were predominantly male, between 22 and 44 years of age, possessed a 12th grade or better education, were economically disadvantaged, and unemployed. The key differences were that in Atlanta the clients were predominantly black, and a larger percentage were offenders and special veterans than was the case nationally. The results demonstrate responsiveness of the Atlanta Manpower Program to the composition of its labor force, but shows that under Title I those most likely to receive training and work experience were

TABLE 1

PERCENT ACHIEVEMENT OF PLANNED OBJECTIVE TO SERVE SIGNIFICANT SEGMENTS OF TITLE II AREAS OF SIGNIFICANT UNEMPLOYMENT, ATLANTA MANPOWER PROGRAM (July 1, 1974-June 30, 1975)

Significant Segments	Persons Served (Number)[a]		Percent Actual Is of Planned
	Planned	Actual	
Unemployed	1,168	704	60
Special veterans	100	83	83
Public assistance recipients	50	52	104
Females	500	225	45
Minorities	930	634	68
Offenders	0	176	—
Vocationally handicapped	0	30	—

a. Figures are nonadditive because of overlapping categories.

SOURCE: Compiled from Atlanta Manpower Program, Cumulative Quarterly Progress Report, July 1, 1974 to June 30, 1975.

TABLE 2

PUBLIC SERVICE EMPLOYMENT, TITLE II, CLIENT CHARACTERISTICS,
ATLANTA MANPOWER PROGRAM AND NATIONAL SUMMARY
(July 1, 1974-June 30, 1975)

Characteristics	Atlanta		National	
	Number	Percent of Total	Number	Percent of Total
Total enrollment	(704)	100.0	(227,000)	100.0
Sex:				
Male	479	68.00	149,336	65.8
Female	225	31.96	77,664	34.2
Age:				
Under 22	201	28.55	53,799	23.7
22-44	464	65.9	142,783	62.9
45 and over	39	5.5	30,418	13.4
Education:				
11 grades or less	215	30.5	62,879	27.7
12 or over	489	69.5	164,121	72.3
Family Income:				
AFDC	51	7.2	13,620	6.6
Public assistance	1	.1	20,884	9.2
Economically disadvantaged	704	100.0	109,641	48.3
Ethnic Group:				
White	70	9.9	147,777	65.1
Black	631	89.8	49,486	21.8
Other	3	.4	29,737	13.1
Labor Force Status:				
Employed	0	—	8,853	3.9
Underemployed	—	—	19,068	8.4
Unemployed	704	100.0	200,668	88.4
Not in labor force	—	—	7,037	3.1
Offenders	176	25.0	6,583	2.9
Veterans:				
Special Vietnam	83	11.7	25,651	11.3
Other	72	10.2	28,602	12.6

SOURCE: Compiled from data reported on Atlanta Manpower Program, Title II Quarterly Progress Report, July 1, 1974-June 30, 1975, and as reported by the U.S. Department of Labor, Office of Administration and Management, September 1975.

females. Both of these types of services are likely to produce lower annual income (wages/allowances) than a classified Title II or work experience job.

THE EMERGENCY JOBS PROGRAM

The bulk of the jobs in Title VI were not a part of the classified service, i.e., part of the city's merit system. Most Title VI jobs were work experience. Job titles were simplified to Work Experience Aide (WEA) I, II, and III, at the respective salaries of $5,967, $7,345, and $9,048. There was no contracting under Title VI, but Title VI employees were assigned to nonprofit private and public agencies as city employees on assignment.

Expenditure and Performance. Between January 10, 1975, and June 30, 1975, one-third of the $3.2 million available to Atlanta was spent to make operational the Title VI program. The city's planned accrued expenditure for those six months was $1.489 million. It spent $1.085 million; only 1.85% of this was spent for administrative cost.

Eighty-eight of the 447 Title VI positions created were treated as classified transitionable slots in the city and were filled through the Personnel Department as Title II PSE positions were filled. Other Title VI persons were placed in city departments as work experience aides. Work experience positions under Titles II and VI consisted of 254 positions; 154 were located in the Department of Environment and Streets. The majority of the remaining positions was distributed among three departments: Administrative Services; Community and Human Development; and Parks, Libraries and Cultural Affairs. Of the remaining 193 positions, 54 were assigned to state and educational institutions, 125 were assigned to work in private nonprofit agencies, and 14 were assigned to special projects.

ADMINISTRATIVE COMPLEXITY

Seldom is the administrative minutiae necessary to get persons off the street and onto the job explored, discussed, or even appreciated by those who design or evaluate public policies at any level in the federal system. Yet administrative minutiae are precisely the elements that can cause a policy or program pursuant to it to succeed, look bad, or even fail. The administrative complexity of PSE and emergency jobs programs includes a number of discrete processes such as access, application, determination of eligibility, establishment of eligibility register, selection, counselling, placement and transitioning.

Access. Access to PSE jobs in Atlanta was gained via the Personnel Department if minimum requirements for filling the vacant position (as a regular city employee) were met and the basic CETA requirements, including residency within the corporate limits of the city of Atlanta and being unemployed for 30 days at the time of application, were satisfied. If qualified, the applicant had to complete both a regular city application form for employment and a set of CETA forms. The CETA forms provided information such as whether the applicant had been unemployed 15 weeks or longer, was currently receiving public assistance or welfare, or had been a former manpower program participant. Five preference points were awarded applicants for possessing either of these statuses. Veterans were automatically awarded 5 points, and 10 points were awarded to disabled veterans. No more than 10 points were awarded to any one applicant. Each applicant was evaluated on the basis of training and experience.[15]

Eligibility Register. Once applicants had been evaluated, their names were ranked in order of score and placed on an eligibility register for a particular job classification. The Personnel Department verified CETA eligibility (a second stage check for ineligibles) through the Planning and Zoning Department, which verified that the applicant's address was within the corporate limits of the city of Atlanta. When eligibility had been verified and ineligibles removed, a revised list of eligibles was sent to the appointing city agency or department. *Only names of qualified applicants* scoring one of the top three scores, listed alphabetically, were sent to departments and agencies rather than names *and* scores, since all applicants in the top three score categories were qualified. Eligible applicants were then advised of their responsibility to contact the appointing agency or department for interviews.

Selection Process. A department had to interview as many applicants as necessary before selection, but had to indicate on the register of eligibles specific action taken regarding *each* applicant on the register. This included an indication of who was hired, why certain persons were not hired, the starting date, position number, date of interview, persons not reporting, and which applicants indicated that they were not interested because of location, required hours, or salary.

Hiring information was passed on to the Title II Office, which contacted the newly hired employee, transmitted the name of the assigned counsellor, and provided explanatory information about the purpose and nature of the Title II PSE program. This is the first point at which the employee got detailed information about the PSE program.

Fringe Benefits. All Title II employees in city government received the same benefits as regular city employees. Benefits for CETA employees hired in noncity agencies by contractors were to be the same as those provided other employees hired by that agency. The cost of these benefits to employees in noncity agencies was incorporated in the contract price negotiated between the agency and the CETA Title II Office and approved by the Atlanta Manpower Program, the city council and the mayor.

Counseling. Each PSE client was assigned to a Title II staff counselor. Counselors assisted clients in solving problems developed during or resulting from long periods of unemployment, especially if they interfered with good work performance. Counselors were responsible for the eventual transition of assigned clients into permanent jobs either inside or outside of the city. This responsibility was carried out by monitoring both the client's performance and the vacancies that might develop at his work site or elsewhere. The client and his department head or supervisor were notified when a position became vacant to which the client was qualified to be transitioned (or transferred). Title VI emergency jobs holders were also assigned to counselors.

Transitioning. When an opening occurred in a department, a PSE client with the same job title and pay scale as the vacant position could be transitioned by a simple departmental request for reassignment. This meant that there would be no change in job title, pay, hours, or fringe benefits, except that the job would now be funded by the city.

When the CETA position was different from the opening that occurred and the CETA employee had been hired in the department for less than six months, the department had to requisition a new position to be approved by the Finance Department. The CETA employee had to apply for the new job through the merit system procedures on an open competitive basis and not on a promotional basis. If the employee's rating fell within the top three scores category, the department might exercise the choice to hire. This procedure assured that competent persons were selected for CETA PSE positions. If the CETA employee had been employed for more than six months the department might request that the position be filled on a promotional basis.

Displacement Potential. The current design of Title II and VI is such that fraud and displacement are greatly minimized. The greatest opportunity for fraud exists at the point of initial enrollment.

CETA AS AN ANTIRECESSIONARY DEVICE

Title VI was used by Atlanta as the key antirecessionary device to quickly reduce the level of unemployment in the city. The effect was that there were more problems encountered under Title VI than under Title II with its administrative complexity. By the end of the first six months of fiscal 1975, 102 persons had been terminated from the emergency jobs program; 73% of these were negative terminations, i.e., terminations which did not result in employment or enrollment in some skill development or training program.

At the end of the first fiscal year, the typical Title VI work experience enrollee in the city of Atlanta was male, between 22 and 44 years old, possessed a high school education or better, was economically disadvantaged, black, and unemployed. Nationally, the typical enrollee was white, possessed a high school education, was unemployed and not previously in the labor market (U.S. Department of Labor, 1975).

CONCLUSIONS

At the beginning of this essay I posed a number of questions about the impact of Public Service Employment under Title II of CETA. Based on the short Atlanta experience, I can tentatively offer *some* answers to *some* of these questions that serve as a summary of findings reported in this essay and an agenda (particularly in the case of unanswered questions) for future research.

First, "Were the jobs made available under Public Service Employment (PSE) in addition to existing jobs in the local jurisdiction?" In the case of Atlanta, yes. During the first year 1,151 new jobs were created in Atlanta, 704 under Title II and 447 under Title VI.

Second, "Did the PSE jobs that were created help to provide otherwise unprovided public services?" The answer to this question is less clear. Whether the services provided by CETA PSE would not have been provided if the PSE positions had not been used was difficult to determine because of the inability to say what services would have existed without CETA funds. However, it is clear that in Atlanta additional services of the type provided before CETA were made possible through the use of PSE positions, and that the "dual service" concept of providing such services (e.g., using PSE positions to generate demand for services by those eligible to receive them on the one hand, and to help expand service delivery on the other) was clearly affected by PSE usage. It is clear that different types of services from those existing prior to CETA did exist during the first year of CETA.

Third, "Were PSE positions integrated into the city's governmental structure in such a manner as to assure continued delivery of city services?" The answer is yes. This is especially true with Title II PSE transitional positions which were integrated into the city governmental structure through the Personnel Department and the various other departments.

Fourth, "Were PSE clients used to replace city employees, thus becoming substitutive in nature with the salary of such employees comprising, in effect, a federal subsidy to local government budgets?" Based on the data collected, interviews with city officials, and the record of dismissals or layoffs, the answer is that PSE positions were *not* used to replace regular city employees. During the first year of CETA, Atlanta *did not* lay off any of its regular employees in order to use PSE in a substitutive fashion. However, it is really difficult to determine both the true existence and the magnitude of a substitutive effect in the first year's operation of a program like PSE under CETA. This is an area which would require much more extensive research on a longitudinal basis. But I am not sure what would be the value of the results. With respect to PSE dollars comprising a federal subsidy to local government budgets, it seems fair to say yes. But comprising a subsidy to the Atlanta budget does not mean necessarily that positions created will be substitutive.

Fifth, "Did the prime sponsor maintain its local effort in level of government service and employment when PSE positions under CETA were authorized?" Yes, Atlanta did maintain its local effort in service provision and employment. Except for the use of PSE positions in the Department of Environment and Streets, where attrition occurred and the number of days of weekly garbage collection was reduced marginally, there was no indication of PSE positions effecting any reduction of local service and employment efforts. Even the case of using PSE positions in the Department of Environment and Streets was not an instance of PSE causing a reduction in local service delivery.

Sixth, "Were jobs created quickly?" In Atlanta, job creation got off to a slow start, partly because of the city's development of procedural mechanisms to reduce the probability of jobs being "loosely" created without maximum integraion into the city structure. This was especially true for Title II PSE and work experience positions both within the city and in private nonprofit organizations under contract. Title VI positions were created quicker than Title II positions because of the simplicity of registering and determination of eligibility and the simplified classification of Work Experience Aides. The research did not determine the exact date of establishing each PSE position, which would be necessary in order to say how quickly jobs were created. But it is reasonably clear that, although the PSE Title II office did not become fully staffed until November 1974 (four months after the operational data of CETA) and Title VI jobs were not authorized to begin until January 1975, there were 447 Title VI positions filled, 422 PSE positions filled in private, nonprofit agencies under Title II, and 100 PSE positions filled within city government by June 30, 1975.

Seventh, "Were PSE jobs created 'loosely' and handed out with no consideration of an applicant's ability or potential ability to do the job?" No. This was not the case in Atlanta. A process for justifying that city PSE jobs were additional and warranted involved certification by the personnel and finance departments as well as by the manpower office. With respect to Title VI work experience jobs, the rigor of job development was less so than for Title II but not without city review, especially because of the contract review and approval process.

Eighth, "Were the jobs that were created meaningful jobs?" For the most part, yes. There were mixed results. Some of the Title VI emergency jobs may have been less meaningful than some of the PSE Title II positions. But the absence of sound criteria of what is "meaningful" makes such evaluative judgments difficult.

Ninth, "Was transition into permanent jobs with city or nonprofit private or public agencies likely?" The Atlanta experience in the first year provided little evidence of either the *fact of* or the *likelihood of* a high percentage of transition from PSE positions to full, unsubsidized city employment status. From the very beginning, the city proposed a very modest transition goal for the first fiscal year. Only about 2% of the planned 1,168 PSE positions to be created were initially planned to be transitioned by the city, a percentage considerably lower than the 50% transition goal initially envisaged by the legislation when formulated in 1973.

Finally, "Did PSE jobs help the economically disadvantaged as well as the unemployed?" In Atlanta, partly because of the population composition, virtually all of the recipients of Title II PSE positions could be place in the economically disadvantaged category.

The tentative answers to these questions suggest that there are gaps in our knowledge with respect to the effects of Public Service Employment and the economy in general. With the passage of the Public Works Employment Act of 1976 and President Jimmy Carter's request for CETA PSE extension for the rest of fiscal 1977 and other antirecessionary public jobs related programs, it seems inescapable that additional empirical research on the effects of PSE on local government budgets, on the holder of the created jobs, and on the types of jobs created, as well as research into the design of the implementation system of local governments, will be required.

The design of the intent and the actual use of public service employment in the city of Atlanta illustrates how the policy model of block grant programs is useful in facilitating greater understanding of how public policies are transformed from simple espoused policy to policy-in-use. The Atlanta experience shows that local jurisdictions do possess some flexibility in how they may use PSE and emergency jobs funds, but also illustrates how a jurisdiction's administrative procedures and mechanisms can be designed to assure that jobs are additive and integrated into the government structure.

The burden of responsibility for performance gaps between what is espoused by local jurisdictions and what outcomes are realized rests clearly with CETA prime sponsors. But resulting outcomes my sometimes only approximate what is intended by the national espoused public policy because of modifications by rules, regulations, and discretion exercised by various actors, including that of regional manpower officers.

The conflicting objectives of trying to (1) improve the fiscal stability of local budgets, (2) improve the quality of skills possessed by the disadvantaged throught work experience, and (c) employ the unemployed and underemployed through public service employment and emergency jobs require numerous administrative complexities to assure that the antirecessionary effects of public service employment are realized. An understanding of these complexities before rendering evaluative judgments about the worth of PSE or CETA seems imperative. These complexities vary from prime sponsor to prime sponsor and from region to region. Often they are lost in aggregate analysis of output figures.

NOTES

1. The notion that public policy considerations are system considerations is not new to the study of national and urban problems. The orientation requires, however, that one recognize that the systems approach requires that public policy be viewed as comprising several interrelated components, all or any one of which may affect the nature of the system's output and performance. For a more thorough discussion of the systems approach and public policy see Dye (1972), Churchman (1968), Van Gigch (1974), and Easton (1972).

2. Although former President Richard M. Nixon had envisaged extensive decategorization, decentralization, and devolution of power to state and local governments to make manpower decisions, considerable federal involvement was retained, especially with regard to the review and approval of prime sponsor plans, the use of discretionary funds by the Secretary of Labor, assessing manpower performance, and in providing technical assistance. These responsibilities are incorporated in the NEPP; see Comprehensive Employment and Training Act of 1973, Public Law 92-203, sec. 105(a) (1).

3. The view of public policy as theory is similiar to that taken by Pressman and Wildavsky (1973). Their position is that "policies imply theories, whether stated explicitly or not, policies point to a chain of causation between initial conditions and future consequences."

4. The concepts "espoused policy" and "policy-in-use" are adaptations of Argyris and Schon's (1974) usage of "espoused theory" and "theory-in-use" at the interpersonal level in order to improve professional effectiveness.

5. As a system, each of these elements possesses an environment of its own. Each is influenced by other factors and tends to influence other factors and elements. Not shown in the simplified model is the state-local espoused policy link and the state espoused policy link. The use of the LEPP in the model in a general way is meant to represent also state espoused public policy (SEPP) and any partnership policy, such as a state-local espoused public policy (S/LEPP). If the model were represented in detail it would show the points of intervention in the system by the judicial system resulting from action brought at points of controversy and conflict over discrepancies between the NEPP and PPRM or PPDM, between the NEPP and the LEPP, between the LEPP and PPIU, and between the NEPP and outcomes.

6. The specific provisions of CETA permit local, state and consortia prime sponsors to formulate their own local/state espoused policy plans for use of CETA funds—given that priority of service is ultimately consistent with overall national goals.

7. The absence of standard decision making by the different federal representatives who monitor CETA operations within the various regions contributes to the range of variance from the original intent incorporated in the NEPP.

8. The gap between the NEPP and the LEPP is one measure of incongruence that results from policy modifications. The gap between either the LEPP or the SEPP and the PPIU is a reflection of changes in state or local preferences from originally intended activities. The source of these changes is often the different preferences that result from activity in the environment of the LEPP and PPIU stages by decision makers, clients, and variation in fiscal resources.

9. The enactment of the Emergency Jobs and Unemployment Assistance Act of 1974, Public Law 93-587, amended CETA by inserting a new Title VI and redesignating the existing Title VI as Title VII.

10. The "parties at interest" are those constituent components which stand to gain from the existence of a manpower policy system. Not limited to community-based organizations, "parties at interest" include recipients of services, contractors, city administrators and politicians, public employee unions, the U.S. Department of Labor, and the state manpower services councils.

11. Reference to the "unemployed" as a priority refers to the unemployed as defined by the NEPP and the federal rules and regulations issued between December 1974 and June 30, 1975. According to the federal regulations, the definition of "unemployed" underwent four changes over this period of time; see, for example, definitions in 40 FR 22685 §94.4 (ggg).

12. Section 205(a) of Title II of the Comprehensive Employment and Training Act of 1973, as amended, requires that eligible prime sponsors submit an application to be approved by the Secretary of Labor in accordance with provisions of Title II (and accompanying regulations). This application, sometimes referred to as the project operation plan, is, in effect, the local espoused public policy plan with respect to what is the intended use of Title II funds. It may be appropriately referred to as the LEPP.

13. Under Title I of CETA, the city of Atlanta planned an expenditure of $994,000 of $4,600,000 for work experience for fiscal year 1975. The actual expenditure for work experience under Title I was $641,538; 64.5% of planned work experience expenditure and 13.9% of all actual expenditures.

14. While the *Quarterly Progress Report* reports the data in Table 3, the Title II Administrator took the position that the figures really did not reflect the real situation because of the unrealistic nature of planned enrollment and expenditure figures.

15. At the time of this research, all paper-and-pencil tests previously used by the city of Atlanta were being revised or validated. Their use was temporarily suspended, except for the performance test for secretaries and the shorthand tests for stenographers.

REFERENCES

ARGYRIS, C., and SCHON, D. (1974). Theory in practice: Increasing professional effectiveness. San Francisco: Jossey-Bass.

CHURCHMAN, C.W. (1968). The systems approach. New York: Delta.

DYE, T. (1972). Understanding public policy. Englewood Cliffs, N.J.: Prentice-Hall.

EASTON, D. (1972). Framework for political analysis. Englewood Cliffs, N.J.: Prentice-Hall.

LEVITAN, S., and ZICKLER, J.K. (1975). The quest for a federal manpower policy. Ann Arbor, Mich.: Institute of Labor and Industrial Relations.

MASSELL, S. (1973). Atlanta area manpower plan: Fiscal year 1974. Atlanta: City of Atlanta.

NIXON, R.M. (1971). "The State of the Union message," Weekly complilation of presidential documents. Washington, D.C.: Office of the Federal Register.

PRESSMAN, J., and WILDAVSKY, A. (1973). Implementation. Berkeley: University of California Press.

Research Atlanta (1973). Which way Atlanta. Atlanta: Author.

SUCHMAN, E.A. (1967). Evaluative research: Principles and practice in public service and social action programs. New York: Russell Sage Foundation.

U.S. Congress (1973). Comprehensive Employment and Training Act. 87 Stat. 839. Pub. L. 93-203.

U.S. Department of Labor (1975). "Comprehensive Employment and Training Act review and oversight; Part II: Public policy issue." Employment and Training Administration, (December).

Van GIGCH, J. (1974). General systems theory. New York: Harper and Row.

THE AUTHORS

ROBERT W. BACKOFF is Assistant Professor of Public Administration and Political Science at Ohio State University. He previously taught at the University of Pennsylvania. He has authored and coauthored articles in *Administration and Society* and the *Public Administration Review*.

ROY W. BAHL is Professor of Economics and Director of the Metropolitan Studies Program at Syracuse University. He is the author of numerous books and journal articles in the areas of public finance and urban economics. Most recently he has coauthored *Fiscal Centralization and Tax Burdens* and *Taxes, Expenditures and Economic Base: A Case Study of New York City*.

DONALD C. BAUMER is a research associate with the Mershon Center of the Ohio State University. He has been involved with the CETA project since its inception and is coauthor of a Department of Labor monograph, *The Implementation of CETA in Ohio* (1977).

JESSE BURKHEAD is Maxwell Professor of Economics at Syracuse University. He has authored books and journal articles in public finance, particularly budgeting, expenditure analysis and state-local finance. He coauthored *Productivity in the Local Government Sector* (1974) and served recently as a consultant to the Detroit Productivity Commission.

JOHN J. DeMARCO is a research assistant at Syracuse University involved in projects related to leadership and collective bargaining. His masters thesis at Wayne State University (1976) concerned the impact of state collective bargaining statutes on unionization.

WILLIAM B. EDDY is Helen Kemper Professor and Director of Public Administration in the School of Administration, University of Missouri-Kansas City. He previously served as Associate Director of the Federal Executive Institute in Charlottesville, Virginia. He is the author of numerous articles and books dealing with the application of behavioral science methods to groups in public sector organizations. He is currently serving as coeditor of *Administration and Society*.

HARLAN HAHN is Professor of Political Science at the University of Southern California. He previously taught at the University of California, Riverside, and the University of Michigan. His most recent books are *People and Politics in Urban Society* (1972), *Ghetto Revolts* (1973), *American Government: Minority Rights Versus Majority Rule* (1976), and *Corruption in the American Political System* (1976). He currently serves as editor of the *American Politics Quarterly*.

[317]

CHARLES H. LEVINE is Associate Professor of Urban Studies at the University of Maryland at College Park. In 1977 he rejoined the Maryland faculty after a four year absence during which time he served as Associate Professor at Syracuse University and as Visiting Research Professor at Cornell University. He is the author of *Racial Conflict and the American Mayor* (1974) and served for three years as coeditor of *Administration and Society*.

EUGENE B. McGREGOR, Jr., is an Associate Professor in the School of Public and Environmental Affairs at Indiana University. He previously served on the faculty of Government and Politics at the University of Maryland, College Park. He published earlier work on civil service structure, manpower management, management simulations, productivity improvement, and decision making at national nominating conventions.

THOMAS P. MURPHY is Director of the Federal Executive Institute in Charlottesville, Virginia. He previously served as Director of Public Administration at the University of Missouri-Kansas City and Director of the Institute of Urban Studies at the University of Maryland. His previous government experience includes service as Deputy Assistant Administrator for Legislative Affairs in NASA. He also has served as the County Manager of Missouri's second largest county (Jackson-Kansas City) and as President of the National Association of Schools of Public Affairs and Administration. Included among his publications are ten books on the Congress, urban management, science policy, internships, and public administration.

FELIX A. NIGRO is Professor of Political Science at the University of Georgia. He is the author of *Management-Employee Relations in the Public Sector* (1969) and coauthor with Lloyd G. Nigro of *Modern Public Administration* (4th ed., 1977) and *The New Public Personnel Administration* (1976). He has been in government and teaching positions since 1937 when he started as an apprentice with Public Administration Clearing House He is a member of the National Labor Panel of the American Arbitration Association and the roster of arbitrators of the Federal Mediation and Conciliation Service, as well as arbitrator and factfinder for the Florida Public Employment Relations Commission.

LLOYD G. NIGRO is Associate Professor of Public Administration at Syracuse University. He previously served on the faculty of the University of Southern California and has also been a Fulbright Senior Lecturer at the University of Madrid, Spain. His publications include *The New Public Personnel Administration* (1976) and *Modern Public Administration* (4th ed., 1977), both coauthored with Felix A. Nigro, and several articles in the areas of personnel management and administrative theory.

JAMES L. PERRY is Assistant Professor of Administration at the Graduate School of Administration, University of California, Irvine. He has authored and coauthored articles on collective bargaining and is currently engaged in studies of innovation in local government and the impact of labor-management relationships on governmental efficiency and effectiveness.

HAL G. RAINEY is Assistant Professor of Public Administration at Florida State University. He has coauthored articles on organization behavior and organization theory in the *Journal of Personality and Social Psychology* and the *Public Administration Review.*

RANDALL B. RIPLEY is Chairman and Professor in the Department of Political Science at Ohio State University. He is author of numerous books and articles on American government and public policy and has directed major studies of the implementation of CETA since mid-1974.

DAVID H. ROSENBLOOM is an Associate Professor of Political Science at the University of Vermont. He previously taught at the Universities of Kansas and Tel Aviv and served as an American Society for Public Administration Fellow with the United States Civil Service Commission. Among his published works are *Federal Service and the Constitution, Federal Equal Employment Opportunity,* and numerous articles appearing in professional journals and law reviews.

DAVID T. STANLEY, of Vienna, Virginia, is a consultant, writer, and lecturer on public administration. From 1961 through 1975 he was a Senior Fellow in the Governmental Studies Program of the Brookings Institution of Washington, D.C. He is the author of seven Brookings books, including *Professional Personnel for the City of New York* (1963) and *Managing Local Government Under Union Pressure* 1972. Previously he served for 22 years in various federal management positions, the last of which was as Director of Management Policy in the Department of Health, Education, and Welfare.

FRANK J. THOMPSON is an Assistant Professor in the Department of Political Science at the University of Georgia. He has served as a Public Administration Fellow with the Department of Health, Education, and Welfare and has also worked for the city government of Oakland, California. He is the author of several articles and a book, *Personnel Policy in the City: The Politics of Jobs in Oakland* (1975).

CHARLES W. WASHINGTON is Assistant Professor in the Department of Public Administration, The George Washington University. He is a former Intern with the Advisory Commission on Intergovernmental Relations and has worked with other research organizations, including the Joint Center for Environmental and Urban Problems, Florida Atlantic University, and the Metropolitan Studies Center of the Maxwell School, Syracuse University.